A First Course In

CONTINUUM MECHANICS

Second Edition

Y. C. FUNG

Professor of Applied Mechanics & Bioengineering
University of California, San Diego

PRENTICE-HALL, INC.,
ENGLEWOOD CLIFFS, NEW JERSEY 07632

Library of Congress Cataloging in Publication Data

FUNG, YUAN-CHENG,
   A first course in continuum mechanics.

   Includes bibliographies and index.
   1. Continuum mechanics.   I. Title.
QA808.2.F85 1977   531   76-41212
ISBN   0-13-318311-4

dedicated to LUNA

PRENTICE-HALL INTERNATIONAL, INC., *London*
PRENTICE-HALL OF AUSTRALIA PTY. LIMITED, *Sidney*
PRENTICE-HALL OF CANADA, LTD., *Toronto*
PRENTICE-HALL OF INDIA PRIVATE LIMITED, *New Delhi*
PRENTICE-HALL OF JAPAN, INC., *Tokyo*
PRENTICE-HALL OF SOUTHEAST ASIA PTE. LTD., *Singapore*
WHITEHALL BOOKS LIMITED, *Wellington, New Zealand*

NOTE: The material contained in this book is prepared for a one-year course or an intensive one-semester course. For a shorter course, the following sections may be omitted: 3.7, 4.6–4.8, 5.6, 5.8–5.10, 6.3, 8.3, 8.4, 8.7, 9.1–9.13, 10.7–10.9, 11.4–11.12, 12.4–12.8 These sections are marked with an asterisk.

Contents

Preface to Second Edition   ix

Preface to First Edition   x

CHAPTER ONE

Introduction   1

1.1   The objective of this course   1
1.2   What is mechanics?   2
1.3   Continuum mechanics   2
1.4   Plan of study   4
1.5   Newton's laws of motion   6
1.6   Equilibrium   8
1.7   Free-body diagram   11
1.8   Ad hoc theory vs. general theory   24

CHAPTER TWO

Vectors and Tensors   37

2.1   Vectors   37
2.2   Vector equations   40
2.3   The summation convention   43

2.4 Translation and rotation of coordinates · 47
2.5 Coordinate transformation in general · 51
2.6 Analytical definitions of scalars, vectors, and Cartesian tensors · 52
2.7 The significance of tensor equations · 55
2.8 Notations for vectors and tensors: Boldface or indices? · 56
2.9 Quotient rule · 56
2.10 Partial derivatives · 57

CHAPTER THREE

Stress · 63

3.1 The idea of stress · 63
3.2 Notations for stress components · 65
3.3 The laws of motion and free-body diagrams · 67
3.4 Cauchy's formula · 70
3.5 Equations of equilibrium · 72
3.6 Change of stress components in transformation of coordinates · 76
3.7 Stress components in orthogonal curvilinear coordinates · 77
3.8 Stress boundary conditions · 79

CHAPTER FOUR

Principal Stresses and Principal Axes · 91

4.1 Introduction · 91
4.2 Plane state of stress · 92
4.3 Mohr's circle for plane stress · 95
4.4 Principal stresses · 98
4.5 Shearing stresses · 101
4.6 Stress-deviation tensor · 103
4.7 Lamé's stress ellipsoid · 106
4.8 Mohr's circles for three-dimensional stress states · 108

CHAPTER FIVE

Analysis of Deformation · 121

5.1 Deformation · 121
5.2 The strain · 125
5.3 Strain components in rectangular Cartesian coordinates · 127

5.4   Geometric interpretation of infinitesimal strain components   129
5.5   Infinitesimal rotation   130
5.6   Finite strain components   132
5.7   Principal strains. Mohr's circle   134
5.8   Infinitesimal strain components in polar coordinates   135
5.9   Direct derivation of the strain-displacement relations in polar
        coordinates   138
5.10 Other strain measures   140

CHAPTER SIX

Velocity Fields and Compatibility Conditions   **153**

6.1   Velocity fields   153
6.2   The so-called compatibility condition   154
6.3   Compatibility of strain components in three dimensions   156

CHAPTER SEVEN

Constitutive Equations   **163**

7.1   Specification of the properties of materials   163
7.2   The nonviscous fluid   164
7.3   Newtonian fluid   165
7.4   Hookean elastic solid   167
7.5   Effect of temperature   169
7.6   Materials with more complex mechanical behavior   170
7.7   Simple beam theory   171

CHAPTER EIGHT

Isotropy   **187**

8.1   The concept of material isotropy   187
8.2   Isotropic tensor   188
8.3   Isotropic tensors of rank 3   191
8.4   Isotropic tensors of rank 4   192
8.5   Isotropic materials   195
8.6   Coincidence of principal axes of stress and of strain   195
8.7   Other methods of characterizing isotropy   196

CHAPTER NINE

Mechanical Properties of Fluids and Solids   **199**

9.1	Fluids	200
9.2	Tensile strength of a liquid	204
9.3	Viscosity	207
9.4	The compressibility of air	212
9.5	The compressibility of liquids	213
9.6	The elasticity of solids	216
9.7	Plasticity of metals	219
9.8	Theoretical strength of metals	222
9.9	Large deformation. Nonlinear elasticity	223
9.10	Viscoelasticity	227
9.11	Non-Newtonian fluids	232
9.12	Visco-plastic materials	233
9.13	Sol-gel transformation. Thixotropy	235

CHAPTER TEN

Derivation of Field Equations   **241**

10.1	Gauss' theorem	242
10.2	Material description of the motion of a continuum	245
10.3	Spatial description of the motion of a continuum	247
10.4	The material derivative of a volume integral	248
10.5	The equation of continuity	250
10.6	The equations of motion	250
10.7	Moment of momentum	252
10.8	The balance of energy	253
10.9	The equations of motion and contimuity in polar coordinates	256

CHAPTER ELEVEN

Field Equations and Boundary Conditions in Fluids   **265**

11.1	The Navier-Stokes equations	265
11.2	Boundary conditions at a solid-fluid interface	268
11.3	Surface tension and the boundary conditions at a free surface	270
11.4	Dynamic similarity and Reynolds number	273

11.5   Laminar flow in a horizontal channel or tube   275
11.6   Boundary layer   279
11.7   Laminar boundary layer over a flat plate   282
11.8   Nonviscous fluid   285
11.9   Vorticity and circulation   287
11.10 Irrotational flow   289
11.11 Compressible nonviscous fluids   291
11.12 Subsonic and supersonic flow   294

CHAPTER TWELVE

Some Simple Problems in Elasticity   307

12.1   Basic equations of elasticity for homogeneous isotropic bodies   307
12.2   Plane elastic waves   310
12.3   Simplifications   312
12.4   The Airy stress function   314
12.5   Torsion of a circular cylindrical shaft   315
12.6   Saint-Venant's principle   320
12.7   Beams   321
12.8   Concluding remarks   324

General Problems   333

Index   335

Preface
to the Second Edition

The focus of revision for the second edition is to improve the usefulness of the book from the point of view of the students. The original objective as set out in the preface to the first edition is unchanged. Students are receptive to our basic approach and can quickly appreciate the importance of the subject; but they demand more examples and more details on how to solve problems. Examples and solutions are, then, the main increments to this edition.

Chapter 1 was completely rewritten. The new emphasis is on the use of the ageless method of a free-body diagram. A number of examples and exercises illustrate the method. The problems of rods and plates lead naturally to the concept of stress tensor, which motivates the study of Chapter 2. The examples on beams introduce the students to one of the most useful practical applications of mechanics. The simple beam theory is introduced in Chapter 7.

A plan of study is outlined in Sec. 1.4. We emphasize that Chapter 10 is important. Mastery of that chapter will make this course useful to the student for his entire engineering and scientific career.

Preface

to the First Edition

This book is intended for students of science and engineering who are beginning a series of courses in mechanics. At this stage, students normally have had courses in calculus, physics, vector analysis, and elementary differential equations. A course in continuum mechanics then provides a fundation for studies in fluid and solid mechanics, material sciences, and other branches of science and engineering.

It is my opinion that, for a beginner, the approach should be physical rather than mathematical. To engineers and physicists who use continuum mechanics constantly, the primary attraction of the subject lies in its simplicity of conception and concreteness in applications. Therefore, the students should be introduced to the applications as soon as possible.

For the scientist or engineer, the important questions he must find answers to are: How shall I formulate the problem? How shall I state the governing field equations and boundary conditions? How shall I choose alternate hypotheses? What kind of experiments would justify or deny or improve my hypotheses? How exhaustive should the investigation be? Where might errors appear? How much time is required to obtain a reasonable solution? At what cost? These are questions which concern active investigators, and are questions of synthesis, which employ analyses as tools. Complete answers to these questions are beyond the scope of this "first course," but we can make a good beginning. In this book, I often ask the reader to formulate problems, regardless of whether he can solve his equations and understand all the mathematical subtleties. I have known many students who have read many books and worked innumerable exercises without ever formulating a problem of their own. I hope they will learn the other way, to generate many problems of their own and then strive to discover the methods and subtleties of solutions. They should be encouraged to observe nature and to think of problems in engineering and then to take the first step to write down a possible set of governing equations and boundary conditions. This "first step"—to derive the basic governing equations—is the object of this book. Perhaps it is justifiable for a "first course" to be concerned only with this first step. But the preparation required for taking this step is extensive. For such a step to be firm, one would have to understand the basic concepts of mechanics and their mathematical expressions. To be able to use these basic equations with confidence one must know their origins and their derivations. Therefore, the discussions of basic ideas must be thorough. It is for this reason that the first ten chapters of this book are rather comprehensive and detailed.

As for the organization of the book: At the outset, the concept of continua is explained. Then a thorough treatment of the concepts of stress and strain follows. The practical techniques of determining the principal stress and strain, and the concept of compatibility, are given emphasis in two separate chapters. The description of motion is considered. In Chapter 7, an idealized specification of fluids and solids is presented. The important concept of isotropy is described in detail in Chapter 8. Data on the mechanical properties of common fluids and solids appear in Chapter 9. In Chapter 10, a thorough treatment of the basic conservation laws of physics is given. Beginning with Chapter 11, some features of perfect fluids, viscous flow, boundary layer theory, linearized theory of elasticity, theories of bending and torsion, and elastic waves are described briefly. The last two chapters provide a glimpse into the rich fields of fluid and solid mechanics; to treat them comprehensively would require many volumes at a more advanced mathematical level. The introduction given here should prepare the student to enter these fields with greater ease.

If the reader obtains clear ideas about the stress, strain, and constitutive equations from this book, I would consider this introductory text a success. Beyond this, only a sketch of some classical problems is provided. Many discussions are given in the exercises, which should be regarded as an integral part of the text.

I have quoted frequently and borrowed heavily from my previous book, *Foundations of Solid Mechanics*, which can be used for a course following the present one. The material for this "first course" was organized for my class at the University of California, San Diego, where the curriculum offers emphasis on general sciences before specialization. The book should be useful for undergraduates and younger graduate students who have a reasonable background in mathematics and physics.

The writing of this book was a pleasant experience. My wife, Luna, cooperated throughout the task. A mathematician, she gave up her teaching career when I came to La Jolla. Willing to learn some mechanics, she worked through the manuscript very thoroughly. Many passages are clearer because of her declaration that she did not understand. My friend, Chia-Shun Yih, Timoshenko Professor at the University of Michigan, read through the manuscript and gave me many valuable comments. I am also grateful to Drs. Pin Tong of the Massachusetts Institute of Technology and Gilbert Hegemier of the University of California, San Diego, for their comments. Finally, I wish to register my thanks to Nicholas Romanelli of Prentice-Hall for editorial assistance, to Mrs. Ling Lin for preparing the index, and to Mrs. Barbara Johnson, whose fast, accurate typing and cheerful good humor made the work a pleasure.

Y. C. FUNG

CHAPTER ONE

Introduction

1.1 THE OBJECTIVE OF THIS COURSE

Our objective is to learn how to formulate problems and how to reduce vague questions and ideas to precise mathematical statements.

Let us consider a few such questions: An airplane is flying above us. The wings must be under strain in order to support the passengers and freight. How much strain are the wings subjected to? If you were flying a glider and an anvil cloud appears, the thermal current would carry the craft higher. Dare you fly into the cloud? Have the wings sufficient strength? Ahead you see the Golden Gate Bridge. Its cables support a tremendous load. How does one design such cable? The cloud contains water and the countryside needs that water. If the cloud were seeded, would that produce the rain? And would it fall where needed? Would the amount of rainfall be adequate and not produce a flood? In the distance there is a nuclear reactor power station. How is the heat transported in the reactor? What kind of thermal stresses are there in the reactor? How does one assess the safety of the power station against earthquake? What happens to the Earth in an earthquake? Thinking about the globe, you may wonder how the continents float, move, or tear apart. And how about ourselves: How do we breathe? What changes take place in our lungs if we do a yoga exercise and stand on our heads?

Interestingly, all these questions can be reduced to certain differential equations and boundary conditions. By solving such equations we obtain precise quantitative information. In this book we deal with the fundamental

1

...ciples that underlie such differential equations and boundary conditions. Although it would be a pleasure to solve these equations once they are formulated, we shall not become involved in discussing their solutions in detail. Our objective is formulation: the formal reduction of general ideas to a mathematical form. These mathematical problems may not be easy to solve. Many scientific and engineering disciplines devise special methods to solve problems quickly and efficiently. A generation ago students of science and engineering spent countless hours learning the techniques needed for solving differential equations. Today the task is made much easier by using computers.

1.2 WHAT IS MECHANICS?

Mechanics is the study of the motion of matter and the forces that cause such motion. Mechanics is based on the concepts of time, space, force, energy, and matter. A knowledge of mechanics is needed for the study of all branches of physics, chemistry, biology, and engineering.

To consider all aspects of mechanics would be too great a task for us. Instead, in this book, we shall study only the mechanics of continua. We shall concern ourselves with the basic principles common to fluids and solids. The details of specialized subjects are beyond the scope of this text.

1.3 CONTINUUM MECHANICS

The concept of a continuum is derived from mathematics. We say that the real number system is a *continuum*. Between any two distinct real numbers there is another distinct real number, and therefore there are infinitely many real numbers between any two distinct real numbers. Intuitively we feel that *time* can be represented by a real number system t and that a three-dimensional *space* can be represented by three real number systems x, y, z. Thus we identify *time* and *space* as a four-dimensional continuum.

Extending the concept of continuum to *matter*, we speak of a continuous distribution of matter in space. This may be best illustrated by considering the concept of *density*. Let the amount of matter be measured by its *mass*, and let us assume that a certain matter permeates a certain space $\mathcal{V}_0$, as in Fig. 1.1. Let us consider a point P in $\mathcal{V}_0$, and a sequence of subspaces $\mathcal{V}_0, \mathcal{V}_1, \mathcal{V}_2, \ldots$, converging on P:

$$(1.3\text{-}1) \qquad \mathcal{V}_n \subset \mathcal{V}_{n-1}, \qquad P \in \mathcal{V}_n, \qquad (n = 1, 2, \ldots).$$

Let the volume of $\mathcal{V}_n$ be V_n and the mass of the matter contained in $\mathcal{V}_n$ be M_n. We form the ratio M_n/V_n. Then if the limit of M_n/V_n exists as $n \longrightarrow \infty$ and $V_n \longrightarrow 0$, the limiting value is defined as the *density* of the mass distribu-

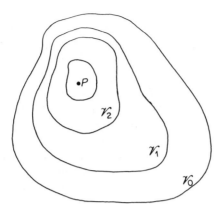

Fig. 1.1 A sequence of spatial domains converging on P.

tion at the point P, and is denoted by $\rho(P)$:

$$(1.3\text{-}2) \qquad\qquad \rho(P) = \lim_{\substack{n\to\infty \\ V_n\to 0}} \frac{M_n}{V_n}.$$

If the density is well defined everywhere in $\mathcal{V}_0$, the mass is said to be *continuously distributed.*

A similar consideration can be used to define the density of momentum, the density of energy, and so on. *A material continuum is a material for which the densities of mass, momentum, and energy exist in the mathematical sense. The mechanics of such a material continuum is continuum mechanics.* It is in this mathematical sense that continuum mechanics will be developed in this book.

Continuum mechanics has wide applications to problems of the physical world. But because it is a mathematical abstraction, the question of applicability deserves a careful consideration. In particular, since matter is conceived in modern physics as structures of elementary particles, there is a need to reconcile the particle point of view with the continuum point of view. Obviously, if water is considered as a structure of elementary particles, then water is not a continuum. If we try to define the density of water according to the definition (1.3-2), we shall encounter an obvious difficulty when the dimension of $\mathcal{V}_n$ is reduced to the order of the atomic radius. Thus, if $\mathcal{V}_n$ and $\mathcal{V}_{n+1}$ differ by a neutron, the ratio M_n/V_n will have a finite difference from M_{n+1}/V_{n+1}. It is easy to see that as the particles move about, the limit M_n/V_n either does not exist or fluctuates with time and space. To rescue this situation, we shall apply a smoothing process as follows. We consider the ratio M_n/V_n. Let $\mathcal{V}_n$ become smaller and smaller but *always remain so large that it contains a large number of particles in it.* If the ratio

M_n/V_n tends to a definite limit $\rho(P)$ *within this added restriction*, then $\rho(P)$ is defined as the density of the matter. In other words, corresponding to the real material, we define a mathematical continuum which has the same density $\rho(P)$ in the strict sense of (1.3-2). Further analysis of the mechanics of the material can then be based on the mathematical model.

In practice, there is little difficulty in applying such a smoothing procedure. The molecular dimension of water is about 1 Å (10^{-8} cm); hence, if we are concerned about the liquid water in a problem in which we never have to consider dimensions less than say 10^{-6} cm, we are safe to treat water as a continuum. The mean free path of the molecules of the air on the surface of the earth at room temperature is about 5×10^{-6} cm; hence, if we consider the flow of air about an airplane, we may treat air as a continuum. The red blood cell in our body has a diameter about 8×10^{-4} cm; hence, we can treat our blood as a continuum if we consider the flow in arteries of diameter, say, 0.5 cm.

These arguments can be made rigorous statistically. The kinetic theory of gases is the best known example. The motion of molecules of a gas is separated into "random" and "systematic" parts. The former contributes zero resultant momentum but has a finite kinetic energy which is identified with thermal energy and is related to the absolute temperature. On the other hand, the systematic part (the local mean value), contributes to the systematic motion of the body of gas as conceived in continuum mechanics.

Thus the concept of a material continuum as a mathematical idealization of the real world is applicable to problems in which the fine structure of matter can be ignored. When the fine structure attracts our attention, we should return to particle physics and statistical mechanics. The duality of continuum and particles helps us to understand the physical world as a whole, in a manner which was made famous by modern optics, in which light is treated sometimes as particles and sometimes as waves.

1.4 PLAN OF STUDY

Since a continuum is an abstraction applied to a large collection of material particles, the purpose of continuum mechanics is to describe the motion (or, in a special case, the equilibrium) of these particles in space and time, in response to whatever forces influencing that motion (or equilibrium). Through the years convenient methods and an accurate language have been developed in continuum mechanics. We shall study these methods and the language.

Our study must start with Newton's laws of motion of particles. Then we must know how to apply Newton's law to a *system* of particles or to a *part of the system*. The last four words of the preceding sentence turn out to be the key to the understanding of mechanics. In order to learn what's going on at a place in a continuum, we cut open (in our minds) that part and examine the interaction of one part of the body on another. The part that is cut out is an entity in itself and is called a *free body*. The uninhibited use of *free-body diagrams* is of utmost importance to students of continuum mechanics. The topic is discussed in this chapter and is illustrated by a number of examples.

Consideration of the force of interaction of one part of a body on another across an imagined surface leads to the concept of *stress*. Stress is a vector. At any given place in a continuum, the stress vector depends on the orientation of the surface on which it acts. The orientation of a surface is specified by its normal vector. Thus stress is a vector that is associated with another vector. Quantities like stresses are called *tensors of rank 2*. The analysis of stresses will be greatly simplified if we use the language of tensors, therefore, in Chapter 2 we digress to a brief introduction to Cartesian tensors.

We then return to a full discussion of the properties of the *stress tensor* in Chapters 3 and 4.

The analysis of deformation of a continuum is considered in Chapter 5. Here again it will be found that the best way of describing deformation is to use a tensor called *strain*. The rate of change of deformation is discussed in Chapter 6.

Each material has a specific mechanical property. The mathematical expression of the mechanical property of a material is called the *constitutive equation* of that material. Out of the milliards of materials in this world, three idealized materials are studied in great detail in engineering science and in physics. These are the nonviscous fluid, the Newtonian viscous fluid, and the Hookean elastic solid. The constitutive equations of these three ideal materials are presented in Chapter 7. A further simplification due to *isotropy* is discussed in Chapter 8. Real materials often behave differently from these ideal materials. Some of the more common and important engineering materials are discussed in Chapter 9.

Now we come to the heart of the subject in Chapter 10, "Derivation of Field Equations." The field equations of continuum mechanics are statements of the balance of forces, the balance of matter, and the balance of energy (i.e., the laws of conservation of momentum, mass, and energy). For a continuum, these are expressed in terms of the stress and strain tensors. Boundary conditions are often subsidiary statements based on the same principles. These equations, plus the constitutive equations, are often all that is necessary to define a problem. *The material in Chapter 10 is the most important in this book.*

In the remaining chapters, some basic features of fluid and solid mechanics are presented.

Sections that may be passed over in first reading are marked with an asterisk.

Learning mechanics is like learning a language. It is absolutely necessary to practice it and to use it. The examples and exercises are, therefore, an integral part of the plan for learning.

In the sections that follow, we shall begin with a reveiw of Newton's laws.

1.5 NEWTON'S LAWS OF MOTION

Newton's laws of motion are abstractions based on experience. In the nearly 300 years since their publication, they have been applied to the analysis of all kinds of problems and have been found to be accurate whenever the velocity of motion is much slower than the velocity of light. There are no exceptions. In this book, we shall consider systems which move at speeds much slower than that of the light. Hence Newton's laws apply, and no relativistic corrections are necessary.

Newton's laws are stated with respect to material particles in a three-dimensional space that obeys Euclidean geometry. A material particle is matter that has a unique, positive measure, the *mass* of the particle. The location of the particle can be described with respect to a rectangular Cartesian frame of reference. In such a space an *inertial frame of reference* exists, with respect to which the Newtonian equations of motion listed below are valid. It can be shown that any frame of reference moving with a uniform velocity with respect to an inertial frame is again inertial. Consider a particle of mass m. Let the position, velocity, and acceleration of this particle be denoted by the vectors $\mathbf{x}$, $\mathbf{v}$, and $\mathbf{a}$, respectively, all defined in an inertial frame of reference. By definition,

$$(1.5\text{-}1) \qquad \mathbf{v} = \frac{d\mathbf{x}}{dt}, \qquad \mathbf{a} = \frac{d\mathbf{v}}{dt}.$$

Let $\mathbf{F}$ be the total force acting on the particle. If $\mathbf{F} = 0$, then *Newton's first law* states that

$$(1.5\text{-}2) \qquad \mathbf{v} = \text{constant}.$$

If $\mathbf{F} \neq 0$, then *Newton's second law* states that

$$(1.5\text{-}3) \qquad \frac{d}{dt} m\mathbf{v} = \mathbf{F}, \quad \text{or} \quad \mathbf{F} = m\mathbf{a}.$$

When Eq. (1.5-3) is written as

$$(1.5\text{-}4) \qquad \mathbf{F} + (-m\mathbf{a}) = 0,$$

it appears as an equation of equilibrium of two forces. The term $-m\mathbf{a}$ is called the *inertia force*. Equation (1.5-4) states that the sum of the external force acting on a particle and the inertia force vanishes; i.e., the inertia force balances the external force. The Newtonian equation of motion stated in this way is called *D'Alembert's principle.*

Now, consider a system of particles that interact with each other. Every particle is influenced by all the other particles in the system. Let an index I denote the Ith particle. Let $\mathbf{F}_{IJ}$ denote the force of interaction exerted by a particle number J on a particle number I, and $\mathbf{F}_{JI}$, that of particle I on particle J. Then *Newton's third law* states that

$$(1.5\text{-}5) \qquad\qquad \mathbf{F}_{IJ} = -\mathbf{F}_{JI} \quad \text{or} \quad \mathbf{F}_{IJ} + \mathbf{F}_{JI} = 0.$$

If $I = J$ we set $\mathbf{F}_{II} = 0$, in agreement with Eq. (1.5-5). Let K be the total number of particles in the system. The force $\mathbf{F}_I$ that acts on the Ith particle consists of an external force $\mathbf{F}_I^{(e)}$, such as gravity, and an internal force that is the resultant of mutual interaction between particles. Thus

$$(1.5\text{-}6) \qquad\qquad \mathbf{F}_I = \mathbf{F}_I^{(e)} + \sum_{J=1}^{K} \mathbf{F}_{IJ}.$$

The equation of motion of the Ith particle is, therefore

$$(1.5\text{-}7) \qquad \frac{d}{dt} m_I \mathbf{v}_I = \mathbf{F}_I^{(e)} + \sum_{J=1}^{K} \mathbf{F}_{IJ}, \qquad (I = 1, 2, \ldots, K).$$

Each particle is described by such an equation. The totality of K equations describes the motion of the system.

To make further progress we must specify how the forces of interaction $\mathbf{F}_{IJ}$ can be computed. Such a specification is a statement of the material property of the system of particles, and is referred to as a *constitutive equation of the material system.*

If the number of particles of a system is small, we may consider every pair of interacting particles being connected with a spring and set about to calculate the forces in the springs as the particles move. A specification of the nature of the springs is a constitutive equation.

If the number of particles is very large, it may become unfeasible to write down each and every equation of the type Eq. (1.5-7) and to account for the interacting forces $\mathbf{F}_{IJ}$ one by one. For example, to account for the motion that takes place in a cloud on a Sunday afternoon, it seems silly to write down an equation of motion for each water droplet in the cloud, because the total number will be so large as to make the equations intractable. In such cases a better approach is needed. In this course we address ourselves to this need. We shall be concerned with the constitutive equations for systems with huge numbers of particles, such as the molecules of a gas or the atoms of a solid. We shall be concerned with the question of how to replace the entire system of ordinary differential equations (1.5-7) with a *single* partial differential equation. A rigorous and comprehensive approach requires some prepara-tory work. In this book the full answer to this objective is given in Chapter 10; nine chapters of preparatory work are necessary.

Within the framework of Newtonian laws the mass of the particles must not change. Equation (1.5-3), although written as $d(m\mathbf{v})/dt = \mathbf{F}$, must not be interpreted as permitting m to be variable. The importance of this remark becomes clear if we consider a rocket that burns fuel at a certain finite rate dm/dt and flies at a velocity $\mathbf{v}$. If we ask how to compute the rocket thrust, we must consider the momentum change of the exhaust gas particles that were in the liquid or solid state before combustion. In applying Newton's law, we must consider a given group of particles. Otherwise we may obtain a wrong answer. Some books erroneously say that Newton's law needs modi-fication to deal with rockets.

1.6 EQUILIBRIUM

A special motion is *equilibrium*; i.e., one in which there is no acceleration for any particles of the system. This is a very important case, especially for engineers. Many engineering constructions are set in equilibrium conditions, e.g., buildings, bridges, automobiles, airplanes. If you think of these products of civilization, you realize that the study of equilibrium, or *statics*, is not trivial.

The condition of equilibrium of a system of particles (we shall call it a "body" for simplicity) can be derived easily. At equilibrium, Eq. (1.5-7) becomes

$$(1.6\text{-}1) \qquad \mathbf{F}_I^{(e)} + \sum_{J=1}^{K} \mathbf{F}_{IJ} = 0, \qquad (I = 1, 2, \ldots, K).$$

Sum over I from 1 to K,

$$(1.6\text{-}2) \qquad \sum_{I=1}^{K} \mathbf{F}_I^{(e)} + \sum_{I=1}^{K} \sum_{J=1}^{K} \mathbf{F}_{IJ} = 0.$$

In the last sum, whenever $\mathbf{F}_{IJ}$ appears, $\mathbf{F}_{JI}$ appears also; they add up to zero according to Eq. (1.5-5). Therefore Eq. (1.6-2) reduces to

$$(1.6\text{-}3) \qquad \sum_{I=1}^{K} \mathbf{F}_I^{(e)} = 0.$$

Put into words, Eq. (1.6-3) states that *for a body in equilibrium, the summation of all external forces acting on the body is zero.* Thus at equilibrium the resul-tant of all external forces vanish.

Next let us consider the tendency of a body to rotate. A body turns if there is a *couple* acting on it. A couple is a pair of forces that are equal in

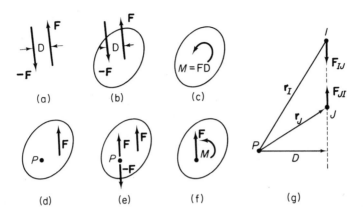

Fig. 1.2 Couples and moments.

magnitude, opposite in direction, and separated by a certain distance. In the example shown in Fig. 1.2(a), the couple is formed by the forces $\mathbf{F}$ and $-\mathbf{F}$, at a distance D apart. The product FD is called the *moment* of the couple, where F is the magnitude of $\mathbf{F}$.

If the couple shown in Fig. 1.2(a) is applied to a free body as shown in Fig. 1.2(b), the body will rotate. Evidently we would have to distinguish the direction of rotation. We therefore define a moment as a vector as shown in Fig. 1.2(c), whose magnitude is FD, whose direction is perpendicular to the plane containing the forces $\mathbf{F}$ and $-\mathbf{F}$, and whose sense is determined by the right-hand screw rule. Finally, consider a body pivoted at a point P as shown in Fig. 1.2(d). When a force $\mathbf{F}$ acts on this body, the body tends to rotate about the pivot P. If we add a pair of equal and opposite forces at P as shown in Fig. 1.2(e), we see that the action of $\mathbf{F}$ is equivalent to a force $\mathbf{F}$ and a couple FD [Fig. 1.2(f)]. Thus the moment FD is the driving cause for rotation about P.

Now consider a body that is acted on by a system of external and internal forces. We shall prove the second condition of equilibrium: *The summation of the moments of all the external forces acting on the body about any point is zero.* Refer to Fig. 1.2(g). Let the mass at the point J be acted on by a force $\mathbf{F}_{JI}$ due to the mass located at point I. The *moment* of the force $\mathbf{F}_{JI}$ about a point P is the product of $\mathbf{F}_{JI}$ and the perpendicular distance D from P to the line of action of $\mathbf{F}_{JI}$. In the meantime the mass I is acted on by a force of interaction $\mathbf{F}_{IJ}$ due to mass J. But $\mathbf{F}_{IJ} = -\mathbf{F}_{JI}$ according to Eq. (1.5-5), and their distance D to P is the same. Thus the two moments cancel each other. Since I and J can refer to any pair of particles in the body, we see that the summation of the moments of all pairs of forces of interaction vanishes.

We can put this in vector notation by writing the moment of $\mathbf{F}_{JI}$ about P as $\mathbf{r}_J \times \mathbf{F}_{JI}$, where $\mathbf{r}_J$ is the position vector from P to J and the $\times$ sign signifies a vector product (see Sec. 2.1). With this notation we form the vector product $\mathbf{r}_I$ with every term in Eq. (1.6-1) and obtain

$$(1.6\text{-}4) \qquad \mathbf{r}_I \times \mathbf{F}_I^{(e)} + \sum_{J=1}^{K} \mathbf{r}_I \times \mathbf{F}_{IJ} = 0,$$

which may be translated by saying that the moment of the external force acting on the mass I about a point P is equal and opposite to the sum of moments of all the forces of interaction acting on the mass I due to all other particles in the system. Now sum (1.6-4) over I; we obtain

$$(1.6\text{-}5) \qquad \sum_{I=1}^{K} \mathbf{r}_I \times \mathbf{F}_I^{(e)} + \sum_{I=1}^{K} \sum_{J=1}^{K} \mathbf{r}_I \times \mathbf{F}_{IJ} = 0.$$

The double sum on the left-hand side represents the sum of the moments of all pairs of interacting forces between the particles and therefore is zero as discussed above. Hence Eq. (1.6-5) becomes

$$(1.6\text{-}6) \qquad \sum_{I=1}^{K} \mathbf{r}_I \times \mathbf{F}_I^{(e)} = 0.$$

This equation states the second necessary condition of equilibrium mentioned above.

The choice of the point P is arbitrary. It can be proved that if Eq. (1.6-3) holds and Eq. (1.6-6) is valid for moments taken about one particular point P, then it is also valid for moments taken about any other point in space. See the problem below.

PROBLEM 1.1 A body is in equilibrium so that Eq. (1.6-3) and (1.6-6) hold. Show that the sum of the moments of all the external forces about every point in space vanishes.

Solution: Equation (1.6-6) expresses the condition that the sum of the moments about the point P vanishes. Let Q be an arbitrary point whose position vector is $\mathbf{a}$ relative to P. Then the vector $\mathbf{r}_I' = \mathbf{r}_I + \mathbf{a}$ is the relative position vector from Q to the particle I. The sum of moments about Q is

$$\sum_{I=1}^{K} \mathbf{r}_I' \times \mathbf{F}_I^{(e)} = \sum (\mathbf{r}_I + \mathbf{a}) \times \mathbf{F}_I^{(e)} = \sum \mathbf{r}_I \times \mathbf{F}_I^{(e)} + \sum \mathbf{a} \times \mathbf{F}_I^{(e)}.$$

The first term on the right-hand side vanishes according to Eq. (1.6-6). The second term vanishes according to Eq. (1.6-3) because

$$\sum \mathbf{a} \times \mathbf{F}_I^{(e)} = \mathbf{a} \times \sum \mathbf{F}_I^{(e)} = 0.$$

Hence $\sum \mathbf{r}_I' \times \mathbf{F}_I^{(e)} = 0$. But $\mathbf{a}$ is completely arbitrary. Hence the moment of all the external forces about the arbitrarily chosen point Q vanishes. Q.E.D.

PROBLEM 1.2 Equations (1.6-3) and (1.6-6) are necessary conditions of equilibrium. Are they sufficient? In other words, do they guarantee zero acceleration of every particle of the system? If your answer is yes, give a proof. If no, give a counter example.

Answer: No. Equations (1.6-3) and (1.6-6) do not guarantee the reduction of Eq. (1.5-7) to Eq. (1.6-1) In other words, we cannot conclude from (1.6-3) and (1.6-6) that $d(m_I v_I)/dt = 0$ for $I = 1, 2, \ldots, K$.

Any system capable of a free vibration without external force is a counter example, e.g., a wire in the piano After the wire is struck by the key, it is free from external forces, but it continues to vibrate (its particles have nonvanishing accelerations).

1.7 FREE-BODY DIAGRAM

The word *body* or the phrase *a system of particles* used in the previous section can be interpreted in the most general way. If a machine is in equilibrium, every part of it is in equilibrium. By a proper selection of the parts to be examined, a variety of information can be obtained. We shall illustrate such applications by a number of examples below.

The method we shall use may be likened to a surgeon's method of exploring a diseased organ. If he wants to know what is going on in the organ, he makes an incision and takes a bit of tissue to examine. We do the same to any material body, except that our "biopsy" is conceptual, or mathematical. Our experiments can be "thought" experiments. With imagined sections, we cut free certain parts of the body and examine their conditions of equilibrium. A diagram of the part with all the external forces acting on it clearly indicated is called a *free-body diagram*. The method we use is therefore called the *free-body method*.

Example 1. Analysis of a Truss

Trusses are frame structures commonly seen in bridges, buildings, lifts in construction sites, TV towers, radio astronomical antennas, etc. Figure 1.3(a) shows a typical truss of a small railway bridge. It is made of steel members $ab, bc, ac, \ldots$, bolted together. The joints at $a, b, c, \ldots$ may be considered as "pin-jointed," meaning that the members are joined together with pins and are free to rotate relative to each other. The whole truss is "simply supported" at the ends a and l, which anchor the truss but impose no moment on the truss. The support at l rests on a roller so that the horizontal reaction from the foundation is eliminated.

Railway trusses are made with slender members. The weights of these members are small compared with the load carried by the truss. Hence as a first approximation we may ignore the weight of the members.

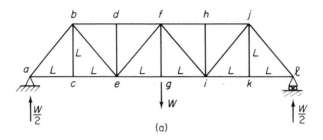

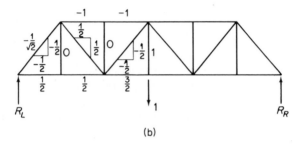

(b)

Fig. 1.3 A pin-jointed truss.

Since each member is pin-jointed and is considered weightless, the condition of equilibrium of the member requires that the pair of forces coming from the joints must be equal and opposite. Hence each member can only transmit forces along its axis.

Let the truss be loaded with a weight W at the center (point g). We would like to know the load acting in various members of the truss.

Let us first compute the reactions at the two supports. Consider the whole truss as a free body. It is subjected to three external forces, W, R_L, and R_R [Fig. 1.3(b)]. The conditions of equilibrium are

(1) Summation of vertical forces is zero:

$$W - R_L - R_R = 0.$$

(2) Summation of moments about the point a is zero:

$$W \cdot 3L - R_R \cdot 6L = 0.$$

The solutions are $R_R = R_L = W/2$.

Next, we wish to know the tension in the members ab and ac. For this purpose we cut through ab and ac with an imaginary plane and consider the portion $ab'c'$ as a free body [see Fig. 1.3(c)]. At the cut and exposed end b' the tension $\mathbf{F}_{ab}$ acts in the member ab. The tension $\mathbf{F}_{ac}$ acts at the cut c'

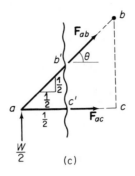

(c)

Fig. 1.3 (Cont.)

of ac. At the support a force $W/2$ acts (the reaction R_L computed above). Now summing all the forces in the vertical direction, and letting F denote the magnitude of **F**, we obtain:

$$\frac{W}{2} + F_{ab} \sin \theta = 0.$$

Since $\theta = 45°$, $\sin \theta = \sqrt{2}/2$, we obtain $F_{ab} = -W/\sqrt{2}$.
Summing all the forces in the horizontal direction yields

$$F_{ab} \cos \theta + F_{ac} = 0.$$

Hence, for $\theta = 45°$ and $F_{ab} = -W/\sqrt{2}$, we obtain $F_{ac} = W/2$.

Next we compute the tensions in the members df, ef, and eg. We pass a cut through these members and consider the left portion of the truss as a free body [see Fig. 1.3(d)]. For convenience we resolve the tension in the member ef into two components: the horizontal H_{ef} and the verticle V_{ef}.

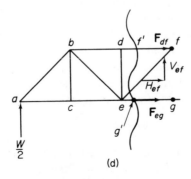

(d)

Fig. 1.3 (Cont.)

All the external forces acting on this free body are shown in Fig. 1.3(d). The equilibrium conditions are

(1) Summation of all horizontal forces vanish:

$$F_{df} + H_{ef} + F_{eg} = 0.$$

(2) Summation of all vertical forces vanish:

$$\frac{W}{2} + V_{ef} = 0.$$

(3) Sum of moments about the point e vanish:

$$\frac{W}{2} \cdot 2L + F_{df} \cdot L + 0 \cdot F_{ef} + 0 \cdot F_{eg} = 0.$$

(4) Sum of moments about the point f vanish:

$$\frac{W}{2} \cdot 3L - F_{eg} \cdot L = 0.$$

Hence, from condition 3 we obtain $F_{df} = -W$; from condition 4 we obtain $F_{eg} = 3W/2$; from condition 2 we obtain $V_{ef} = -W/2$; and, finally, from condition 1 we obtain $H_{ef} = -W/2$. A similar calculation can be done for other members of the truss.

The results can be presented as in Fig. 1.3(b). Since the load in every member is proportional to W, we may express the load in each member in units of W and set W equal to 1. For the truss design it is important to know whether a member is subjected to tension or compression (a rod pushed at both ends is said to subject to *compression*; a rod pulled at both ends is said to subject to *tension*). The design of a steel member in tension is different from that in compression. A tension member may fail by plastic yielding; a compression member may fail by elastic buckling. Whereas the signs of F_{ac}, F_{ab}, V_{ef}, H_{ef}, etc., in the equations above depend on the directions of the vectors we draw on the free-body diagram (which is done arbitrarily), the tension-compression character of the stress in each member is fixed by the load W. We present the final result in Fig. 1.3(b) with the *convention* that *if a member is in tension, we give the load a positive sign; if the member is in compression, we give the load a negative sign.* Thus, in Fig. 1.3(b) we see that the member ab is in compression; the members be, ac, and eg are in tension; and the member ef is in compression.

Example 2. A Simply Supported Beam

A beam is a solid member that resists lateral load by bending. Figure 1.4 illustrates a simply supported beam. Its function is similar to the truss discussed in Example 1. Whereas the truss resists the load by tension or com-

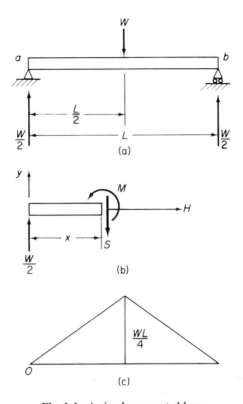

Fig. 1.4 A simply supported beam.

pression in the members, the beam resists it by continuously distributed tensile and compressive stresses.

The ends of the beam shown in Fig. 1.4 are supported on pins that do not resist moment. The reactions at the supports are obviously $W/2$.

Let us ask how the beam resists the external load. For this purpose let us make a cut with an imaginary plane perpendicular to the beam at a distance x from the left end [Fig. 1.4(b)]. Consider the free-body diagram of the left portion of the beam as shown in Fig. 1.4(b). At the cut surface there acts a "shear force" S tangential to the cut, an "axial force" H perpendicular to the cut, and a couple M, called the *bending moment* in the beam. The conditions of equilibrium are

(1) Sum of all forces in the horizontal direction vanish:

$$H = 0.$$

(2) Sum of all forces in the vertical direction vanish:

$$S = \frac{W}{2}.$$

(3) Sum of the moments of all forces about the left end support vanish:

$$M = Sx = \left(\frac{W}{2}\right)x.$$

Thus in a cross section at a distance x from the left end, the stresses in the beam are equipolent to a shear force $S = W/2$ and a moment $M = Wx/2$.

As x varies, the moment varies as shown in Fig. 1.4(c). Such a figure is called a *bending moment diagram* of a beam subjected to a specific loading. Knowing the bending moment, we can compute the stresses acting in a beam. See Sec. 7. 7 , 12.7, and 12.8. Beams are generally designed on the basis of the maximum bending moments they have to resist.

Example 3. Stresses in a Block

Consider a block of solid material compressed by a load W acting on its end as shown in Fig. 1.5. The block is a rectangular parallelopiped. We wish to know the forces of interaction among the material particles of this block. To shorten the verbal expression, we speak of "stresses" when we mean the "forces of interaction between particles." Thus our question is: What are the stresses in the block?

Let us assume that at a distance sufficiently far away from the ends the stresses are uniform in the block, i.e., everywhere the same. Let us erect a set of rectangular Cartesian coordinates x, y, z as shown in the figure with the z-axis parallel to the axis of the block. Let us pass an imaginary plane $z = 0$ to cut the block and consider the free-body diagram of the upper half of the block. The stresses acting on the surface $z = 0$ must have a resultant force and a resultant moment. Applying the conditions of equilibrium as before, we find at once that the horizontal component of the resultant force vanishes, that the vertical component of the resultant force is W, and that the resultant moment is zero. In this case, we say that the *stress* acting on a plane $z = 0$ is a *compressive*, *normal* stress with a magnitude

$$\sigma = -\frac{W}{A},$$

where A is the normal cross-sectional area of the block (cut by the plane $z = 0$ normal to the axis of the block). The stress is *compressive* because the material is compressed in that direction. It is *normal* because σ is a force (per unit area) perpendicular (normal) to the surface $z = 0$. We indicate the compressive nature of the stress by giving it a negative value.

Next, let us make a cut with a plane that is inclined at an angle θ to the xy-plane. The simplest way to express the orientation of a plane is to specify the normal vector of the plane. Let $\mathbf{v}$ be the *unit* normal vector (of unit length) of the inclined plane; then $\mathbf{v} \cdot \mathbf{z} = \cos(\mathbf{v}, \mathbf{z}) = \cos\theta$. Consider the

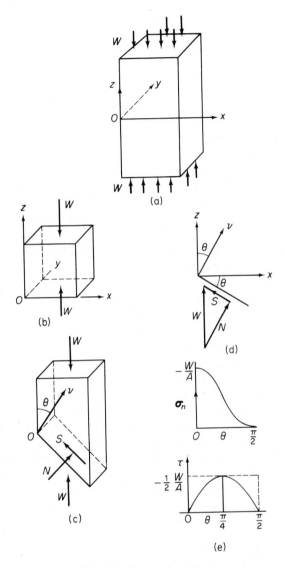

Fig. 1.5 Stresses in a block.

upper half of the block as a free body as shown in Fig. 1.5(c). The balance of forces requires that the resultant force acting on the plane **v** (a plane whose unit normal vector is **v**) is exactly equal to $-W$. This resultant can be resolved into two components, one normal and one tangential to the surface, as shown in Fig. 1.5(c). Let these be N and S, respectively; then (See Fig. 1.5(d)):

$$N = -W \cos \theta, \qquad S = -W \sin \theta.$$

The cross-sectional area of the block cut by the plane $\mathbf{v}$ is $A/\cos\theta$, where A is the normal cross-sectional area. Dividing N and S by the area of the surface on which they act and denoting the results by σ_n and τ, we obtain

$$\sigma_n = -\cos^2\theta\, W/A, \qquad \tau = -\sin\theta\cos\theta\, W/A.$$

These are the *normal stress* and *shear stress*, respectively, acting on the inclined surface $\mathbf{v}$. We give the normal stress σ_n a negative value to indicate that it is a *compressive* stress. If the load W is reversed so that the block is *pulled*, then the material on the two sides of the plane $\mathbf{v}$ tends to be pulled apart. We say in that case that the stress is in *tension* and indicate that fact by assigning σ_n a positive numerical value.

The sign convention of the shear stress will be discussed in Sec. 3.2.

The normal and shear stresses σ_n and τ vary with the angle θ. If we plot them as a function of θ, we obtain curves as shown in Fig. 1.5(e). We see that σ_n is a maximum when $\theta = 0$, whereas the shear τ reaches a maximum when $\theta = 45°$, and that the maximum shear is $\tau_{\max} = \frac{1}{2}W/A$.

The principal lesson that we learn from this example is that there are two components of stress, normal and shear, whose values at any given point in a body depend on the direction of the surface on which the stress acts. Thus stress is a vector (σ_n, τ) associated with another vector $(\mathbf{v})$. To specify a stress we have to specify two vectors. To specify fully the *state of stress at a given point in a continuum* we must know the stresses acting on all possible planes $\mathbf{v}$ (i.e., sections oriented in all possible directions). A quantity such as the state of stress is called a *tensor*. Thus this example tells us that the *stress is a tensorial quantity*.

In the International System of Units (SI Units), the basic unit of force is the *newton* (n) and that of length is the *meter* (m). Thus the basic unit of stress is *newtons per square meter* (n/m²) or *pascals* (Pa, in honor of Pascal). 1 MPa = 1 n/mm². A force of 1 n can accelerate a body of mass 1 kg to 1 m/sec². (A newton is the approximate weight of a small apple.) A force of 1 dyne can accelerate a body of mass 1 gram to 1 cm/sec². Hence, 1 dyne = 10^{-5} newton. Some of the conversion factors are listed below:

> 1 pound force $\doteq$ 4.448 newtons
>
> 1 pound mass avoirdupois $\doteq$ 0.4535 kg
>
> 1 pound per square inch (psi) $\doteq$ 6894 n/m² $\sim$ 7 kPa
>
> 1 dyne/cm² = 0.1 n/m²
>
> 1 atmosphere $\doteq$ 1.013 × 10⁵ n/m² = 1.013 bar
>
> 1 mm Hg at 0°C $\doteq$ 133 n/m² = 1 torr $\sim$ $\frac{1}{8}$ kPa
>
> 1 cm H₂O at 4°C $\doteq$ 98 n/m², 1 inch H₂O $\sim$ $\frac{1}{4}$ kPa
>
> 1 poise (viscosity) = 0.1 newton sec/m².

The notion of stress has practical value. If you have large blocks and small blocks of the same material, obviously the large ones can take larger loads, the small ones can take smaller loads; but both will break at the same critical state of stress. Hence engineers look at stresses.

Example 4. Stresses in a Plate

Consider a thin rectangular plate of uniform thickness and homogeneous material. As shown in Fig. 1.6(a) and (b), the plate is subjected to a uniformly distributed load acting on the surfaces $x = \pm a$ and $y = \pm b$, and no load on the surface $z = \pm h/2$. In Fig. 1.6(b) it is shown that the stress acting on the

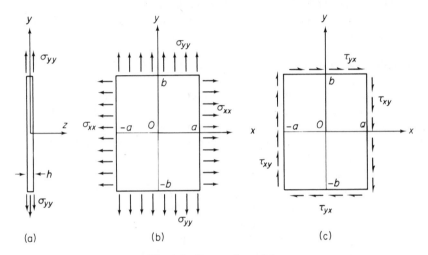

Fig. 1.6 Stresses in a plate.

edge $x = a$ is of a magnitude σ_{xx} per unit area (σ_{xx} is equal to the total load on the edge $x = a$ divided by the cross-sectional area of the plate cut by the plane $x = a$). The stress acting on the edge $y = b$ is of a magnitude σ_{yy} per unit area. In Fig. 1.6(c), it is shown that the plate is subjected to a shear stress τ_{xy} on the edge $x = a$ (τ_{xy} is equal to the total shear load acting on the edge $x = a$ in the direction of the y-axis divided by the cross-sectional area of the section $x = a$) and a shear stress τ_{yx} on the edge $y = b$. σ_{xx}, σ_{yy}, τ_{xy}, τ_{yx} are called stresses because they are all in units of force per unit area.

Applying the equations of equilibrium to the plate shown in Fig. 1.6(b), we see that the stress σ_{xx} acting on the edge $x = -a$ is equal to that acting on the edge $x = a$. Applying the equations of equilibrium to the plate shown in Fig. 1.6(c), we see that τ_{xy} on $x = -a$ is equal to τ_{xy} on $x = a$, that τ_{yx} on $y = b$ and $y = -b$ are also equal, and further that by taking the moment of all forces (stresses × cross-sectional area) about the origin 0 we obtain

$$2a \cdot \tau_{xy} \cdot 2bh - 2b \cdot \tau_{yx} \cdot 2ah = 0, \quad \text{or} \quad \tau_{xy} = \tau_{yx}.$$

The state of stress in the plate shown in Fig. 1.6(b) is specified by σ_{xx} and σ_{yy}. The state of stress in the plate shown in Fig. 1.6(c) is specified by $\tau_{xy} = \tau_{yx}$. If a plate is subjected to both the *normal stresses* σ_{xx}, σ_{yy} and the *shear stresses* τ_{xy}, τ_{yx} [a superposition of the condition shown in Fig. 1.6(b) and (c)], then the state of stress is specified by the four numbers $\sigma_{xx}, \sigma_{yy}, \tau_{xy}, \tau_{yx}$ ($\tau_{yx} = \tau_{xy}$). To clarify this double subscript notation, we specify the rule that *the first index of the stress denotes the plane on which the stress acts, whereas the second index denotes the direction in which the force acts.* Thus the tensorial character of stress mentioned at the end of Example 3 becomes even clearer in the present example.

Example 5. A Pressurized Spherical Shell (Fig. 1.7)

The wall of an inflated balloon is in tension. We would like to know the tensile stress in the wall. For this purpose it is simplest to cut the sphere with a diametrical plane and consider the hemisphere as a free body as shown in Fig. 1.7(b). Let the inner radius of the shell be r_i, the outer radius be r_o, and

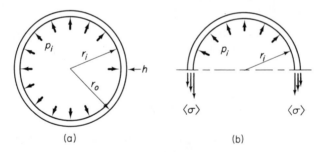

(a)   (b)

Fig. 1.7   A pressurized spherical shell.

the thickness of the wall be $h = r_o - r_i$. The internal pressure p_i acts on the inner wall. The resultant pressure force acting on a hemisphere is $\pi r_i^2 p_i$. The normal stress in the wall of the shell is not uniform; the calculation of this must await a general formulation (Chapter 10, et seq.) but it is easy to calculate the average tensile stress in the wall. Let $\langle \sigma \rangle$ be the average normal stress acting on a surface normal to the wall (i.e., passing through the center of the sphere). The area of the wall on the diametrical plane is $\pi r_o^2 - \pi r_i^2$. The resultant tensile force due to wall stress is $\pi(r_o^2 - r_i^2)\langle \sigma \rangle$. The balance of the forces in equilibrium requires, therefore,

$$\pi(r_o^2 - r_i^2)\langle \sigma \rangle = \pi r_i^2 p_i$$

or

$$\langle\sigma\rangle = p_i \frac{r_i^2}{r_o^2 - r_i^2} = \frac{r_i^2 p_i}{h(r_o + r_i)}.$$

This is a useful formula that is valid for thick-walled, as well as thin-walled, spherical shells.

If a pressure p_o acts on the outside of the shell, (Fig. 1.8), the resulting normal stress in the wall will be

$$\langle\sigma\rangle = -\frac{r_o^2 p_o}{h(r_o + r_i)}.$$

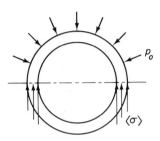

Fig. 1.8 A spherical shell subjected to external pressure.

If the walls of the sphere are very thin, then $r_o - r_i = h$, but $r_o \cong r_i = r$, and the equations above are reduced to

$$\langle\sigma\rangle = \frac{rp_i}{2h} - \frac{rp_o}{2h}.$$

Example 6. Pressurized Circular Cylindrical Tanks

Consider a cylindrical shell subjected to an internal pressure p_i as shown in Fig. 1.9(a). Let us pass two planes perpendicular to the axis of the cylinder to cut the shell into a ring, pass another plane through the axis of the cylinder to cut the ring into two halves, and isolate the semicircular ring as a free body as shown in Fig. 1.9(b). The stress acting on the radial cut CD is normal to the surface and is directed in the direction of increasing polar angle θ in a polar coordinates; hence it will be denoted by σ_θ. As in Example 5, we do not know the exact distribution of σ_θ in the cross section, but if $\langle\sigma_\theta\rangle$ denotes the average value of σ_θ over the cross section, then $\langle\sigma_\theta\rangle$ multiplied by the area $(r_o - r_i)L$ is the resultant force acting in the cross section CD. Similarly, the tensile force in the section EF [Fig. 1.9(b)] is also $\langle\sigma_\theta\rangle(r_o - r_i)L$. The resultant of pressure acting on the inside is $2r_i Lp_i$. The balance of forces

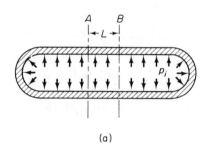

(a)

(b)   (c)

Fig. 1.9 A pressurized cylindrical tank.

acting on the semicircular ring in the vertical direction requires that

$$2\langle\sigma_\theta\rangle(r_o - r_i)L = 2r_iLp_i.$$

Hence

$$\langle\sigma_\theta\rangle = \frac{r_ip_i}{r_o - r_i},$$

which is another very useful exact formula.

   If we make a cut of the cylinder by a plane perpendicular to the cylinder axis and consider the left half of the tank as a free body as shown in Fig. 1.9(c), we can examine the average value of the axial stress σ_x that acts in the axial direction x on a cross section perpendicular to x. We note that the area on which σ_x acts is $\pi(r_o^2 - r_i^2)$. On the other hand, the surface on which the internal pressure p_i acts has a projected area in the axial direction equal to πr_i^2. Hence the balance of forces in the axial direction yields

$$\pi r_i^2 p_i = \langle\sigma_x\rangle\pi(r_o^2 - r_i^2)$$

or

$$\langle\sigma_x\rangle = \frac{r_i^2 p_i}{r_o^2 - r_i^2}.$$

If the shell wall is very thin, so that $r_o - r_i = h$, $r_o \cong r_i = r$, then these equations are simplified to

$$\langle \sigma_\theta \rangle = \frac{rp_i}{h}, \qquad \langle \sigma_x \rangle = \frac{rp_i}{2h}.$$

Example 7. An Airplane

Figure 1.10(a) illustrates an airplane in flight. Consider the airplane as a free body: It is subjected to aerodynamic pressure, skin friction, the thrust of the engine, and gravitation. The motion of the airplane must be determined by the balance of the lift, drag, thrust, weight, and control forces. Following the methods of general physics, we can write the equations of motion of the center-of-mass of the plane, and the equations of rotation about the center-of-mass. Through these equations we can study airplane dynamics.

Figure 1.10(b) shows a free-body diagram for the wing, with the aerodynamic forces acting on it indicated. The left boundary is an imagined surface introduced to construct the free-body diagram. The resultant shear force S and the resultant bending moment M must balance the aerodynamic and inertia forces. Solution of the equations of equilibrium yields the desired information about the bending moment and transverse shear distributions in the wing.

Figure 1.10(c) shows the free-body diagram for a segment of the wing structure. The forces acting on the cross-sectional surfaces are shown. Analysis of such a segment leads to a differential equation relating M, S, and the load.

Figure 1.10(d) shows the free-body diagram for a small element of the wing's skin. The conditions of equilibrium for such an element yield the differential equation governing the stress distribution in the sheet-metal skin.

Figure 1.10(e) shows the free-body diagram for a volume of air surrounding the airplane. The upper and lower surfaces are stream surfaces comprised of streamlines. The velocity profiles at the front and rear sections are indicated. If the diagram represents the condition in a wind tunnel, the velocity profile can be measured and the momentum flow into the domain calculated. In a steady flow, the net momentum flow per unit time equals the sum of forces acting on the boundaries, and the measurement of velocity profile can be used as a means to measure aerodynamic drag.

The usefulness of free-body diagrams is evident!

1.8 AD HOC THEORY VS. GENERAL THEORY

The analysis presented above is typical of ad hoc theories. Special cases are investigated one by one. Simplifying assumptions are introduced whenever expedient, and generalizations are made whenever practical. Such an

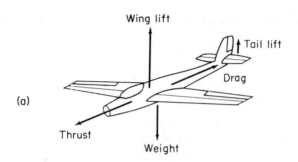

(a)

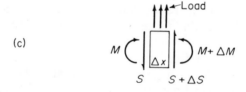

(b)

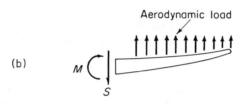

(c)

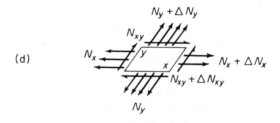

(d)

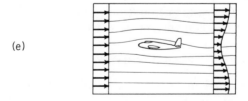

(e)

Fig. 1.10 Free-body diagrams of an airplane. (old 3.5)

approach is attractive to most scientists and engineers; most of our scientific knowledge was accumulated using the ad hoc approach. It is probably also the simplest and clearest way various subjects can be taught and learned.

However, a quicker way to describe a subject is to use a *general* theory. With suitable mathematical preparation, a general theory can lay out the entire subject and list the assumptions, the complete set of equations, and the permissible boundary conditions. Specialization to various simpler cases can be introduced in an orderly fashion, and different ad hoc theories can be put in clear perspective. These are the important merits of the general approach.

The objective of this book is to introduce the reader to the formulation of physical problems in continuum mechanics. In order to be able to deal with a wide variety of problems that occur in the physical world, we must provide a reasonably broad treatment of the basic concepts of stress; strain; constitutive equations; equations of conservation of mass, momentum, and energy; and various possible boundary conditions. Hence our approach tends to be the general one. However, we hope that numerous examples and exercises will keep us in close contact with the many interesting special problems in mechanics.

PROBLEMS

1.3 Figure 9.3 on p. 206 shows a photomicrograph of a stainless steel specimen enlarged 560 times. The specimen was polished and etched. The picture shows that it is very nonuniform. However, if we use the stainless steel in designing a rocket case, it is a common practice to treat the steel as a homogeneous material. How can this be justified?

1.4 The water molecule is made of one oxygen atom and two hydrogen atoms. The oxygen atom is larger than the hydrogen atom. The distance between the center of the oxygen atom to that of the hydrogen atoms is 0.957 Å (1 Å = 10^{-8} cm). The lines joining the center of the oxygen atom to those of the hydrogen atoms form an angle of 105.3°.

How large must the minimum size of a body of water be if its motion is to be treated by the method of continuum mechanics?

1.5 For a gas, the length of mean free path between collisions is a measure of the average distance between molecules. For air at 1 km above ground, the mean free path is 8×10^{-6} cm; at 100 km it is 9.5 cm; at 200 km it is 3×10^4 cm. To analyze the flow of air around a spacecraft reentering the atmosphere, would it be permissible to analyze the air flow by the method of continuum mechanics?

1.6 Look at the skin on your hands. The stress and strain in it may be very important to a plastic surgeon. At what kind of typical dimension (1 cm?, 1 mm?, . . .) would you stop considering the skin as homogeneous? At what dimension would you refuse to think of it as a continuum? Inhomogeneity has many levels: macroscopic, microscopic to the extent of being visible in an ordinary light microscope, or to the limit of an electron microscope. Can you think of various phenomena concerning the skin in which continuum mechanics is useful?

1.7 Find the loads in the rods a, b, c of the simple pin-jointed truss loaded by W as shown in Fig P1.7(a).

Note: Answers to problems like this are best given graphically as shown in Fig. P1.7(b). If all forces are expressed in units of W, we may take $W = 1$.

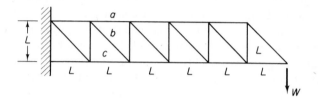

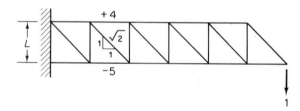

Fig. P1.7   A pin-jointed truss.

1.8 Compare three designs of a truss to hoist a heavy load W, as shown in Fig. P1.8(a).

The slopes of the roofs are shown in the figure. The lengths of the base are one meter in every case. The cross-sectional area of each member should be made proportional to the load (tension or compression acting in that member). The same material will be used for all members. Which one of the three designs is the best from the point of view of economy on the total amount of material of the truss?

Note that the truss in the middle is drawn to a larger scale for clarity.

Answer: The loads in the members of the three trusses are shown in Fig. P1.8(b). Since the same material is used and since the cross-sectional area is proportional to the load, the volume of each member is proportional to the product of load and length. This is shown in Fig. P1.8(c). The total volume of material in the three cases are, respectively, 4, 5, 5. Hence the first design is the most economical.

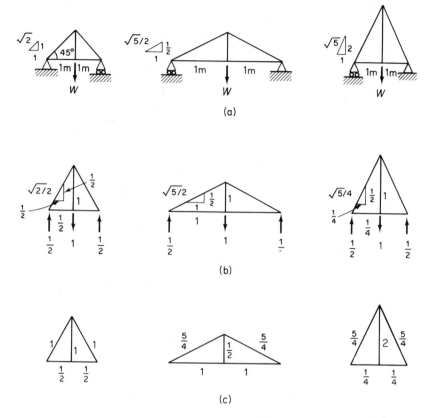

Fig. P1.8 Three truss designs.

1.9 A simple pin-jointed truss is loaded by a force W as shown in Fig. P1.9(a). Find the load in every member. Then another rod AB is added as shown in Fig. P1.9(b). The members AB, CD are not connected where they cross. Can you find the loads in all the members? Why can't you?

Note: The truss in Fig. P1.9(b) is called a *statically indeterminate* structure.

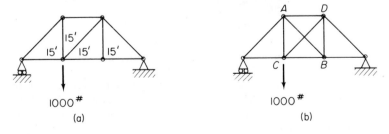

Fig. P1.9 A statically determinate truss and a statically indeterminate truss.

1.10 Consider a heavy-weight lift used on a building construction site as shown in Fig. P1.10(a). It is a pin-jointed truss. The left end is enlarged in Fig. P1.10(b) and two members *AB, AC* are added. They do not make any difference to the stresses in the rest of the truss. Why?

Determine the loads (tension or compression) in the members *BC, CD, CE, DE,* and *DF.*

Determine the loads in members *PQ, PR, RQ,* and *RS.*

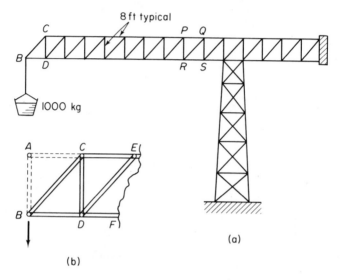

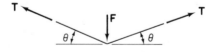

Fig. P1.10   A pin-jointed truss used as a lift.

1.11 It is pretty easy to demonstrate that the tension in an Achilles tendon is considerable when we stand on tiptoe or when we poise for a jump. A tension gauge can be built using the same principle as we would have used to measure the tension in the string of a bow, or a rubber slingshot. Design such a gauge.

Hint: If we pull on a bow as shown in Fig. P1.11, and if a lateral force *F* induces a deflection angle θ, show that $T = F/(2 \sin \theta)$.

Fig. P1.11   Measurement of tension in a string.

1.12 A man worked hard shoveling dirt in his garden. If the dirt and shovel weighs 10 kg and has a center of gravity located at a distance of 1 m from the lumbar vertebra of his backbone, what is the moment about that vertebra?

The construction of a man's backbone is sketched in Fig. P1.12(b). The disks between the vertebrae serve as the pivots of rotation. It may be assumed

Fig. P1.12   Loads in the spine when a man shoves weight.

that the intervertebral disks cannot resist moments. Hence the weight of the dirt and shovel has to be resisted by the vertebral column and the muscle. Estimate the loads in his back muscle, vertebrae, and disks.

Lower back pain is such a common affliction that the loads acting on the disks of patients were measured with strain gauges in some cases. It was found that no agreement can be obtained if we do not take into account the fact that when one lifts a heavy weight, one tenses up abdominal muscle so that the pressure in the abdomen is increased. A free-body diagram of the upper body of a man is shown in Fig. P1.12(c). Show that it helps to have a large abdomen and strong abdominal muscles.

1.13 Compare the bending moment acting on the spinal column at the level of a lumbar vertebra for the following cases:

(a) A secretary bends down to pick up a book on the floor (i) with her knees straight and (ii) with her knees bent.

(b) A water skier skiis (i) with his arms straight and (ii) with elbows hugging to his waist.

Discuss these cases quantitatively with proper free-body diagrams.

1.14 A small balloon is inflated by air at an internal pressure that is p pascals larger than the external pressure, which is atmospheric. I wish to have a feel about what that p is by pressing my finger on the balloon. How far down do I need to press? I'd assume that the bending resistance of the wall is negligible.

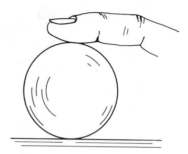

Fig. P1.14   Feeling the pressure in a balloon.

Hint: Consider the free-body diagram of a small piece of balloon under the finger. Consider the condition when the spot under the finger becomes flattened into a plane surface.

1.15 One man is twice as tall as another man. Assume that they are completely similar and doing exactly the same stationary gymnastic maneuver. Are they subjected to the same stresses in their bones and muscles?

Answer: The ratio of the linear dimensions is 2. The ratio of the mass of the corresponding organs is 8. The ratio of the corresponding areas is 4. The ratio of stresses is 2.

1.16 Gas pressure in a soap bubble is related to the surface tension σ and the radius R by the equation $P = 4\sigma/R$. Derive this equation.

Take a pippette, put a valve in the middle, close it, and blow two bubbles, one at each end: One bubble is large and one is small. Now open the middle valve so that the gas in the two bubbles can move into each other. In which way will the bubble diameters change? Explain in detail.

Answer: The small one will disappear.

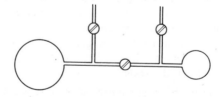

Fig. P1.16   Collapsing of a small soap bubble into a large one.

1.17 Consider two beams of equal length but supported differently as shown in Fig. P1.17. Determine the bending moment distribution in the beams.

1.18 The beams in Prob. 1.17 are loaded by their own weight, which is w lb per unit length (Fig. P1.18). Determine the bending moments at the middle points of the beams. Which one is larger?

1.19 Find the moment distribution in the beams shown in Fig. P1.19.

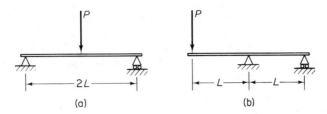

Fig. P1.17 Two beams carrying the same load but supported differently.

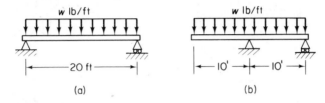

Fig. P1.18 Two beams carrying their own weight.

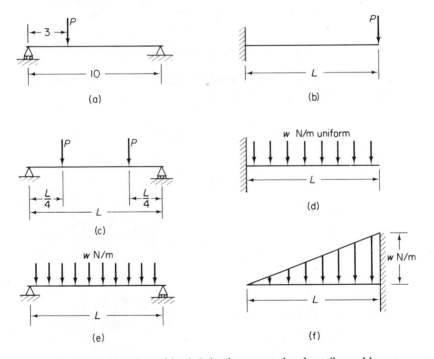

Fig. P1.19 Several loaded simply supported and cantilevered beams.

1.20 Two beams of the same uniform cross section and equal span are designed to support a uniformly distributed load. One of the beams is simply supported at the ends. The other is cantilevered at the center. See Fig. P1.20(a). The beams are to be designed with respect to the bending moment. Work out the moment distribution in these beams. Plot the moment diagrams. Which one of the beams is the stronger?

Answer: The moment diagrams are shown in Fig. P1.20(b). The two beams are equally strong.

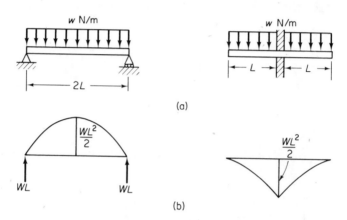

Fig. P1.20   Comparison of two designs.

1.21 Let M denote bending moment in a beam; S, the shear; and w, the load. Show that, according to the free-body diagram in Fig. P1.21,

$$S = \frac{dM}{dx}, \qquad w = \frac{dS}{dx};$$

hence

$$\frac{d^2M}{dx^2} = w.$$

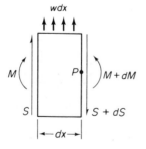

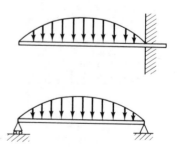

Fig. P1.21   Equilibrium of a beam segment.

Fig. P1.22   Bending of beams under a sinusoidally distributed load.

1.22 Using the differential equation derived in Prob. 1.21, find the bending moment distribution in the beams shown in Fig. P1.22(a) and (b) under the loading w per unit length:

$$w = a \sin \frac{\pi x}{L}.$$

1.23 A person weighing W pounds tried to walk over a plank that was simply supported at two ends across a river. The plank will break whenever the bending moment exceeds M_{cr}. At what place (x) would the plank break and the person fall into the river?

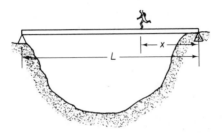

Fig. P1.23   A man walking over a plank.

Answer: $x = \frac{1}{2}[L \pm (L^2 - 4K)^{1/2}]$, where $K = LM_{cr}/W$.

1.24 A vehicle moves on a simply supported girder bridge (a beam) at a velocity v. The bridge has a dip that has a profile of a cosine curve, $y = -a[1 + \cos(2\pi x/L)]$, as shown in Fig. P1.24 for x in $-L/2$ and $L/2$, and 0 elsewhere. Assume that the vehicle is very short compared with L (unlike the drawing). Compute:

    (a) The maximum vertical acceleration of the vehicle.

    (b) The centrifugal force experienced by the passenger in the car, i.e., the load acting on his seat.

    (c) The maximum bending moment in the bridge due to the moving vehicle.

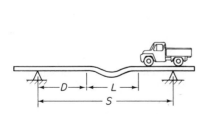

Fig. P1.24   A vehicle moving over a simply supported bridge.

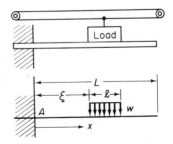

Fig. P1.25   A moving load over a cantilever beam.

1.25 Consider a cantilever beam with a moving load acting on it (Fig. P1.25). The load extends over a length l. Determine the bending moment in the beam as a function of x, and, in particular, at the clamped edge A as a function of the location of the load. The load is w per unit length.

Solution: It is algebraically messy to write different expressions for different ranges of x. Hence we use a trick as follows:

$$\frac{d^2M}{dx^2} = w[\mathbf{1}(x - \xi) - \mathbf{1}(x - \xi - l)] \tag{1}$$

when $\mathbf{1}(x)$ is a unit step function, which is defined as 1 when the argument is greater than zero and zero when the argument is negative. On integrating Eq. (1) and introducing the integration constants c_1, c_2, c_3, we obtain

$$\frac{dM}{dx} = (xw + c_1)[\mathbf{1}(x - \xi) - \mathbf{1}(x - \xi - l)] + c_2[\mathbf{1}(x) - \mathbf{1}(x - \xi)]$$
$$+ c_3[\mathbf{1}(x - \xi - l) - \mathbf{1}(x - L)]. \tag{2}$$

The physical meaning of dM/dx is the shear in the beam and it is a continuous function. When x is slightly greater than ξ, the first term gives $dM/dx = w\xi + c_1$, whereas the second and third terms are zero. When x is slightly less than ξ, only the second term is nonvanishing and gives $dM/dx = +c_2$. The continuity thus requires $c_2 = w\xi + c_1$. Similarly, $c_3 = (\xi + l)w + c_1$. Finally, the boundary condition that the shear is zero at the free end $x = L$ gives us the relation $c_3 = 0$. Hence we obtain

$$c_1 = -(\xi + l)w, \qquad c_2 = -lw, \qquad c_3 = 0. \tag{3}$$

Substituting these into Eq. (2) and integrating once more, we obtain

$$M = \left(w\frac{x^2}{2} + c_1x + c_4\right)[\mathbf{1}(x - \xi) - \mathbf{1}(x - \xi - l)]$$
$$+ (c_2x + c_5)[\mathbf{1}(x) - \mathbf{1}(x - \xi)] + c_6[\mathbf{1}(x - \xi - l) - \mathbf{1}(x - L)]. \tag{4}$$

Again, M is a continuous function of x. By the same method as above, we obtain

$$\tfrac{1}{2}w\xi^2 + c_1\xi + c_4 = c_2\xi + c_5,$$
$$\tfrac{1}{2}w(\xi + l)^2 + c_1(\xi + l) + c_4 = +c_6. \tag{5}$$

Again, the boundary condition that the bending moment is zero at the free end $x = L$ implies that $c_6 = 0$. Hence by Eqs. (5) and (3),

$$c_4 = -\tfrac{1}{2}w(\xi + l)^2 + (\xi + l)^2w = \tfrac{1}{2}w(\xi + l)^2. \tag{6}$$

Thus, by solving for c_4, c_5, c_6 and substituting into Eq. (4), we obtain, finally, the bending moment distribution in the beam. In particular, the bending moment at the clamped end, point A, where $x = 0$, is

$$M(0) = c_5 = \frac{w\xi^2}{2} + c_1\xi + c_4 - c_2\xi = wl\left(\xi + \frac{l}{2}\right). \qquad (7)$$

Note: The last result, Eq. (7), could have been obtained very easily by treating the distributed load by its static equivalent, a resultant force wl acting through the centroid of the distributed load.

1.26 One of the most beautiful results in the aerodynamic theory says that the best design for minimum induced drag of an airplane (the air resistance to forward motion due to lifting the weight) is one with elliptic loading. By loading is meant the aerodynamic lift force per unit span. By elliptic loading is meant that the lift distribution from wing tip to wing tip is an ellipse. Let x be the distance along the wing span measured from the centerline of the airplane. Let b be the semi-span of the wing (distance from the center line to the wing tip.) Then the theorem says that the minimum induced drag is achieved if the lift is distributed according to the formula

$$\ell(x) = k\sqrt{1 - \frac{x^2}{b^2}}.$$

where k is a constant. This lift distribution yields the best fuel economy.

Assume that an airplane has an elliptic loading. Consider the wing as a cantilever beam. Compute the bending moment in the wing, $M(x)$, at x. Plot the bending moment diagram to show the moment at every station in the wing due to aerodynamic load.

CHAPTER TWO

Vectors and Tensors

The examples on pp. 16–20 show that the stress acting on a surface passing through a point in a body depends on the orientation of that surface. Thus stress is a vector that depends on another vector. Quantities like stresses are called tensors, or more precisely, tensors of rank 2. The analysis of stresses in a body requires analysis of tensors. Later we shall show that strains are also tensors and that the derivatives of tensors are tensors of higher ranks. Thus we shall be very much involved with tensors.

The study of mechanics will be simplified if we familiarize ourselves with the notations, algebra, and calculus of tensors. In this chapter the definitions and operations of vectors and tensors are given. It is simple, but it is necessary that the reader be thoroughly familiar with it.

2.1 VECTORS

We assume that you are already familiar with the elements of vector analysis, so only a brief review is given here.

For our purposes a vector in a Euclidean space is best defined as a directed line segment with a given magnitude and a given direction. We shall denote vectors by $\overrightarrow{AB}$, $\overrightarrow{PQ}$, . . . , or by boldface letters **u, v, F, T,** . . . .

Two vectors are *equal* if they have the same direction and same magnitude. A *unit vector* is a vector of magnitude one. The *zero vector*, denoted by **0,** is a vector of zero magnitude. We use the symbols $|\overrightarrow{AB}|$ $|\mathbf{u}|$, or v to represent the magnitudes of $\overrightarrow{AB}$, $\mathbf{u}$, or $\mathbf{v}$ respectively.

The sum of two vectors is another vector obtained by the "parallelogram law," and we write, for example, $\overrightarrow{AB} + \overrightarrow{BC} = \overrightarrow{AC}$. Vector addition is commutative and associative.

A vector multiplied by a number yields another vector. If k is a positive real number, $k\mathbf{a}$ represents a vector having the same direction as $\mathbf{a}$ and a magnitude k times as large. If k is negative, $k\mathbf{a}$ is a vector whose magnitude is $|k|$ times as large and whose direction is opposite to $\mathbf{a}$. If $k = 0$, we have $0 \cdot \mathbf{a} = \mathbf{0}$.

The subtraction of vectors can be defined by

$$\mathbf{a} - \mathbf{b} = \mathbf{a} + (-\mathbf{b}).$$

If we let $\mathbf{i}, \mathbf{j}, \mathbf{k}$ be the unit vectors in the directions of positive x, y, z-axes, respectively, one can show that every vector in space may be represented by a linear combination of $\mathbf{i}, \mathbf{j}$, and $\mathbf{k}$. Furthermore, if the end points of a given vector $\mathbf{u}$ have the coordinates (x_1, y_1, z_1) and (x_2, y_2, z_2), we can write

(2.1-1) $\mathbf{u} = (x_2 - x_1)\mathbf{i} + (y_2 - y_1)\mathbf{j} + (z_2 - z_1)\mathbf{k}$
            $= u_x\mathbf{i} + u_y\mathbf{j} + u_z\mathbf{k},$

where u_x, u_y, u_z are called the *components* of $\mathbf{u}$ in the directions of $\mathbf{i}, \mathbf{j}, \mathbf{k}$, respectively.

The magnitude $|\mathbf{u}|$ is then given by

(2.1-2) $$|\mathbf{u}| = \sqrt{u_x^2 + u_y^2 + u_z^2},$$

and therefore $\mathbf{u} = \mathbf{0}$ if and only if $u_x = u_y = u_z = 0$.

The *scalar* (or *dot*) *product* of $\mathbf{u}$ and $\mathbf{v}$, denoted by $\mathbf{u} \cdot \mathbf{v}$, is defined by the formula

(2.1-3) $$\mathbf{u} \cdot \mathbf{v} = |\mathbf{u}||\mathbf{v}| \cos \theta, \qquad (0 \leq \theta \leq \pi)$$

where θ is the angle between the given vectors. This represents the product of the magnitude of one vector and the component of the second vector in the direction of the first; that is,

(2.1-4) $\mathbf{u} \cdot \mathbf{v} = $ (magnitude of $\mathbf{u}$)(component of $\mathbf{v}$ along $\mathbf{u}$).

If

$$\mathbf{u} = u_x\mathbf{i} + u_y\mathbf{j} + u_z\mathbf{k}, \qquad \mathbf{v} = v_x\mathbf{i} + v_y\mathbf{j} + v_z\mathbf{k},$$

the scalar product of these two vectors can also be expressed in terms of the components:

$$\text{(2.1-5)} \qquad \mathbf{u \cdot v} = u_x v_x + u_y v_y + u_z v_z.$$

Whereas the scalar product of two vectors is a scalar quantity, the *vector* (or *cross*) *product* of two vectors $\mathbf{u}$ and $\mathbf{v}$ produces another vector $\mathbf{w}$; and we write $\mathbf{w} = \mathbf{u} \times \mathbf{v}$. The magnitude of $\mathbf{w}$ is defined as

$$\text{(2.1-6)} \qquad |\mathbf{w}| = |\mathbf{u}||\mathbf{v}| \sin \theta, \qquad (0 \le \theta \le \pi)$$

where θ is the angle between $\mathbf{u}$ and $\mathbf{v}$, and the direction of $\mathbf{w}$ is defined as perpendicular to the plane determined by $\mathbf{u}$ and $\mathbf{v}$, in such a way that $\mathbf{u}, \mathbf{v}, \mathbf{w}$ form a right-handed system. The vector products satisfy the following relations:

$$\text{(2.1-7)} \qquad \begin{aligned} &\mathbf{u \times v} = -(\mathbf{v \times u}) \\ &\mathbf{u \times (v + w)} = \mathbf{u \times v} + \mathbf{u \times w} \\ &\mathbf{u \times u} = 0 \\ &\mathbf{i \times i} = \mathbf{j \times j} = \mathbf{k \times k} = 0 \\ &\mathbf{i \times j} = \mathbf{k}, \qquad \mathbf{j \times k} = \mathbf{i}, \qquad \mathbf{k \times i} = \mathbf{j} \\ &k\mathbf{u \times v} = \mathbf{u} \times k\mathbf{v} = k(\mathbf{u \times v}). \end{aligned}$$

Using these relations, the vector product can be expressed in terms of the components as follows:

$$\text{(2.1-8)} \quad \mathbf{u \times v} = (u_y v_z - u_z v_y)\mathbf{i} + (u_z v_x - u_x v_z)\mathbf{j} + (u_x v_y - u_y v_x)\mathbf{k},$$

or, symbolically,

$$\text{(2.1-9)} \qquad \mathbf{u \times v} = \begin{vmatrix} \mathbf{i} & \mathbf{j} & \mathbf{k} \\ u_x & u_y & u_z \\ v_x & v_y & v_z \end{vmatrix}.$$

PROBLEMS

2.1 Given vector $\mathbf{u} = -3\mathbf{i} + 4\mathbf{j} + 5\mathbf{k}$, find a unit vector in the direction of $\mathbf{u}$.

Answer: $(\sqrt{2}/10)\mathbf{u}$.

2.2 If $\overrightarrow{AB} = -2\mathbf{i} + 3\mathbf{j}$ and the midpoint of the segment $\overrightarrow{AB}$ has coordinates $(-4, 2)$, find the coordinates of A and B.

Answer: $(-3, \frac{1}{2}), (-5, \frac{7}{2})$.

2.3 Prove, for any two vectors $\mathbf{u}$, $\mathbf{v}$, $|\mathbf{u} - \mathbf{v}|^2 + |\mathbf{u} + \mathbf{v}|^2 = 2(|\mathbf{u}|^2 + |\mathbf{v}|^2)$.

2.4 Find the magnitude and direction of the resultant force of three coplanar forces of 10 lb each acting outward on a body at the origin, and making angles of 60°, 120°, and 270°, respectively, with the x-axis.

Answer: $10(\sqrt{3} - 1)$, $\perp x$.

2.5 Find the angle between $\mathbf{u} = 6\mathbf{i} + 2\mathbf{j} - 3\mathbf{k}$ and $\mathbf{v} = -\mathbf{i} + 8\mathbf{j} + 4\mathbf{k}$.

Answer: $\cos^{-1}(-\frac{2}{63})$.

2.6 Given $\mathbf{u} = 3\mathbf{i} + 4\mathbf{j} - \mathbf{k}$, $\mathbf{v} = 2\mathbf{i} + 5\mathbf{k}$, find the value of α so that $\mathbf{u} + \alpha\mathbf{v}$ is orthogonal to $\mathbf{v}$.

Answer: $-\frac{1}{29}$.

2.7 Given $\mathbf{u} = 2\mathbf{i} + 3\mathbf{j}$, $\mathbf{v} = \mathbf{i} - \mathbf{j} + 2\mathbf{k}$, $\mathbf{w} = \mathbf{i} - 2\mathbf{k}$, evaluate $\mathbf{u} \cdot (\mathbf{v} \times \mathbf{w})$ and $(\mathbf{u} \times \mathbf{v}) \cdot \mathbf{w}$.

Answer: 16.

2.8 $(\mathbf{u} \times \mathbf{v}) \cdot \mathbf{w}$ is called the scalar triple product of $\mathbf{u}, \mathbf{v}, \mathbf{w}$. Show that $(\mathbf{u} \times \mathbf{v}) \cdot \mathbf{w} = \mathbf{u} \cdot (\mathbf{v} \times \mathbf{w})$.

2.9 Find the equation of the plane through $A(1, 0, 2)$, $B(0, 1, -1)$, and $C(2, 2, 3)$.

Answer: $7x - 2y - 3z - 1 = 0$.

2.10 Find the area of $\triangle ABC$ in Prob. 2.9.

Answer: $\sqrt{62}/2$.

2.11 Find a vector perpendicular to both $\mathbf{u} = 2\mathbf{i} + 3\mathbf{j} - \mathbf{k}$ and $\mathbf{v} = \mathbf{i} - 2\mathbf{j} + 3\mathbf{k}$.

Answer: $7\mathbf{i} - 7\mathbf{j} - 7\mathbf{k}$.

2.2 VECTOR EQUATIONS

The spirit of vector analysis is to use symbols to represent physical or geometric quantities and to express a physical relationship or a geometric fact by an equation. For example, if we have a particle on which act the forces $\mathbf{F}^{(1)}$, $\mathbf{F}^{(2)}, \ldots, \mathbf{F}^{(n)}$, then we say that the condition of equilibrium for this particle is

$$(2.2\text{-}1) \qquad\qquad \mathbf{F}^{(1)} + \mathbf{F}^{(2)} + \ldots + \mathbf{F}^{(n)} = \mathbf{0}.$$

The meaning of this elegant equation is self-evident. No further explanation is needed as to where the observer is located, or how the components of the force vectors are to be resolved. As another example, we say that the following equation for the vector $\mathbf{r}$ represents a plane, if $\mathbf{n}$ is a unit vector and p is a constant:

$$(2.2\text{-}2) \qquad\qquad\qquad \mathbf{r} \cdot \mathbf{n} = p.$$

By this statement we mean that the locus of the end point of a radius vector **r** satisfying the equation above is a plane. The geometric meaning is again clear. The vector **n**, called the unit normal vector of the plane, is specified. The scalar product **r** · **n** represents the scalar projection of **r** on **n**. Equation (2.2-2) then states that if we consider all radius vectors **r** whose component on **n** is a constant p we shall obtain a plane. See Fig. 2.1.

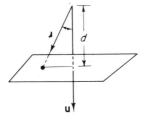

Fig. 2.1   Equation of a plane, **r·n** $= p$.

On the other hand, elegant as they are, vector equations are not always convenient. Indeed it was a great contribution of Descartes, who introduced analytic geometry in which vectors are expressed by their components with respect to a fixed frame of reference. Thus, with reference to a set of rectangular Cartesian coordinate axes O-xyz, Eqs. (2.2-1) and (2.2-2) may be written, respectively, as

$$(2.2\text{-}3) \qquad \sum_{i=1}^{n} F_x^{(i)} = 0, \qquad \sum_{i=1}^{n} F_y^{(i)} = 0, \qquad \sum_{i=1}^{n} F_z^{(i)} = 0,$$

$$(2.2\text{-}4) \qquad ax + by + cz = p,$$

where $F_x^{(i)}$, $F_y^{(i)}$, $F_z^{(i)}$ represent the components of the vector $\mathbf{F}^{(i)}$ with respect to the frame of reference O-xyz; x, y, z represent the components of **r**; a, b, c represent those of the unit normal vector **n**.

Why is the analytic form preferred? Why are we willing to sacrifice the elegance of the vector notation? The answer is a compelling one: We like to express physical quantities in numbers. To specify a radius vector, it *is* convenient to specify a triple of numbers (x, y, z). To specify a force **F**, it *is* convenient to define the three components F_x, F_y, F_z. In fact, in practical calculations we use Eqs. (2.2-3), (2.2-4) much more frequently than Eqs. (2.2-1) and (2.2-2).

PROBLEM 2.12 Express the basic laws in elementary physics in the form of vector equations, e.g., Newton's law of motion, Coulomb's law for the attraction or repulsion between electric charges, Maxwell's equation for electromagnetic field, etc.

For example, to express Newton's law of gravitation in vector form, let m_1, m_2 be the masses of two particles. Let the position vector from particle 1 to particle 2 be $\mathbf{r}_{12}$. Then the force produced on particle 1 due to the gravitational attraction between 1 and 2 is

$$\mathbf{F}_{12} = \frac{m_1 m_2}{|\mathbf{r}_{12}|^2} \frac{\mathbf{r}_{12}}{|\mathbf{r}_{12}|}.$$

PROBLEM 2.13 Consider a particle constrained to move in a circular orbit at a constant speed. Let $\mathbf{v}$ be the velocity at any instant. What is the acceleration of the particle; i.e., what is the vector $d\mathbf{v}/dt$?

Answer: The velocity vector $\mathbf{v}$ may be represented in polar coordinates as follows. Let $\hat{\mathbf{r}}, \hat{\boldsymbol{\theta}}, \hat{\mathbf{z}}$, be, respectively, the unit vectors with origin at P in the directions of the radius, the tangent, and the polar axis perpendicular to the plane of the orbit. See Fig. P2.13. Then $\mathbf{v} = v\hat{\boldsymbol{\theta}}$ where v is the absolute value of $\mathbf{v}$. Hence, by differentiation,

$$\frac{d\mathbf{v}}{dt} = v \frac{d\hat{\boldsymbol{\theta}}}{dt} + \frac{dv}{dt} \hat{\boldsymbol{\theta}}.$$

The last term vanishes because v is a constant. To evaluate $d\hat{\boldsymbol{\theta}}/dt$, we note that $\hat{\boldsymbol{\theta}}$ is a unit vector; hence it can only change direction. $d\hat{\boldsymbol{\theta}}/dt$ is, therefore, perpendicular to the vector $\hat{\boldsymbol{\theta}}$, i.e., parallel to $\hat{\mathbf{r}}$. Let ω be the angular velocity of the particle about the center of the orbit. Obviously $\hat{\boldsymbol{\theta}}$ is turning at a rate of $\omega = v/a$. Hence $d\hat{\boldsymbol{\theta}}/dt = (v/a)\hat{\mathbf{r}}$, and $d\mathbf{v}/dt = (v^2/a)\hat{\mathbf{r}}$.

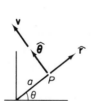

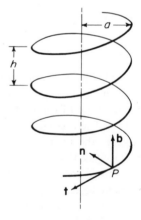

Fig. P2.13 Velocity vector of a particle moving in a circular orbit.

Fig. P2.14 A helical orbit.

PROBLEM 2.14 A particle is constrained to move along a circular helix of radius a and pitch h at a constant speed v. What is the acceleration of the particle? If the particle is located at P as shown in Fig. P2.14, express the velocity and acceleration vectors in terms of unit vectors **t**, **n**, and **b** which are, respectively, tangent, normal, and binormal to the helix at P.

Answer: The velocity vector is parallel to **t** and has a magnitude v. Hence **v** $= v$**t**. By differentiation, and noting that v is a constant, we have d**v**$/dt = v\, d$**t**$/dt$. But since **t** has a constant length of unity, d**t**$/dt$ must be perpendicular to **t** and hence must be a combination of **n** and **b**:

$$\frac{d\mathbf{t}}{dt} = \kappa\mathbf{n} + \tau\mathbf{b}$$

where κ and τ are constants. If the particle moves with unit velocity, the constants κ and τ are called the *curvature* and the *torsion* of the space curve, respectively.

It is convenient to use polar coordinates for this problem. Let the unit vectors in the direction of the radial, circumferential, and axial directions be $\hat{\mathbf{r}}$, $\hat{\boldsymbol{\theta}}$, and $\hat{\mathbf{z}}$, respectively. Then

$$\mathbf{v} = u\hat{\boldsymbol{\theta}} + w\hat{\mathbf{z}}$$

where u and w are the circumferential and axial velocities, respectively. Hence $d\mathbf{v}/dt = (du/dt)\,\hat{\boldsymbol{\theta}} + u\, d\hat{\boldsymbol{\theta}}/dt + (dw/dt)\,\hat{\mathbf{z}} + w\,(d\hat{\mathbf{z}}/dt) = u\, d\hat{\boldsymbol{\theta}}/dt = -(u^2/a)\hat{\mathbf{r}}$. The velocities u and w are related to v as follows: In the time interval $\Delta t = 2\pi a/u$, the axial position z is changed by h. Hence $w = h/\Delta t = hu/2\pi a$, and $v = u[1 + h^2/(4\pi^2a^2)]^{1/2}$.

2.3 THE SUMMATION CONVENTION

For further development an important matter of notation must be mastered.

A set of n variables $x_1, x_2, \ldots, x_n$ is usually denoted as x_i, $i = 1, \ldots, n$. When written singly, the symbol x_i stands for *any one* of the variables x_1, $x_2, \ldots, x_n$. The *range* of i must be indicated in every case; the simplest way is to write, as illustrated here, $i = 1, 2, \ldots, n$. The symbol i is an *index*. An index may be either a subscript or a superscript. A system of notations using indices is said to be an *indicial notation*.

Consider an equation describing a plane in a three-dimensional space referred to a rectangular Cartesian frame of reference with axes x_1, x_2, x_3,

$$(2.3\text{-}1) \qquad a_1x_1 + a_2x_2 + a_3x_3 = p,$$

where a_i, and p are constants. This equation can be written as

$$(2.3\text{-}2) \qquad \sum_{i=1}^{3} a_ix_i = p.$$

However, we shall introduce the *summation convention* and write the equation above in the simple form

(2.3-3) $a_i x_i = p.$

The convention is as follows: *The repetition of an index in a term will denote a summation with respect to that index over its range.* The *range* of an index i is the set of n integer values 1 to n. An index that is summed over is called a *dummy index*; one that is not summed is called a *free index.*

Since a dummy index just indicates summation, it is immaterial which symbol is used. Thus, $a_i x_i$ may be replaced by $a_j x_j$, etc. This is analogous to the dummy variable in an integral

$$\int_a^b f(x)\, dx = \int_a^b f(y)\, dy.$$

The use of the index and summation convention may be illustrated by other examples. Consider a unit vector $\mathbf{v}$ in a three-dimensional Euclidean space with rectangular Cartesian coordinates x, y, and z. Let the direction cosines α_i be defined as

$$\alpha_1 = \cos(\mathbf{v}, x), \qquad \alpha_2 = \cos(\mathbf{v}, y), \qquad \alpha_3 = \cos(\mathbf{v}, z),$$

where $(\mathbf{v}, x)$ denotes the angle between $\mathbf{v}$ and the x-axis, and so forth. The set of numbers $\alpha_i (i = 1, 2, 3)$ represents the components of the unit vector on the coordinates axes. The fact that the length of the vector is unity is expressed by the equation

$$(\alpha_1)^2 + (\alpha_2)^2 + (\alpha_3)^2 = 1,$$

or, simply,

(2.3-4) $\alpha_i \alpha_i = 1.$

As another illustration, consider a line element with components dx, dy, dz in a three-dimensional Euclidean space with rectangular Cartesian coordinates x, y, and z. The square of the length of the line element is

(2.3-5) $ds^2 = dx^2 + dy^2 + dz^2.$

If we define

(2.3-6) $dx_1 = dx, \qquad dx_2 = dy, \qquad dx_3 = dz,$

and

(2.3-7)
$$\delta_{11} = \delta_{22} = \delta_{33} = 1,$$
$$\delta_{12} = \delta_{21} = \delta_{13} = \delta_{31} = \delta_{23} = \delta_{32} = 0,$$

then (2.3-5) may be written as

(2.3-8) ▲ $$ds^2 = \delta_{ij}\, dx_i\, dx_j,$$

with the understanding that the range of the indices i and j is 1 to 3. Note that there are two summations in the expression above, one over i and one over j. The symbol δ_{ij} as defined in (2.3-7) is called the *Kronecker delta*.

The following determinant illustrates another application:

$$\begin{vmatrix} a_{11} & a_{12} & a_{13} \\ a_{21} & a_{22} & a_{23} \\ a_{31} & a_{32} & a_{33} \end{vmatrix} = \begin{aligned} & a_{11}a_{22}a_{33} + a_{21}a_{32}a_{13} + a_{31}a_{12}a_{23} \\ & - a_{11}a_{32}a_{23} - a_{21}a_{12}a_{33} - a_{31}a_{22}a_{13}. \end{aligned}$$

If we denote the general term in the determinant by a_{ij} and write the determinal as $|a_{ij}|$, then the equation above can be written as

(2.3-9) ▲ $$|a_{ij}| = e_{rst} a_{r1} a_{s2} a_{t3}$$

where e_{rst}, the *permutation symbol*, is defined by the equations

(2.3-10)
$$e_{111} = e_{222} = e_{333} = e_{112} = e_{121} = e_{211} = e_{221} = e_{331} = \ldots = 0,$$
$$e_{123} = e_{231} = e_{312} = 1,$$
$$e_{213} = e_{321} = e_{132} = -1.$$

In other words, e_{ijk} vanishes whenever the values of any two indices coincide; $e_{ijk} = 1$ when the subscripts permute like 1, 2, 3; and $e_{ijk} = -1$ otherwise.

The Kronecker delta and the permutation symbol are very important quantities which will appear again and again in this book. They are connected by the identity

(2.3-11) ▲ $$e_{ijk} e_{ist} = \delta_{js}\delta_{kt} - \delta_{jt}\delta_{ks}.$$

This e-δ identity is used frequently enough to warrant special attention here. It can be verified by actual trial.

Finally, we shall extend the summation convention to differentiation formulas. Let $f(x_1, x_2, \ldots, x_n)$ be a function of n variables $x_1, x_2, \ldots, x_n$. Then its differential shall be written as

(2.3-12) $$df = \frac{\partial f}{\partial x_1} dx_1 + \frac{\partial f}{\partial x_2} dx_2 + \cdots + \frac{\partial f}{\partial x_n} dx_n = \frac{\partial f}{\partial x_i} dx_i.$$

PROBLEMS

2.15 Write Eq. (2.2-1) or (2.2-3) in the index form. Let the components of $\mathbf{F}^{(i)}$ be written as $F_k^{(i)}$, $k = 1, 2, 3$; i.e., $F_x = F_1$, etc.

Answer: $\sum_{i=1}^{n} F_k^{(i)} = 0$.

2.16 Show that
(a) $\delta_{ii} = 3$
(b) $\delta_{ij}\delta_{ij} = 3$
(c) $e_{ijk}e_{jki} = 6$
(d) $e_{ijk}A_jA_k = 0$
(e) $\delta_{ij}\delta_{jk} = \delta_{ik}$
(f) $\delta_{ij}e_{ijk} = 0$

2.17 Write Eqs. (2.1-1), (2.1-5) in the index form, e.g., $\mathbf{u} \cdot \mathbf{v} = u_i v_i$.

Note: For Eq. (2.1-1), we may do the following: Define three unit vectors $\mathbf{v}^{(1)} = \mathbf{i}$, $\mathbf{v}^{(2)} = \mathbf{j}$, $\mathbf{v}^{(3)} = \mathbf{k}$; then $\mathbf{u} = u_i \cdot \mathbf{v}^{(i)}$.

Answer: $\mathbf{u} = u_i \mathbf{v}^{(i)}$, $\mathbf{u} \cdot \mathbf{v} = u_i v_i$.

2.18 Use the index form of vector equations to solve prob. 2.5 through 2.9.

2.19 The vector product of two vectors $\mathbf{u} = (u_1, u_2, u_3)$ and $\mathbf{v} = (v_1, v_2, v_3)$ is the vector $\mathbf{w} = \mathbf{u} \times \mathbf{v}$ whose components are

$$w_1 = u_2v_3 - u_3v_2, \qquad w_2 = u_3v_1 - u_1v_3, \qquad w_3 = u_1v_2 - u_2v_1.$$

Show that this can be shortened by writing

$$w_i = e_{ijk}u_jv_k.$$

2.20 Express Eq. (2.1-7) in the index form.

2.21 Derive the vector identity connecting three arbitrary vectors $\mathbf{A}, \mathbf{B}, \mathbf{C}$ by the method of vector analysis:

$$\mathbf{A} \times (\mathbf{B} \times \mathbf{C}) = (\mathbf{A} \cdot \mathbf{C})\mathbf{B} - (\mathbf{A} \cdot \mathbf{B})\mathbf{C}.$$

Solution: Since $\mathbf{A} \times (\mathbf{B} \times \mathbf{C})$ is perpendicular to $\mathbf{B} \times \mathbf{C}$, it must lie in the plane of $\mathbf{B}$ and $\mathbf{C}$. Hence we may write $\mathbf{A} \times (\mathbf{B} \times \mathbf{C}) = a\mathbf{B} + b\mathbf{C}$ where a, b are scalar quantities. But $\mathbf{A} \times (\mathbf{B} \times \mathbf{C})$ is a linear function of $\mathbf{A}, \mathbf{B},$ and $\mathbf{C}$; hence a must be a linear scalar combination of $\mathbf{A}$ and $\mathbf{C}$, b must be one of $\mathbf{A}$ and $\mathbf{B}$. Hence a, b are proportional to $\mathbf{A} \cdot \mathbf{C}$ and $\mathbf{A} \cdot \mathbf{B}$, respectively, and we may write

$$\mathbf{A} \times (\mathbf{B} \times \mathbf{C}) = \lambda(\mathbf{A} \cdot \mathbf{C})\mathbf{B} + \mu(\mathbf{A} \cdot \mathbf{B})\mathbf{C}$$

where λ, μ are pure numbers, independent of $\mathbf{A}, \mathbf{B},$ and $\mathbf{C}$. We can, therefore, evaluate λ, μ by special cases, e.g., if $\mathbf{i}, \mathbf{j}, \mathbf{k}$ are the unit vectors in the directions of x-, y-, z-axes (right-handed rectangular Cartesian), respectively, we may put $\mathbf{B} = \mathbf{i}$, $\mathbf{C} = \mathbf{j}$, $\mathbf{A} = \mathbf{i}$ to show that $\mu = -1$; and $\mathbf{B} = \mathbf{i}$, $\mathbf{C} = \mathbf{j}$, $\mathbf{A} = \mathbf{j}$ to show that $\lambda = 1$.

2.22 Write the equation in Prob. 2.21 in the index form, and prove its validity by means of the e-δ identity (2.3-11).

Note: Since the equation in Prob. 2.21 is valid for arbitrary vectors $\mathbf{A}$, $\mathbf{B}$, $\mathbf{C}$, this verification may be regarded as a proof for the e-δ identity.

Solution: $[\mathbf{A} \times (\mathbf{B} \times \mathbf{C})]_l = e_{lmn}a_m(\mathbf{B} \times \mathbf{C})_n = e_{lmn}a_m e_{njk}b_j c_k = e_{nlm}e_{njk}a_m b_j c_k$. By the e-δ identity, Eq. (2.3-11), this becomes $(\delta_{lj}\delta_{mk} - \delta_{lk}\delta_{mj})a_m b_j c_k$. Hence it is $\delta_{lj}a_m c_m b_j - \delta_{lk}a_m b_m c_k = a_m c_m b_l - a_m b_m c_l = (\mathbf{A} \cdot \mathbf{C})(\mathbf{B})_l - (\mathbf{A} \cdot \mathbf{B})(\mathbf{C})_l$.

2.4 TRANSLATION AND ROTATION OF COORDINATES

Consider two sets of rectangular Cartesian frames of reference O-xy and O'-$x'y'$ on a plane. If the frame of reference O'-$x'y'$ is obtained from O-xy by a shift of origin without a change in orientation, then the transformation is a *translation*. If a point P has coordinates (x, y) and (x', y') with respect to the old and new frames of reference, respectively, then

(2.4-1)
$$\begin{cases} x = x' + h \\ y = y' + k \end{cases} \text{ or } \begin{cases} x' = x - h \\ y' = y - k. \end{cases}$$

If the origin remains fixed, and the new axes are obtained by rotating Ox and Oy through an angle θ in the counterclockwise direction, then the transformation of axes is a *rotation*. Let P have coordinates (x, y), (x', y') relative to the old and new frames of reference, respectively. Then (see Fig. 2.2),

(2.4-2)
$$x = x' \cos \theta - y' \sin \theta$$
$$y = x' \sin \theta + y' \cos \theta.$$

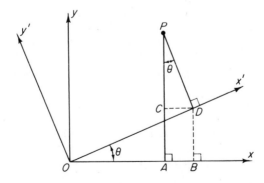

Fig. 2.2   Rotation of coordinates.

$$x' = x \cos \theta + y \sin \theta$$
$$(2.4\text{-}3)$$
$$y' = -x \sin \theta + y \cos \theta.$$

Using the index notion, we let x_1, x_2 replace x, y, and x'_1, x'_2 replace x', y'. Then obviously a rotation specified by (2.4-3) can be represented by the equation

$$(2.4\text{-}4) \qquad x'_i = \beta_{ij} x_j, \qquad (i = 1, 2)$$

where β_{ij} are elements of the square matrix (β_{ij}):

$$(2.4\text{-}5) \qquad \begin{pmatrix} \beta_{11} & \beta_{12} \\ \beta_{21} & \beta_{22} \end{pmatrix} = \begin{pmatrix} \cos \theta & \sin \theta \\ -\sin \theta & \cos \theta \end{pmatrix}.$$

The inverse transform of (2.4-4) is

$$(2.4\text{-}6) \qquad x_i = \beta_{ji} x'_j, \qquad (i = 1, 2)$$

where, according to (2.4-2), β_{ji} is the element in the jth row and ith column of the (β_{ij}) matrix. It is clear that the matrix (β_{ji}) is the *transpose* of the matrix (β_{ij}),

$$(2.4\text{-}7) \qquad (\beta_{ji}) = (\beta_{ij})^T.$$

On the other hand, from the point of view of the solution of a set of simultaneous linear equations (2.4-4), the matrix (β_{ji}) in (2.4-6) must be identified as the *inverse* of the matrix (β_{ij}):

$$(2.4\text{-}8) \qquad (\beta_{ji}) = (\beta_{ij})^{-1}.$$

Thus we obtain here a fundamental property of the transformation matrix (β_{ij}) which defines a rotation of rectangular Cartesian coordinates:

$$(2.4\text{-}9) \qquad (\beta_{ij})^T = (\beta_{ij})^{-1}.$$

A matrix (β_{ij}), $i, j = 1, 2, \ldots, n$, which satisfies Eq. (2.4-9) is called an *orthogonal* matrix. A transformation is said to be orthogonal if the associated matrix is orthogonal. The matrix (2.4-5) defining a rotation of coordinates is orthogonal.

For an orthogonal matrix we have

$$(\beta_{ij})(\beta_{ij})^T = (\beta_{ij})(\beta_{ij})^{-1} = (\delta_{ij}),$$

where δ_{ij} is the Kronecker delta. Hence,

$$(2.4\text{-}10) \qquad \beta_{ik} \beta_{jk} = \delta_{ij}.$$

To clarify the geometric meaning of this important equation, we rederive it directly for the rotation transformation as follows. A unit vector issued from the origin along the x_i'-axis has direction cosines β_{i1}, β_{i2} with respect to the x_1-, x_2-axes, respectively. The fact that its length is unity is expressed by the equation

$$(2.4\text{-}11) \qquad (\beta_{i1})^2 + (\beta_{i2})^2 = 1, \qquad (i = 1, 2).$$

The fact that a unit vector along the x_i'-axis is perpendicular to a unit vector along the x_j'-axis, if $j \neq i$, is expressed by the equation

$$(2.4\text{-}12) \qquad \beta_{i1}\beta_{j1} + \beta_{i2}\beta_{j2} = 0, \qquad (i \neq j).$$

Combining (2.4-11) and (2.4-12), we obtain (2.4-10).

Note: Alternatively, since we know what the β_{ij}'s are from (2.4-5), we can verify Eq. (2.4-10) by direct computation.

Obviously, the discussion above can be extended to three dimensions without much ado. The range of indices i, j can be extended to 1, 2, 3. Thus, consider two right-handed rectangular Cartesian coordinate systems x_1, x_2, x_3 and x_1', x_2', x_3', with the same origin O. Let $\mathbf{x}$ denote the position vector of a point P with components x_1, x_2, x_3 or x_1', x_2', x_3'. Let $\mathbf{e}_1$, $\mathbf{e}_2$, $\mathbf{e}_3$ be unit vectors in the directions of the positive x_1, x_2, x_3 axes. They are called *base vectors* of the x_1, x_2, x_3 coordinates system. [See Eq. (2.1-1) in which $\mathbf{e}_1$, $\mathbf{e}_2$, $\mathbf{e}_3$ are written as $\mathbf{i}$, $\mathbf{j}$, $\mathbf{k}$.] Let $\mathbf{e}_1'$, $\mathbf{e}_2'$, $\mathbf{e}_3'$ be the base vectors of the x_1', x_2', x_3' coordinates. Note that since the coordinates are orthogonal, we have

$$(2.4\text{-}13) \qquad \mathbf{e}_i \cdot \mathbf{e}_j = \delta_{ij}, \qquad \mathbf{e}_i' \cdot \mathbf{e}_j' = \delta_{ij}.$$

In terms of the base vectors, the vector $\mathbf{x}$ may be expressed as follows:

$$(2.4\text{-}14) \qquad \mathbf{x} = x_j \mathbf{e}_j = x_j' \mathbf{e}_j', \qquad (j = 1, 2, 3).$$

A scalar product of both sides of (2.4-14) with $\mathbf{e}_i$ gives

$$(2.4\text{-}15) \qquad x_j(\mathbf{e}_j \cdot \mathbf{e}_i) = x_j'(\mathbf{e}_j' \cdot \mathbf{e}_i).$$

But

$$x_j(\mathbf{e}_j \cdot \mathbf{e}_i) = x_j \delta_{ij} = x_i;$$

therefore,

$$(2.4\text{-}16) \qquad x_i = (\mathbf{e}_j' \cdot \mathbf{e}_i) x_j'.$$

Now, define

$$(2.4\text{-}17) \qquad (\mathbf{e}_j' \cdot \mathbf{e}_i) \equiv \beta_{ji};$$

then,

(2.4-18) $$x_i = \beta_{ji}x'_j, \qquad (j = 1, 2, 3).$$

Next, dot both sides of (2.4-14) with e'_i. This gives

$$x_j(e_j \cdot e_i) = x'_j(e'_j \cdot e'_i).$$

But $(e'_j \cdot e'_i) = \delta_{ij}$ and $(e_j \cdot e'_i) = \beta_{ij}$; therefore, we obtain

(2.4-19) $$x'_i = \beta_{ij}x_j, \qquad (i = 1, 2, 3).$$

Equations (2.4-18) and (2.4-19) are generalizations of Eqs. (2.4-4) and (2.4-6) to the three-dimensional case.

Equation (2.4-17) shows the geometric meaning of the coefficient β_{ij}. That (2.4-7) and (2.4-8) hold for $i, j = 1, 2, 3$ is clear because Eqs. (2.4-18) and (2.4-19) are inverse transformations of each other. Then, Eqs. (2.4-9) and (2.4-10) follow.

Now the numbers x_1, x_2, x_3, which represent the coordinates of the point P in Fig. 2.3, are also the components of the radius vector **A**. A recognition

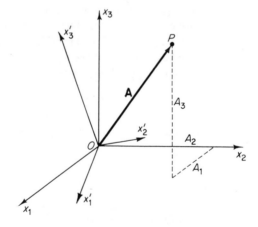

Fig. 2.3   Radius vector and coordinates.

of this fact gives us immediately the law of transformation of the components of a vector in rectangular Cartesian coordinates:

(2.4-20) $$A'_i = \beta_{ij}A_j, \qquad A_i = \beta_{ji}A'_j,$$

in which β_{ij} represents the cosine of the angle between the axes Ox'_i and Ox_j.

Finally, let us point out that the three unit vectors along x'_1, x'_2, x'_3 form the edges of a cube with volume 1. The volume of a parallelepiped having any three vectors **u, v, w** as edges is given either by the triple product **u**·(**v** × **w**), or by its negative; the sign is determined by whether the three vectors **u, v, w**, in this order, form a right-handed screw system or not. If they are right-handed, then the volume is equal to the determinant of their components:

$$(2.4\text{-}21) \qquad \text{Volume} = (\mathbf{u} \times \mathbf{v})\cdot\mathbf{w} = \begin{vmatrix} u_1 & u_2 & u_3 \\ v_1 & v_2 & v_3 \\ w_1 & w_2 & w_3 \end{vmatrix}.$$

Let us assume that x_1, x_2, x_3 and x'_1, x'_2, x'_3 are right-handed. Then it is clear that the determinant of β_{ij} represents the volume of a unit cube and hence has the value 1:

$$(2.4\text{-}22) \qquad |\beta_{ij}| \equiv \begin{vmatrix} \beta_{11} & \beta_{12} & \beta_{13} \\ \beta_{21} & \beta_{22} & \beta_{23} \\ \beta_{31} & \beta_{32} & \beta_{33} \end{vmatrix} = 1.$$

PROBLEM 2.23 Write out Eq. (2.4-10) in extenso and interpret the geometric meaning of the six resulting equations. $i = 1, 2, 3$.

Solution: Let the index i stand for 1, 2, 3.

$$\text{If } i = 1, j = 1: \text{ then } \beta_{11}\beta_{11} + \beta_{12}\beta_{12} + \beta_{13}\beta_{13} = 1. \qquad (1)$$

$$\text{If } i = 1, j = 2: \text{ then } \beta_{11}\beta_{21} + \beta_{12}\beta_{22} + \beta_{13}\beta_{23} = 0. \qquad (2)$$

Equation (1) means that the length of the vector $(\beta_{11}, \beta_{12}, \beta_{13})$ is one. Equation (2) means that the vectors $(\beta_{11}, \beta_{21}, \beta_{13})$, $(\beta_{21}, \beta_{22}, \beta_{23})$ are orthogonal to each other.

Other combinations of i, j are similar.

PROBLEM 2.24 Derive Eq. (2.4-10) by the following alternative procedure. From Eq. (2.4-4), differentiate both sides of the equation with respect to x_j. Then use (2.4-6) and the fact that $\partial x_i/\partial x_j = \delta_{ij}$ to simplify the results.

Solution: Differentiating Eq. (2.4-4) with respect to x'_j, we obtain $\delta_{ij} = \beta_{ik}\,\partial x_k/\partial x'_j$. But $x_i = \beta_{ji}x'_j$. On changing the index i to k, and differentiating, we have $\partial x_k/\partial x'_j = \beta_{jk}$. Combining these results yields $\delta_{ij} = \beta_{ik}\beta_{jk}$.

2.5 COORDINATE TRANSFORMATION IN GENERAL

A set of independent variables x_1, x_2, x_3 may be thought of as specifying the coordinates of a point in a frame of reference. A transformation from x_1, x_2, x_3 to a set of new variables $\bar{x}_1, \bar{x}_2, \bar{x}_3$ through the equation

$$(2.5\text{-}1) \qquad \bar{x}_i = f_i(x_1, x_2, x_3), \qquad (i = 1, 2, 3)$$

specifies a transformation of coordinates. The inverse transformation

$$(2.5\text{-}2) \qquad x_i = g_i(\bar{x}_1, \bar{x}_2, \bar{x}_3), \qquad (i = 1, 2, 3)$$

proceeds in the reverse direction. In order to insure that such a transformation is reversible and in one-to-one correspondence in a certain region R of the variables (x_1, x_2, x_3); i.e., in order that each set of numbers $(\bar{x}_1, \bar{x}_2, \bar{x}_3)$ defines a unique set of numbers (x_1, x_2, x_3), for (x_1, x_2, x_3) in the region R, and vice versa, it is sufficient that

(1) The functions f_i are single-valued, continuous, and possess continuous first partial derivatives in the region R.

(2) The *Jacobian determinant* $J = |\partial \bar{x}_i / \partial x_j|$ does not vanish at any point of the region R.

$$(2.5\text{-}3) \qquad J = \left| \frac{\partial \bar{x}_i}{\partial x_j} \right| \equiv \begin{vmatrix} \dfrac{\partial \bar{x}_1}{\partial x_1} & \dfrac{\partial \bar{x}_1}{\partial x_2} & \dfrac{\partial \bar{x}_1}{\partial x_3} \\[2mm] \dfrac{\partial \bar{x}_2}{\partial x_1} & \dfrac{\partial \bar{x}_2}{\partial x_2} & \dfrac{\partial \bar{x}_2}{\partial x_3} \\[2mm] \dfrac{\partial \bar{x}_3}{\partial x_1} & \dfrac{\partial \bar{x}_3}{\partial x_2} & \dfrac{\partial \bar{x}_3}{\partial x_3} \end{vmatrix} \neq 0.$$

Coordinate transformations with the properties 1 and 2 named above are called *admissible transformations*. If the Jacobian is positive everywhere, then a right-hand set of coordinates is transformed into another right-hand set, and the transformation is said to be *proper*. If the Jacobian is negative everywhere, a right-hand set of coordinates is transformed into a left-hand one, and the transformation is said to be *improper*. *In this book, we shall* tacitly assume *that our transformations are admissible and proper.*

PROBLEM 2.25 Do you understand the meaning of Jacobian thoroughly? Can you relate it to the *implicit function* theorem? Prove the theorem named above; i.e., the conditions 1 and 2 insure a one-to-one transformation (2.5.1). [A detailed discussion can be found in many books. For example, (1) W. Kaplan: *Advanced Calculus*, Sec. 2-8, p. 90, Addison-Wesley, 1952; (2) M. H. Protter and C. B. Morrey: *Modern Mathematical Analysis*, Chap. 11, Sec. 3, p. 498, Addison-Wesley, 1964; (3) E. Goursat (translated by E. R. Hedrick): *A Course in Mathematical Analysis*, Vol. 1, Sec. 25, p. 45, Dover Publications.]

2.6 ANALYTICAL DEFINITIONS OF SCALARS, VECTORS, AND CARTESIAN TENSORS

Let (x_1, x_2, x_3) and $(\bar{x}_1, \bar{x}_2, \bar{x}_3)$ be two fixed sets of rectangular Cartesian coordinates of reference related by the transformation law

$$(2.6\text{-}1) \qquad \bar{x}_i = \beta_{ij} x_j$$

where β_{ij} is the direction cosine of the angle between unit vectors along the coordinate axes $\bar{x}_i$ and x_j. Thus

$$(2.6\text{-}2) \qquad \beta_{21} = \cos(\bar{x}_2, x_1),$$

and so forth. The inverse transform is

$$(2.6\text{-}3) \qquad x_i = \beta_{ji}\bar{x}_j.$$

A system of quantities is called a *scalar*, a *vector*, or a *tensor* depending upon how the components of the system are defined in the variables x_1, x_2, x_3 and how they are transformed when the variables x_1, x_2, x_3 are changed to $\bar{x}_1$, $\bar{x}_2$, $\bar{x}_3$.

A system is called a *scalar* if it has only a single component Φ in the variables x_i and a single component $\bar{\Phi}$ in the variables $\bar{x}_i$ and if Φ and $\bar{\Phi}$ are numerically equal at the corresponding points,

$$(2.6\text{-}4) \qquad \Phi(x_1, x_2, x_3) = \bar{\Phi}(\bar{x}_1, \bar{x}_2, \bar{x}_3).$$

A system is called a *vector field* or a *tensor field of rank one* if it has three components ξ_i in the variables x_i and three components $\bar{\xi}_i$ in the variables $\bar{x}_i$ and if the components are related by the characteristic law

$$(2.6\text{-}5) \qquad \begin{aligned} \bar{\xi}_i(\bar{x}_1, \bar{x}_2, \bar{x}_3) &= \xi_k(x_1, x_2, x_3)\beta_{ik}, \\ \xi_i(x_1, x_2, x_3) &= \bar{\xi}_k(\bar{x}_1, \bar{x}_2, \bar{x}_3)\beta_{ki}. \end{aligned}$$

Generalizing these definitions to a system which has nine components when i and j range over 1, 2, 3, we define a *tensor field of rank 2* if it is a system which has nine components t_{ij} in the variables x_1, x_2, x_3 and nine components $\bar{t}_{ij}$ in the variables $\bar{x}_1$, $\bar{x}_2$, $\bar{x}_3$, and if the components are related by the characteristic law

$$(2.6\text{-}6) \qquad \begin{aligned} \bar{t}_{ij}(\bar{x}_1, \bar{x}_2, \bar{x}_3) &= t_{mn}(x_1, x_2, x_3)\beta_{im}\beta_{jn}, \\ t_{ij}(x_1, x_2, x_3) &= \bar{t}_{mn}(\bar{x}_1, \bar{x}_2, \bar{x}_3)\beta_{mi}\beta_{nj}. \end{aligned}$$

Further generalization to tensor fields of higher ranks is immediate. These definitions can obviously be modified to two dimensions if the indices range over 1, 2, or to n-dimensions if the range of the indices is 1, 2, ..., n. Since our definitions are based on transformations from one rectangular Cartesian frame of reference to another, the systems so defined are called *Cartesian tensors*. For simplicity, only Cartesian tensor equations will be used in this book.

Elaboration on Why Vectors and Tensors Are Defined in this Manner

The analytical definition of vectors is designed to follow the idea of a radius vector. We all know that the radius vector, a vector joining the origin $(0, 0, 0)$ to a point (x_1, x_2, x_3), embodies our idea of a vector and expresses it numerically in terms of the components (x_1-0, x_2-0, x_3-0), i.e., (x_1, x_2, x_3). When this vector is viewed from another frame of reference, the components referred to the new frame can be computed from the old according to Eq. (2.6-1), which is the *law of transformation of the components of a radius vector. Our generalization of Eq. (2.6-1) into Eq. (2.6-5), which defines all vectors, is equivalent to saying that we can call an entity a vector if it behaves like a radius vecotr, namely, if it has a fixed direction and a fixed magnitude.*

These remarks are intended to differentiate a matrix from a vector. We can list the components of a vector in the form of a column matrix; but not all column matrices are vectors. For example, to identify myself, I list my age, social security number, street address, and zip code, in a column matrix. What can you say about this matrix? Nothing very interesting! It is certainly not a vector.

The mathematical steps we took in generalizing the definition (2.6-5) for a vector to Eq. (2.6-6) for a tensor are natural enough. These equations are so similar that if we call a vector a tensor of rank 1, we cannot help but call the others tensors of rank 2 or 3, etc. What is the physical significance of these higher order tensors? The most effective way to answer this question is to consider some concrete examples, the best example being the stress tensor. (In French, the word *tension* means "stress"; hence the word *tensor* to indicate a system of quantities which transforms like stress under coordinate transformation.) However, before we turn our attention to specific examples, we shall first discuss the significance of tensor equations through the following problems.

PROBLEM 2.26 Show that, *if all components of a Cartesian tensor vanish in one coordinate system, then they vanish in all other Cartesian coordinate systems.* This is perhaps the most important property of tensor fields.

Proof: This follows immediately from Eq. (2.6-6). If every component of t_{mn} vanishes, then the right-hand side vanishes and $\bar{t}_{ij} = 0$ for all i, j.

PROBLEM 2.27 Prove the theorem: *The sum or difference of two Cartesian tensors of the same rank is again a tensor of the same rank.* Thus, any linear combination of tensors of the same rank is again a tensor of the same rank.

Proof: Let A_{ij}, B_{ij} be two tensors. Under coordinate transformation (2.6-1) we have the new components

$$\bar{A}_{ij} = A_{mn}\beta_{im}\beta_{jn}, \qquad \bar{B}_{ij} = B_{mn}\beta_{im}\beta_{jn}.$$

Adding or subtracting, we obtain

$$\bar{A}_{ij} \pm \bar{B}_{ij} = \beta_{im}\beta_{jn}(A_{mn} \pm B_{mn})$$

and the theorem is proved.

PROBLEM 2.28 Prove the theorem: *Let $A_{\alpha_1 \ldots \alpha_r}$, $B_{\alpha_1 \ldots \alpha_r}$ be tensors. Then the equation*

$$A_{\alpha_1 \ldots \alpha_r}(x_1, x_2, \ldots, x_n) = B_{\alpha_1 \ldots \alpha_r}(x_1, x_2, \ldots, x_n)$$

is a tensor equation; i.e., if this equation is true in one Cartesian coordinate system, then it is true in all Cartesian systems.

Proof: Multiplying both sides of the equation by

$$\beta_{i\alpha_1}\beta_{j\alpha_2} \ldots \beta_{k\alpha_r}$$

and summing over the repeated indices yields the equation

$$\bar{A}_{ij \ldots k}(\bar{x}_1, \bar{x}_2, \ldots, \bar{x}_n) = \bar{B}_{ij \ldots k}(\bar{x}_1, \bar{x}_2, \ldots, \bar{x}_n).$$

Alternatively, write the equation as $\mathbf{A} - \mathbf{B} = 0$. Then every component of $\mathbf{A} - \mathbf{B}$ vanishes. Then apply the results of Prob. 2.27 and 2.26.

Note: Proofs of these statements are so simple as to be trivial. The statements, however, are most important.

2.7 THE SIGNIFICANCE OF TENSOR EQUATIONS

The theorems stated in the problems at the end of the previous section contain the most important property of tensor fields: *If all the components of a tensor field vanish in one coordinate system, they vanish likewise in all coordinate systems which can be obtained by admissible transformations.* Since the sum and difference of tensor fields of a given type are tensors of the same type, we deduce that *if a tensor equation can be established in one coordinate system, then it must hold for all coordinate systems obtained by admissible transformations.*

Thus the importance of tensor analysis may be summarized by the following statement. The form of an equation can have general validity with respect to any frame of reference only if every term in the equation has the same tensor characteristics. If this condition is not satisfied, a simple change of the system of reference will destroy the form of the relationship, and the form would, therefore, be merely fortuitous.

We see that tensor analysis is as important as dimensional analysis in any formulation of physical relations. In dimensional analysis, we study the changes a physical quantity undergoes with particular choices of fundamental units. Two physical quantities cannot be equal unless they have the same dimensions. An equation describing a physical relation cannot be correct unless it is invariant with respect to a change of fundamental units.

Because of the design of the tensor transformation laws, the tensorial equations are in harmony with physics. For example, a tensor of rank 1 is defined in accordance with the physical idea of a vector as discussed in Sec. 2.6.

2.8 NOTATIONS FOR VECTORS AND TENSORS: BOLDFACE OR INDICES?

In continuum mechanics we are concerned with vectors describing displacements, velocities, forces, etc., and with tensors which are generalizations of the concept of vectors. For vectors the usual notation of boldface letters or an arrow, such as $\mathbf{u}$ or $\vec{u}$, is agreeable to all; but for tensors there are differences of opinion. A tensor of rank 2 may be printed as a boldface letter or with a double arrow or with a pair of braces. Thus if T is a tensor of rank 2, it may be printed as $\mathbf{T}$, $\vec{\overrightarrow{T}}$ or $\{T\}$. The first is the simplest, but then you have to remember what the symbol represents; it may be a vector or it may be a tensor. The other notations are cumbersome. More important objections to the simple notation arise when several vectors and tensors are associated together. In vector analysis we have to distinguish scalar products from vector products. How about tensors? Shall we define many kinds of tensor products? We have to, because there is a variety of ways tensors can be associated. The matter becomes complicated. For this reason, in most theoretical works which require extensive use of tensors, such as the general theory of relativity, or the mechanics of solid continua, an index notation is used. In this notation, vectors and tensors are resolved into their components with respect to a frame of reference and denoted by symbols such as u_i, u_{ij}, etc. These components are real numbers. Mathematical operations on these components follow the usual rules of arithmetic. No special rules of combination need to be introduced. Thus it gains a measure of simplicity. Furthermore, the index notation exhibits the rank and the range of a tensor clearly. It displays the role of the frame of reference explicitly.

The last mentioned advantage of the index notation, however, is also a weakness: It draws the attention of the reader away from the physical entity.

On balancing the advantages and disadvantages of various notations, we come to the conclusion that the index notation is indispensable. Its clarity in algebraic and analytical operations far outweighs any drawbacks.

2.9 QUOTIENT RULE

Consider a set of n^3 functions $A(1, 1, 1)$, $A(1, 1, 2)$, $A(1, 2, 3)$, etc., or $A(i, j, k)$ for short, with each of the indices i, j, k, ranging over $1, 2, \ldots, n$. Although the set of functions $A(i, j, k)$ has the right number of components, we do not know whether it is a tensor or not. Now suppose we know something about the nature of the product of $A(i, j, k)$ with an arbitrary tensor. Then there is a method which enables us to establish whether $A(i, j, k)$ is a tensor without going to the trouble of determining the law of transformation directly.

For example, let $\xi_i(x)$ be a vector. Let us suppose that the product $A(i, j, k)\,\xi_i$ (summation convention used over i) is known to yield a tensor of the type $A_{jk}(x)$,

2.9-1) $$A(i, j, k)\xi_i = A_{jk},$$

then we can prove that $A(i, j, k)$ is a tensor of the type $A_{ijk}(x)$.

The proof is very simple. Since $A(i, j, k)\,\xi_i$ is of the type A_{jk}, it is transformed into $\bar{x}$-coordinates as

(2.9-2) $$\bar{A}(i, j, k)\bar{\xi}_i = \bar{A}_{jk} = \beta_{jr}\beta_{ks}A_{rs} = \beta_{jr}\beta_{ks}[A(m, r, s)\xi_m].$$

But $\xi_m = \beta_{im}\bar{\xi}_i$. Inserting this in the right-hand side of the equation above and transposing all terms to one side of the equation, we obtain

(2.9-3) $$[\bar{A}(i, j, k) - \beta_{jr}\beta_{ks}\beta_{im}A(m, r, s)]\bar{\xi}_i = 0.$$

Now $\bar{\xi}_i$ is an arbitraty vector. Hence, the quantity within the brackets must vanish and we have

(2.9-4) $$\bar{A}(i, j, k) = \beta_{im}\beta_{jr}\beta_{ks}A(m, r, s),$$

which is precisely the law of transformation of the tensor of the type A_{ijk}.

The pattern of the preceding example can be generalized to higher order tensors.

2.10 PARTIAL DERIVATIVES

When only Cartesian coordinates are considered, the partial derivatives of any tensor field behave like the components of a Cartesian tensor. To show this, let us consider two sets of Cartesian coordinates (x_1, x_2, x_3) and $(\bar{x}_1, \bar{x}_2, \bar{x}_3)$ related by

(2.10-1) $$\bar{x}_i = \beta_{ij}x_j + \alpha_i,$$

where β_{ij} and α_i are constants.

Now, if $\xi_i(x_1, x_2, x_3)$ is a tensor, so that

(2.10-2) $$\bar{\xi}_i(\bar{x}_1, \bar{x}_2, \bar{x}_3) = \xi_k(x_1, x_2, x_3)\beta_{ik},$$

then, on differentiating both sides of the equation, one obtains

(2.10-3)
$$\frac{\partial \bar{\xi}_i}{\partial \bar{x}_j} = \beta_{ik} \frac{\partial \xi_k}{\partial x_m} \frac{\partial x_m}{\partial \bar{x}_j} = \beta_{ik} \beta_{jm} \frac{\partial \xi_k}{\partial x_m}$$

which verifies the statement.

It is a common practice to *use a comma to denote partial differentiation.* Thus,

$$\xi_{i,j} \equiv \frac{\partial \xi_i}{\partial x_j}, \qquad \Phi_{,i} \equiv \frac{\partial \Phi}{\partial x_i}, \qquad \sigma_{ij,k} \equiv \frac{\partial \sigma_{ij}}{\partial x_k}.$$

When we restrict ouselves to Cartesian coordinates, $\Phi_{,i}$, $\xi_{i,j}$, and $\sigma_{ij,k}$ are tensors of rank 1, 2, 3, respectively, provided that Φ, ξ, and σ_{ij} are tensors.

PROBLEMS

2.29 In any tensor $A_{ijk...m}$, equating two indices and summing over that index is called a *contraction*. Thus for a tensor A_{ijk}, a contraction over i and j $(i, j = 1, 2, 3)$ results in a vector $A_{iik} = A_{11k} + A_{22k} + A_{33k}$. Prove that the contraction of any two indices in a Cartesian tensor of rank n results in a tensor of rank $n - 2$.

Solution: The only significant part of the statement is that the result of contraction is a *tensor*. Let $A_{ijk...n}$ be a tensor of rank n. Then $A_{iik...n}$ has only $(n - 2)$ indices. To show that it is a tensor, consider the definition:

$$\bar{A}_{ijk...n} = A_{\alpha_1 \alpha_2 \alpha_3 ... \alpha_n} \beta_{i\alpha_1} \beta_{j\alpha_2} \beta_{k\alpha_3} ... \beta_{n\alpha_n}.$$

A contraction over i and j yields

$$\bar{A}_{iik...n} = A_{\alpha_1 \alpha_2 \alpha_3 ... \alpha_n} \beta_{i\alpha_1} \beta_{i\alpha_2} \beta_{k\alpha_3} ... \beta_{n\alpha_n}.$$

But we know from (2.4-10) that

$$\beta_{i\alpha_1} \beta_{i\alpha_2} = \delta_{\alpha_1 \alpha_2}.$$

Hence

$$\begin{aligned}
\bar{A}_{iik...n} &= A_{\alpha_1 \alpha_2 \alpha_3 ... \alpha_n} \delta_{\alpha_1 \alpha_2} \beta_{k\alpha_3} ... \beta_{n\alpha_n} \\
&= A_{\alpha_1 \alpha_1 \alpha_3 ... \alpha_n} \beta_{k\alpha_3} ... \beta_{n\alpha_n}.
\end{aligned}$$

Thus $A_{\alpha_1 \alpha_1 \alpha_3 ... \alpha_n}$ obeys the transformation law for a tensor of rank $(n - 2)$. Hence the theorem.

2.30 If A_{ij} is a Cartesian tensor of rank 2, show that A_{ii} is a scalar.

Solution: From Prob. 2.29, A_{ii} is a tensor of rank 0 and hence is a scalar. More directly, we have

$$\bar{A}_{ij} = A_{mn}\beta_{im}\beta_{jn}$$

$$\bar{A}_{ii} = A_{mn}\beta_{im}\beta_{in} = \delta_{mn}A_{mn} = A_{mm},$$

which obeys the definition of a scalar, Eq. (2.6-4).

2.31 Use the index notation and summation convention to prove the following relations (See table of notations below.)

(a) $\mathbf{u} \times \mathbf{v} = -\mathbf{v} \times \mathbf{u}$

(b) $(\mathbf{s} \times \mathbf{t}) \cdot (\mathbf{u} \times \mathbf{v}) = (\mathbf{s} \cdot \mathbf{u})(\mathbf{t} \cdot \mathbf{v}) - (\mathbf{s} \cdot \mathbf{v})(\mathbf{t} \cdot \mathbf{u})$

(c) curl curl $\mathbf{v}$ = grad div $\mathbf{v} - \Delta\mathbf{v}$

Example of solution:

(c) curl curl $\mathbf{v} = e_{ijk} \dfrac{\partial}{\partial x_j}\left(e_{klm}\dfrac{\partial v_m}{\partial x_l}\right)$

$$= e_{ijk}e_{lmk}\dfrac{\partial^2 v_m}{\partial x_j\,\partial x_l}$$

$$= (\delta_{il}\,\delta_{jm} - \delta_{im}\,\delta_{jl})\dfrac{\partial^2 v_m}{\partial x_j\,\partial x_l}$$

$$= \dfrac{\partial^2 v_j}{\partial x_j\,\partial x_i} - \dfrac{\partial^2 v_i}{\partial x_j\,\partial x_j} = \dfrac{\partial}{\partial x_i}\left(\dfrac{\partial v_j}{\partial x_j}\right) - \dfrac{\partial}{\partial x_j}\left(\dfrac{\partial v_i}{\partial x_j}\right)$$

$$= \nabla(\nabla \cdot \mathbf{v}) - \nabla \cdot \nabla\mathbf{v} = \text{grad div } \mathbf{v} - \Delta\mathbf{v}.$$

2.32 Let $\mathbf{r}$ be the radius vector of a typical point in the field and r be the magnitude of $\mathbf{r}$. Prove that, with the notations defined in the table below,

Vector Notation	Index Notation	Rank of Tensor
$\mathbf{v}$ (vector)	v_i	1
$\lambda = \mathbf{u} \cdot \mathbf{v}$ (dot, scalar, or inner product)	$\lambda = u_i v_i$	0
$\mathbf{w} = \mathbf{u} \times \mathbf{v}$ (cross or vector product)	$w_i = e_{ijk}u_j v_k$	1
grad $\phi = \nabla\phi$ (gradient of scalar field)	$\dfrac{\partial\phi}{\partial x_i}$	1
grad $\mathbf{v} = \nabla\mathbf{v}$ (vector gradient)	$\dfrac{\partial v_i}{\partial x_j}$	2
div $\mathbf{v} = \nabla \cdot \mathbf{v}$ (divergence)	$\dfrac{\partial v_i}{\partial x_i}$	0
curl $\mathbf{v} = \nabla \times \mathbf{v}$ (curl)	$e_{ijk}\dfrac{\partial v_k}{\partial x_j}$	1
$\nabla^2\mathbf{v} = \nabla \cdot \nabla\mathbf{v} = \Delta\mathbf{v}$ (Laplacian)	$\dfrac{\partial}{\partial x_i}\left(\dfrac{\partial v_j}{\partial x_i}\right) = \dfrac{\partial^2 v_j}{\partial x_i\,\partial x_i}$	1

(a) $\operatorname{div}(r^n\mathbf{r}) = (n + 3)r^n$

(b) $\operatorname{curl}(r^n\mathbf{r}) = 0$

(c) $\Delta(r^n) = n(n + 1)r^{n-2}$

Example of solution:

(a) Let the components of $\mathbf{r}$ be x_i, $(i = 1, 2, 3)$.

$$\operatorname{div}\mathbf{r} = \nabla\cdot\mathbf{r} = \frac{\partial x_i}{\partial x_i} = 3$$

$$r^2 = x_i x_i; \qquad r\frac{\partial r}{\partial x_i} = x_i; \qquad \therefore \frac{\partial r}{\partial x_i} = \frac{x_i}{r}$$

$$\operatorname{div}(r^n\mathbf{r}) = \nabla\cdot(r^n\mathbf{r}) = \frac{\partial}{\partial x_i}(r^n x_i) = r^n\frac{\partial x_i}{\partial x_i} + x_i\frac{\partial r^n}{\partial x_i}$$

$$= 3r^n + n_i\left(nr^{n-1}\frac{\partial r}{\partial x_i}\right) = 3r^n + nr^{n-2}x_i x_i = (n + 3)r^n.$$

2.33 A matrix-valued quantity a_{ij} $(i, j = 1, 2, 3)$ is given as follows:

$$\begin{pmatrix} a_{11} & a_{12} & a_{13} \\ a_{21} & a_{22} & a_{23} \\ a_{31} & a_{32} & a_{33} \end{pmatrix} = \begin{pmatrix} 1 & 1 & 0 \\ 1 & 2 & 2 \\ 0 & 2 & 3 \end{pmatrix}$$

What are the values of (a) a_{ii}, (b) $a_{ij}a_{ij}$, (c) $a_{ij}a_{jk}$ when $i = 1$, $k = 1$ and $i = 1$, $k = 2$.

Answer: 6, 24, 2, 3.

2.34 It is well-known that rigid-body rotation is noncommutative. For example, take a book, fix a frame of reference with x-, y-, z-axes directed along the edges of the book. First rotate it 90° about y; then rotate it 90° about z. We obtain a certain configuration. But a reversal of the order of rotation yields a different result.

The rotation of coordinates is also noncommutative; i.e., the transformation matrices (β_{ij}) are noncommutative. Demonstrate this in a special case that is analogous to the rigid-body rotation of the book named above. First transform x, y, z to x', y', z' by a rotation of 90° about the y-axis. Then transform x', y', z' to x'', y'', z'' by a 90° rotation about z'. Thus

$$\begin{pmatrix} x' \\ y' \\ z' \end{pmatrix} = \begin{pmatrix} 0 & 0 & -1 \\ 0 & 1 & 0 \\ 1 & 0 & 0 \end{pmatrix}\begin{pmatrix} x \\ y \\ z \end{pmatrix}, \qquad \begin{pmatrix} x'' \\ y'' \\ z'' \end{pmatrix} = \begin{pmatrix} 0 & 1 & 0 \\ -1 & 0 & 0 \\ 0 & 0 & 1 \end{pmatrix}\begin{pmatrix} x' \\ y' \\ z' \end{pmatrix}.$$

Derive the transformation matrix from x, y, z to x'', y'', z''. Now reverse the order of rotation. Show that a different result is obtained.

2.35 Infinitesimal rotations, however, are commutative. Demonstrate this by considering an infinitesimal rotation by an angle θ about y followed by another infinitesimal rotation ψ about z. Compare the results with the case in which the order of rotations is reversed.

2.36 Express the following set of equations in a single equation using index notation

$$e_{xx} = \frac{1}{E}[\sigma_{xx} - \nu(\sigma_{yy} + \sigma_{zz})], \qquad e_{xy} = \frac{1+\nu}{E}\sigma_{xy}$$

$$e_{yy} = \frac{1}{E}[\sigma_{yy} - \nu(\sigma_{xx} + \sigma_{zz})], \qquad e_{yz} = \frac{1+\nu}{E}\sigma_{yz}$$

$$e_{zz} = \frac{1}{E}[\sigma_{zz} - \nu(\sigma_{xx} + \sigma_{yy})], \qquad e_{xz} = \frac{1+\nu}{E}\sigma_{xz}$$

2.37 Write out in longhand, in unabridged form, the following equation:

$$G\left(u_{i,kk} + \frac{1}{1-2\nu}u_{k,ki}\right) + X_i = \rho\frac{\partial^2 u_i}{\partial t^2}.$$

Let

$$x_1 = x, x_2 = y, x_3 = z; \qquad u_1 = u, u_2 = v, u_3 = w.$$

2.38 Show that $e_{ijk}\sigma_{jk} = 0$, where e_{ijk} is the permutation symbol and σ_{jk} is a symmetric tensor, $\sigma_{jk} = \sigma_{kj}$.

FURTHER READING

ARIS, R., *Vectors, Tensors, and the Basic Equations of Fluid Mechanics*, Englewood Cliffs, N.J.: Prentice-Hall (1962).

BRILLOUIN, LÉON, *Tensors in Mechanics and Elasticity*, New York: Academic Press (1964).

MICHAL, A. D., *Matrix and Tensor Calculus, with Applications to Mechanics, Elasticity and Aeronautics*, New York: Wiley (1947).

SOKOLNIKOFF, I. S., *Tensor Analysis, Theory and Applications*, New York: Wiley (1951, 1958).

CHAPTER THREE

Stress

The concept of stress is the heart of our subject. It is the unique way continuum mechanics has for specifying the interaction between one part of a material body and another. We shall examine the idea of stress and the method of describing it. It will be shown that the specification of the state of stress at any point in a body requires nine numbers, called the components of stress. These nine numbers can be arranged to form a matrix. Under the assumption that the body-moment and the couple-stress do not exist, it is shown that the stress-matrix is symmetric, so that three pairs of stress components are equal, and six independent components fully describe the state of stress at any point. A change in the frames of reference changes the values of the stress components. Examination of the rules of the change of stress components under rotation of the frame of reference shows that the rule is a tensor-transformation rule. Therefore the stress is a tensor.

One must be able to compute the stress vector acting on any surface when the stress tensor is known. This relation is furnished by Cauchy's formula, Eq. (3.4-2).

3.1 THE IDEA OF STRESS

In particle mechanics, we study two types of interaction between particles: by collision and by action at a distance. In considering a system of particles

63

we must specify exactly the manner in which one particle is influenced by all the others. In a similar way, in continuum mechanics, we have to consider the interaction between one part of the body and another. However, since a continuum is a mathematical abstraction of the real material, for which even the smallest volume contains a very large number of particles, it would be futile to approach the interaction problem through the particle concept. A new method of description is necessary.

The new concept that evolved is the *stress*. Consider a material continuum *B* occupying a spatial region *V* at some time (Fig. 3.1). Imagine a closed surface *S* within *B*. We would like to express the interaction between the

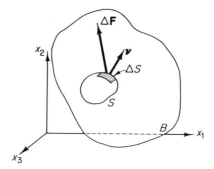

Fig. 3.1   Stress principle.

material outside *S* and that in the interior. This interaction can be divided into two kinds: one, due to the action-at-a-distance type of forces such as the gravitation and the electromagnetic force, which can be expressed as force per unit mass, and are called the *body forces*; and another, due to the action across the boundary surface *S*, called the *surface force*. To express the surface force, let us consider a small surface element of area ΔS on our imagined surface *S*. Let us draw, from a point on ΔS, a unit vector **v** normal to ΔS, with its direction outward from the interior of *S*. Then we can distinguish the two sides of ΔS according to the direction of **v**. Let the side to which this normal vector points be called the positive side. Consider the part of material lying on the positive side. This part exerts a force $\Delta \mathbf{F}$ on the other part which is situated on the negative side of the normal. The force $\Delta \mathbf{F}$ depends on the location and size of the area and the orientation of the normal. We introduce the assumption that *as ΔS tends to zero, the ratio $\Delta \mathbf{F}/\Delta S$ tends to a definite limit $d\mathbf{F}/dS$, and that the moment of the force acting on the surface ΔS about any point within the area vanishes in the limit.* The limiting vector will be written as

(3.1-1)
$$\overset{v}{\mathbf{T}} = \frac{d\mathbf{F}}{dS},$$

where a superscript v is introduced to denote the direction of the normal $\mathbf{v}$ of the surface ΔS. The limiting vector $\overset{v}{\mathbf{T}}$ is called the *traction*, or the *stress vector*, and represents the force per unit area acting on the surface.

The assertion that there is defined upon any imagined closed surface S in the interior of a continuum a stress vector field whose action on the material occupying the space interior to S is equipollent to the action of the exterior material upon it, is the *stress principle of Euler and Cauchy*. This principle is well accepted, and it seems to meet all the needs of conventional fluid and solid mechanics. However, this statement is no more than a basic simplification. For example, there is no *a priori* justification why the interaction of the material on the two sides of the surface element ΔS must be momentless. Indeed, some people who do not like the restrictive idea ". . . that the moment of the forces acting on the surface ΔS about any point within the area vanishes in the limit" have proposed a generalization of the stress principle of Euler and Cauchy to say that "across any infinitesimal surface element in a material the action of the exterior material upon the interior is equipollent to a force *and a couple*." The resulting theory requires the concept of *couple-stress* and is much more complex than the conventional theory. So far no real application has been found for the couple-stress theory, hence we shall not discuss it further in this book.

3.2 NOTATIONS FOR STRESS COMPONENTS

Consider a special case in which the surface ΔS_k, $k = 1$, 2, or 3, is parallel to one of the coordinate planes. Let the normal of ΔS_k be in the positive direction of the x_k-axis. Let the stress vector acting on ΔS_k be denoted by $\overset{k}{\mathbf{T}}$, with three components $\overset{k}{T_1}, \overset{k}{T_2}, \overset{k}{T_3}$ along the direction of the coordinate axes x_1, x_2, x_3, respectively; the index i of $\overset{k}{T_i}$ denoting the components of the force, and the symbol k indicating the normal (x_k-axis) to the surface on which the force acts. In this special case, we introduce a new set of symbols for the stress components,

$$(3.2\text{-}1) \qquad \overset{k}{T_1} = \tau_{k1}, \qquad \overset{k}{T_2} = \tau_{k2}, \qquad \overset{k}{T_3} = \tau_{k3}.$$

If we arrange the components of tractions acting on the surfaces ΔS_k, $k = 1, 2$, and 3, in a square matrix, we obtain

| | Components of Stresses | | |
	1	2	3
(3.2-2) Surface normal to x_1	τ_{11}	τ_{12}	τ_{13}
Surface normal to x_2	τ_{21}	τ_{22}	τ_{23}
Surface normal to x_3	τ_{31}	τ_{32}	τ_{33}

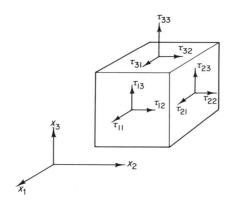

Fig. 3.2   Notations of stress components.

This is illustrated in Fig. 3.2. The components τ_{11}, τ_{22}, τ_{33} are called *normal stresses*, and the remaining components τ_{12}, τ_{13}, etc., are called *shearing stresses*. Each of these components has the dimension of force per unit area, or M/LT^2.

A great diversity in notations for stress components exists in the literature. The most widely used notations in American literature are, in reference to a system of rectangular Cartesian coordinates $x, y, z,$

$$(3.2\text{-}3) \qquad \begin{matrix} \sigma_x & \tau_{xy} & \tau_{xz} \\ \tau_{yx} & \sigma_y & \tau_{yz} \\ \tau_{zx} & \tau_{zy} & \sigma_z \end{matrix}$$

or

$$(3.2\text{-}4) \qquad \begin{matrix} \sigma_{xx} & \sigma_{xy} & \sigma_{xz} \\ \sigma_{yx} & \sigma_{yy} & \sigma_{yz} \\ \sigma_{zx} & \sigma_{zy} & \sigma_{zz} \end{matrix}$$

Love writes X_x, Y_x for σ_x and τ_{xy}, and Todhunter and Pearson use $\widehat{xx}$, $\widehat{xy}$. Since the reader is likely to encounter all these notations in the literature, we shall not insist on uniformity and would use (3.2-2) or (3.2-3) or (3.2-4), whichever happens to be convenient. There should be no confusion.

It is important to emphasize again that a stress will always be understood to be the force (per unit area) that the part lying on the positive side of a surface element (the side on the positive side of the outer normal) exerts on the part lying on the negative side. Thus, if the outer normal of a surface element points in the positive direction of the x_2-axis and τ_{22} is positive, the vector representing the component of normal stress acting on the surface element will point in the positive x_2-direction. But if τ_{22} is positive while the outer normal points in the negative x_2-axis direction, then the stress vector acting on the element also points to the negative x_2-axis direction (see Fig. 3.3).

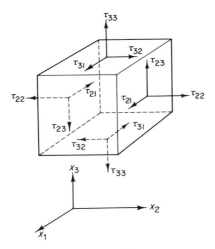

Fig. 3.3 Senses of positive stress components.

Similarly, positive values of τ_{21}, τ_{23} will imply shearing stress vectors pointing to the positive x_1-, x_3-axes if the outer normal agrees in sense with x_2-axis, whereas the stress vectors point to the negative x_1-, x_3-directions if the outer normal disagrees in sense with the x_2-axis, as illustrated in Fig. 3.3. A careful study of the figure is essential. Naturally, these rules agree with the usual notions of tension, compression, and shear.

3.3 THE LAWS OF MOTION AND FREE-BODY
DIAGRAMS

The continuum mechanics is founded on Newton's laws of motion. Let the coordinate system x_1, x_2, x_3 be an inertial frame of reference. Let the space occupied by a material body at any time t be denoted by $B(t)$. Let $\mathbf{r}$ be the position vector of a particle with respect to the origin of the coordinate system, ρ be the density of the material, and $\mathbf{V}$ be the velocity vector of the particle which is located at the point (x_1, x_2, x_3) and has a volume dv and mass $\rho\, dv$. Let the integral of the linear momentum $(\rho\, dv)\mathbf{V}$ of the particles over the domain $B(t)$

$$(3.3\text{-}1) \qquad \mathcal{P} = \int_{B(t)} \mathbf{V}\rho\, dv$$

be called the *linear momentum* of the body in the configuration $B(t)$, and let the integral of the moment of momentum of the particles about the origin, $\mathbf{r} \times \mathbf{V}\rho\, dv$, over the domain $B(t)$

$$(3.3\text{-}2) \qquad \mathcal{K} = \int_{B(t)} \mathbf{r} \times \mathbf{V}\rho\, dv$$

be called the *moment of momentum* of the body. *Newton's laws*, as stated by Euler for a continuum, assert that *the rate of change of the linear momentum is equal to the total applied force* $\mathscr{F}$ *acting on the body,*

(3.3-3)
$$\dot{\mathscr{P}} = \mathscr{F},$$

and that the rate of change of moment of momentum is equal to the total applied torque $\mathscr{L}$ *about the origin,*

(3.3-4)
$$\dot{\mathscr{H}} = \mathscr{L}.$$

It is easy to verify that if Eq. (3.3-3) holds, then when Eq. (3.3-4) is valid for one choice of origin, it is valid for all choices of origin.†

As we have mentioned before, there are two types of external forces acting on material bodies considered in mechanics of continuous media:

(1) Body forces, acting on elements of volume of the body.

(2) Surface forces, or stresses, acting on surface elements.

Examples of body forces are gravitational forces and electromagnetic forces. Examples of surface forces are aerodynamic pressure acting on a body and pressure due to mechanical contact of two bodies.

To specify a body force, we consider a volume bounded by an arbitrary surface S (Fig. 3.4). The resultant force vector contributed by the body force is assumed to be representable in the form of a volume integral taken over the domain B enclosed in S,

$$\int_B \mathbf{X}\, dv.$$

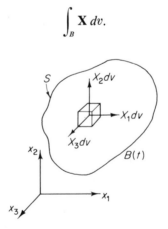

Fig. 3.4   Body forces.

†The derivatives $\dot{\mathscr{P}}$ and $\dot{\mathscr{H}}$ refer to the time rate of change of $\mathscr{P}$ and $\mathscr{H}$ of a fixed set of material particles. Later we shall denote them by $D\mathscr{P}/Dt$, $D\mathscr{H}/Dt$, respectively (see Sec. 10.3).

The vector $\mathbf{X}$, with three components X_1, X_2, X_3, all of dimensions force per unit volume $M(LT)^{-2}$, is called the body force per unit volume. For example, in a gravitational field,

$$X_i = \rho g_i,$$

where g_i are components of a gravitational acceleration field and ρ is the density (mass per unit volume) of the material.

The surface force acting on an imagined surface in the interior of a body is the stress vector conceived in Euler and Cauchy's stress principle. According to this concept, the total force acting upon the material occupying the region B interior to a closed surface S is

$$(3.3\text{-}5) \qquad \mathfrak{F} = \oint_S \overset{v}{\mathbf{T}} \, dS + \int_B \mathbf{X} \, dv,$$

where $\overset{v}{\mathbf{T}}$ is the stress vector acting on dS whose outer normal vector is $\mathbf{v}$. Similarly, the torque about the origin is given by the expression

$$(3.3\text{-}6) \qquad \mathfrak{L} = \oint_S \mathbf{r} \times \overset{v}{\mathbf{T}} \, dS + \int_B \mathbf{r} \times \mathbf{X} \, dv.$$

Combining these equations, we have the equations of motion

$$(3.3\text{-}7) \quad \blacktriangle \qquad \oint_S \overset{v}{\mathbf{T}} \, dS + \int_B \mathbf{X} \, dv = \frac{D}{Dt} \int_B \mathbf{V} \rho \, dv,$$

$$(3.3\text{-}8) \quad \blacktriangle \qquad \oint_S \mathbf{r} \times \overset{v}{\mathbf{T}} \, dS + \int_B \mathbf{r} \times \mathbf{X} \, dv = \frac{D}{Dt} \int_B \mathbf{r} \times \mathbf{V} \rho \, dv.$$

It should be noted that no demand was made on the domain $B(t)$ other than that it must consist of the same material particles at all times. No special rule was made about the choice of the particles, other than that of continuity, i.e., that they form a continuum. Equations (3.3-7) and (3.3-8) are applicable to any material bodies. They can be applied to an ocean, but they are also applicable to a spoonful of water. The boundary surface of $B(t)$ may coincide with the external boundary of an elastic solid, but it may also include only a small portion.

This freedom of choice of the domain $B(t)$ is of great importance. In mechanics it is often necessary to look at various choices of subdomains in order to visualize the whole picture. Each specific choice of subdomain is called a *free body*. A diagram in which all the forces and moments acting on a free body are indicated is called a *free-body diagram*. An intelligent use of free-body diagrams is essential to the success of an analysis.

3.4 CAUCHY'S FORMULA

From the equations of motion, we shall first derive a simple result which states that *the stress vector* $T^{(+)}$ *representing the action of material exterior to a surface element on the interior is equal in magnitude and opposite in direction to the stress vector* $T^{(-)}$ *which represents the action of the interior material on the exterior across the same surface element:*

(3.4-1)   ▲ $$T^{(-)} = -T^{(+)}.$$

To prove this, we consider a small "pillbox" with two parallel surfaces of area ΔS and thickness δ, as shown in Fig. 3.5. When δ shrinks to zero, while ΔS remains small but finite, the volume forces and the linear momentum and its rate of change with time vanish, as well as the contribution of surface forces on the sides of the pillbox. The equation of motion (3.3-3) implies, therefore, for small ΔS,

$$T^{(+)} \, \Delta S + T^{(-)} \, \Delta S = 0.$$

Equation (3.4-1) then follows.

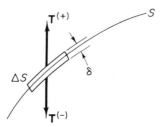

Fig. 3.5   Equilibrium of a "pill box" across a surface S.

Another way of stating this result is that the stress vector is a function of the normal vector to a surface. When the sense of direction of the normal vector reverses, the stress vector reverses also.

Now we shall show that *knowing the components* τ_{ij}, *we can write down at once the stress vector acting on any surface with unit outer normal vector* v *whose components are* v_1, v_2, v_3. *This stress vector is denoted by* $\overset{v}{T}$, *with components* $\overset{v}{T}_1$, $\overset{v}{T}_2$, $\overset{v}{T}_3$ *given by Cauchy's formula,*

(3.4-2)   ▲ $$\overset{v}{T}_i = v_j \tau_{ji},$$

which can be derived in several ways. We shall give first an elementary derivation.

Let us consider an infinitesimal tetrahedron formed by three surfaces parallel to the coordinate planes and one normal to the unit vector $\mathbf{v}$ (see Fig. 3.6). Let the area of the surface normal to $\mathbf{v}$ be dS. Then the areas of the other three surfaces are

$$dS_1 = dS \cos(\mathbf{v}, \mathbf{x}_1)$$
$$= v_1\, dS = \text{area of surface parallel to the } x_2 x_3\text{-plane,}$$
$$dS_2 = v_2\, dS = \text{area of surface parallel to the } x_3 x_1\text{-plane,}$$
$$dS_3 = v_3\, dS = \text{area of surface parallel to the } x_1 x_3\text{-plane,}$$

and the volume of the tetrahedron is

$$dv = \tfrac{1}{3}h\, dS,$$

where h is the height of the vertex P from the base dS. The forces in the positive direction of x_1, acting on the three coordinate surfaces, can be written as

$$(-\tau_{11} + \epsilon_1)\, dS_1, \qquad (-\tau_{21} + \epsilon_2)\, dS_2, \qquad (-\tau_{31} + \epsilon_3)\, dS_3,$$

where $\tau_{11}, \tau_{21}, \tau_{31}$ are the stresses at the vertex P opposite to dS. The negative sign is obtained because the outer normals to the three surfaces are opposite in sense with respect to the coordinate axes, and the ϵ's are inserted because the tractions act at points slightly different from P. If we assume that the stress field is continuous, then $\epsilon_1, \epsilon_2, \epsilon_3$ are infinitesimal eqantities. On the

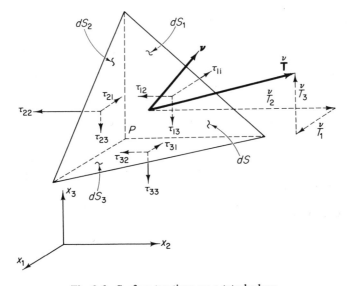

Fig. 3.6 Surface tractions on a tetrahedron.

other hand, the force acting on the triangle normal to $\mathbf{v}$ has a component $(\overset{v}{T}_1 + \epsilon)\, dS$ in the positive x_1-axis direction, the body force has an x_1-component equal to $(X_1 + \epsilon')\, dv$, and the rate of change of linear momentum has a component $\rho \dot{V}_1\, dv$. Here $\overset{v}{T}_1$ and X_1 refer to the point P, and ϵ, ϵ' are again infinitesimal. The first equation of motion is thus

$$(-\tau_{11} + \epsilon_1)v_1\, dS + (-\tau_{21} + \epsilon_2)v_2\, dS + (-\tau_{31} + \epsilon_3)v_3\, dS$$
$$+ (\overset{v}{T}_1 + \epsilon)\, dS + (X_1 + \epsilon')\tfrac{1}{3}h\, dS = \rho \dot{V}_1 \tfrac{1}{3}h\, dS.$$

Dividing through by dS, taking the limit $h \to 0$, and noting that $\epsilon_1, \epsilon_2, \epsilon_3, \epsilon, \epsilon'$ vanish with h and dS, one obtains

$$(3.4\text{-}3) \qquad \overset{v}{T}_1 = \tau_{11}v_1 + \tau_{21}v_2 + \tau_{31}v_3,$$

which is the first component of Eq. (3.4-2). Other components follow similarly.

   Cauchy's formula assures us that the nine components of stresses τ_{ij} are necessary and sufficient to define the traction across any surface element in a body. Hence the stress state in a body is characterized completely by the set of quantities τ_{ij}. Since $\overset{v}{T}_i$ is a vector and Eq. (3.4-2) is valid for an arbitrary vector v_j, it follows that τ_{ij} is a tensor. Henceforth τ_{ij} will be called a stress tensor.

3.5 EQUATIONS OF EQUILIBRIUM

   We shall now transform the equations of motion (3.3-7) and (3.3-8) into differential equations. This can be done elegantly by means of Gauss' theorem and Cauchy's formula as is shown in Chapter 10, but here we shall pursue an elementary course to assure physical clarity.

   Consider the static equilibrium state of an infinitesimal parallelepiped with surfaces parallel to the coordinate planes. The stresses acting on the various surfaces are shown in Fig. 3.7. The force $\tau_{11}\, dx_2\, dx_3$ acts on the left-hand side, the force $[\tau_{11} + (\partial \tau_{11}/\partial x_1)\, dx_1]\, dx_2\, dx_3$ acts on the right-hand side, etc. These expressions are based on the assumption of continuity of the stresses. The body force is $X_i\, dx_1\, dx_2\, dx_3$.

   The stresses indicated in Fig. 3.7 may be explained as follows. We are concerned with a nonuniform stress field. Every stress component is a function of position. Thus the stress component τ_{11} is a function of x_1, x_2, x_3: $\tau_{11}(x_1, x_2, x_3)$. At a point slightly to the right of the point (x_1, x_2, x_3), namely,

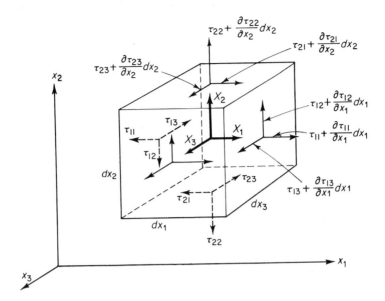

Fig. 3.7 Equilibrating stress components on an infinitesimal parallelo-piped.

at $(x_1 + dx_1, x_2, x_3)$, the value of the stress τ_{11} is $\tau_{11}(x_1 + dx_1, x_2, x_3)$. But if τ_{11} is a continuously differentiable function of x_1, x_2, x_3, then, according to Taylor's theorem with a remainder, we have

$$\tau_{11}(x_1 + dx_1, x_2, x_3) = \tau_{11}(x_1, x_2, x_3) + dx_1 \frac{\partial \tau_{11}}{\partial x_1}(x_1, x_2, x_3)$$
$$+ dx_1^2 \frac{1}{2} \frac{\partial^2 \tau_{11}}{\partial x_1^2}(x_1 + \alpha\, dx_1, x_2, x_3)$$

where $0 \leq \alpha \leq 1$. If $\partial^2 \tau_{11}/\partial x_1^2$ is finite, then the last term can be made arbitrarily small compared with the other terms by choosing dx_1 sufficiently small. With such a choice, we have

$$\tau_{11}(x_1 + dx_1, x_2, x_3) = \tau_{11}(x_1, x_2, x_3) + \frac{\partial \tau_{11}}{\partial x_1}(x_1, x_2, x_3)\, dx_1.$$

In Fig. 3.7 we write, in short, τ_{11} and $\tau_{11} + (\partial \tau_{11}/\partial x_1)\, dx_1$ on the surfaces where the stresses act. The left, bottom, and rear surfaces are located at x_1, x_2, x_3. The edges of the element have lengths dx_1, dx_2, dx_3.

All stresses and their derivatives are evaluated at (x_1, x_2, x_3). Equilibrium of the body demands that the resultant forces vanish. Consider the forces in the x_1-direction. As shown in Fig. 3.8 we have six components of surface

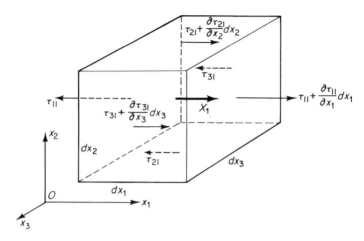

Fig. 3.8   Components of tractions in x_1-direction.

force and one component of body force. The sum is

$$\left(\tau_{11} + \frac{\partial \tau_{11}}{\partial x_1}\, dx_1\right) dx_2\, dx_3 - \tau_{11}\, dx_2\, dx_3 + \left(\tau_{21} + \frac{\partial \tau_{21}}{\partial x_2}\, dx_2\right) dx_3\, dx_1$$

$$- \tau_{21}\, dx_3\, dx_1 + \left(\tau_{31} + \frac{\partial \tau_{31}}{\partial x_3}\, dx_3\right) dx_1\, dx_2$$

$$- \tau_{31}\, dx_1\, dx_2 + X_1\, dx_1\, dx_2\, dx_3 = 0.$$

Dividing by $dx_1\, dx_2\, dx_3$, we obtain

$$(3.5\text{-}1) \qquad\qquad \frac{\partial \tau_{11}}{\partial x_1} + \frac{\partial \tau_{21}}{\partial x_2} + \frac{\partial \tau_{31}}{\partial x_3} + X_1 = 0.$$

A cyclic permutation of symbols leads to similar equations of equilibrium of forces in x_2-, x_3-directions. The whole set, written concisely, is

$$(3.5\text{-}2) \quad \blacktriangle \qquad\qquad \frac{\partial \tau_{ji}}{\partial x_j} + X_i = 0.$$

This is an important result. A shorter derivation will be given later in Sec. 10.6.

The equilibrium of an element requires also that the resultant moment vanish. If there do not exist external moments proportional to a volume, the consideration of moments will lead to the important conclusion that *the stress tensor is symmetric,*

$$(3.5\text{-}3) \quad \blacktriangle \qquad\qquad \tau_{ij} = \tau_{ji}.$$

This is demonstrated as follows. Referring to Fig. 3.9 and considering the

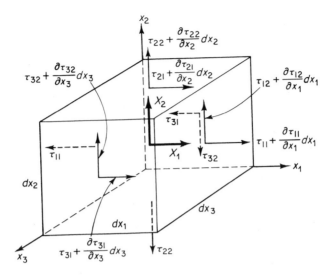

Fig. 3.9 Components of tractions that contribute moment about Ox_3-axis.

moment of all the forces about the x_3-axis, we see that those components of forces parallel to Ox_3 or lying in planes containing Ox_3 do not contribute any moment. The components that do contribute a moment about the x_3-axis are shown in Fig. 3.9. Therefore, properly taking care of the moment arm, we have

$$-\left(\tau_{11} + \frac{\partial \tau_{11}}{\partial x_1} dx_1\right) dx_2\, dx_3 \frac{dx_2}{2} + \tau_{11}\, dx_2\, dx_3 \frac{dx_2}{2}$$

$$+ \left(\tau_{11} + \frac{\partial \tau_{12}}{\partial x_1} dx_1\right) dx_2\, dx_3\, dx_1 - \left(\tau_{21} + \frac{\partial \tau_{21}}{\partial x_2} dx_2\right) dx_1\, dx_3\, dx_2$$

$$+ \left(\tau_{22} + \frac{\partial \tau_{22}}{\partial x_2} dx_2\right) dx_1\, dx_3 \frac{dx_1}{2} - \tau_{22}\, dx_1\, dx_3 \frac{dx_1}{2}$$

$$+ \left(\tau_{32} + \frac{\partial \tau_{32}}{\partial x_3} dx_3\right) dx_1\, dx_2 \frac{dx_1}{2} - \tau_{32}\, dx_1\, dx_2 \frac{dx_1}{2}$$

$$- \left(\tau_{31} + \frac{\partial \tau_{31}}{\partial x_3} dx_3\right) dx_1\, dx_2 \frac{dx_2}{2} + \tau_{31}\, dx_1\, dx_2 \frac{dx_2}{2}$$

$$- X_1\, dx_1\, dx_2\, dx_3 \frac{dx_2}{2} + X_2\, dx_1\, dx_2\, dx_3 \frac{dx_1}{2} = 0.$$

On dividing through by $dx_1\, dx_2\, dx_3$ and passing to the limit $dx_1 \rightarrow 0$, $dx_2 \rightarrow 0$, $dx_3 \rightarrow 0$, we obtain

(3.5-4) $$\tau_{12} = \tau_{21}.$$

Similar considerations of resultant moments about Ox_2, Ox_1 lead to the general result given by Eq. (3.5-3). A shorter derivation will be given later in Sec. 10.7.

So far we have considered the condition of equilibrium. If it is desired to derive the equation of motion instead of equilibrium, it is only necessary to apply the D'Alembert principle to our cubical element. According to the D'Alembert principle a particle in motion may be considered as in equilibrium if the negative of the product of mass and acceleration of the particle is applied as an external force on the particle. This is the inertia force. For a system of particles, D'Alembert's principle applies if the resultant of the inertia forces on all particles is applied to the center-of-mass of the system.

For the element considered in this section, if $\mathbf{a}$ (with components a_1, a_2, a_3) represents the acceleration vector of the particle referred to an inertial frame of reference, then since the mass of the element is $\rho \, dx_1 \, dx_2 \, dx_3$, the inertia force is $-\rho a_i \, dx_1 \, dx_2 \, dx_3$. An addition of this to Eq. (3.5-1) or (3.5-2) leads to the equation of motion,

$$\rho a_1 = \frac{\partial \tau_{11}}{\partial x_1} + \frac{\partial \tau_{21}}{\partial x_2} + \frac{\partial \tau_{31}}{\partial x_3} + X_1, \text{ etc., i.e., } \rho a_i = \frac{\partial \tau_{ij}}{\partial x_j} + X_i.$$

3.6 CHANGE OF STRESS COMPONENTS IN
TRANSFORMATION OF COORDINATES

In the previous section, the components of stress τ_{ij} are defined with respect to a rectangular Cartesian system x_1, x_2, x_3. Let us now take a second set of rectangular Cartesian coordinates x'_1, x'_2, x'_3 with the same origin but oriented differently, and consider the stress components in the new reference system (Fig. 3.10). Let these coordinates be connected by the linear relations

(3.6-1) $x'_k = \beta_{ki} x_i,$ $(k = 1, 2, 3)$

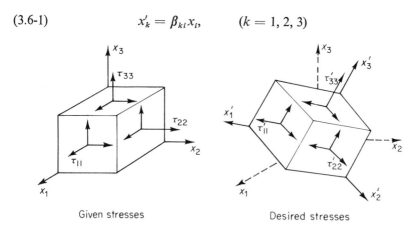

Given stresses   Desired stresses

Fig. 3.10 Transformation of stress components under rotation of coordinate systems.

where β_{ki} are the direction cosines of the x'_k-axis with respect to the x_i-axis. Since τ_{ij} is a tensor (Sec. 3.4) we can write down the transformation law at once. However, in order to emphasize the importance of the result we shall insert an elementary derivation based on Cauchy's formula (derived in Sec. 3.4), which states that if dS is a surface element whose unit outer normal vector $\mathbf{v}$ has components v_i, then the force per unit area acting on dS is a vector $\overset{v}{\mathbf{T}}$ with components

$$(3.6\text{-}2) \qquad \overset{v}{T}_i = \tau_{ji} v_j.$$

If the normal $\mathbf{v}$ is chosen to be parallel to the axis x'_k, so that

$$v_1 = \beta_{k1}, \qquad v_2 = \beta_{k2}, \qquad v_3 = \beta_{k3},$$

then the stress vector $\overset{k}{\mathbf{T}}'$ has components

$$\overset{k}{T}'_i = \tau_{ji} \beta_{kj}.$$

The component of the vector $\overset{k}{\mathbf{T}}'$ in the direction of the axis x'_m is given by the product of $\overset{k}{T}'_i$ and β_{mi}. Hence, the stress component

$$\tau'_{km} = \text{projection of } \overset{k}{\mathbf{T}}' \text{ on the } x'_m\text{-axis}$$
$$= \overset{k}{T}'_1 \beta_{m1} + \overset{k}{T}'_2 \beta_{m2} + \overset{k}{T}'_3 \beta_{m3}$$
$$= \tau_{j1} \beta_{kj} \beta_{m1} + \tau_{j2} \beta_{kj} \beta_{m2} + \tau_{j3} \beta_{kj} \beta_{m3};$$

i.e.,

$$(3.6\text{-}3) \qquad \tau'_{km} = \tau_{ji} \beta_{kj} \beta_{mi}.$$

If we compare Eq. (3.6-3) with Eq. (2.6-6) we see that the stress components transform like a Cartesian tensor of rank 2. Thus, the physical concept of stress which is described by τ_{ij} agrees with the mathematical definition of a tensor of rank 2 in a Euclidean space.

3.7* STRESS COMPONENTS IN ORTHOGONAL CURVILINEAR COORDINATES

Orthogonal curvilinear coordinates are often introduced in continuum mechanics if the boundary conditions are simplified by such a frame of reference. For example, if we want to study the flow in a circular cylindrical tube or the torsion of a circular shaft, it is natural to use the cylindrical coordinates. If we wish to study the stress distribution in a sphere, it is natural to use the spherical coordinates. In fact, if we want to study the explosive forming of a flat sheet of metal into a spherical cap, it may be useful to use a rectangular frame of reference for the original state of the plate and a spherical-polar frame of reference for the deformed state.

It is appropriate to resolve the components of stress in the directions of the curvilinear coordinates and denote them by corresponding subscripts. For example, in a set of cylindrical coordinates, r, θ, z, which are related to the rectangular Cartesian coordinates x, y, z by

$$(3.7\text{-}1) \qquad \begin{cases} x = r\cos\theta, \\ y = r\sin\theta, \\ z = z, \end{cases} \qquad \begin{cases} \theta = \tan^{-1}\dfrac{y}{x}, \\ r^2 = x^2 + y^2, \\ z = z, \end{cases}$$

it is natural to denote the components of stress tensor at a point (r, θ, z) by

$$(3.7\text{-}2) \qquad \begin{pmatrix} \tau_{rr} & \tau_{r\theta} & \tau_{rz} \\ \tau_{\theta r} & \tau_{\theta\theta} & \tau_{\theta z} \\ \tau_{zr} & \tau_{z\theta} & \tau_{zz} \end{pmatrix} \text{ or } \begin{pmatrix} \sigma_r & \tau_{r\theta} & \tau_{rz} \\ \tau_{\theta r} & \sigma_\theta & \tau_{\theta z} \\ \tau_{zr} & \tau_{z\theta} & \sigma_z \end{pmatrix}.$$

To relate these stress components to σ_x, τ_{xy}, etc., let us erect a local rectangular Cartesian frame of reference $x'y'z'$ at the point (r, θ, z), with the origin located at the point (r, θ, z), the x'-axis in the direction of increasing r, the y'-axis in the direction of increasing θ, and the z'-axis parallel to z (see Fig. 3.11). Then, in conventional notation, the stresses $\tau_{x'x'}$, $\tau_{y'y'}$, ...,

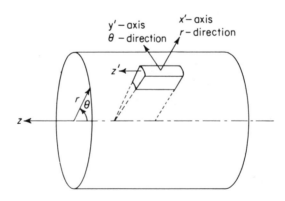

Fig. 3.11   Stress components in cylindrical polar coordinate.

are well defined. Now we can define the stress components listed in (3.7-2) by identifying r, θ, z with x', y', z':

(3.7-3) $\tau_{rr} = \tau_{x'x'}$, $\tau_{r\theta} = \tau_{x'y'}$, $\tau_{\theta\theta} = \tau_{y'y'}$,

etc. Since the coordinated systems x', y', z' and x, y, z are both Cartesian, we can apply the transformation law (3.6-3). The direction cosines of the axes x', y', z' relative to x, y, z are (see Fig. 2.2, and Eq. 2.4-3)

(3.7-4) $$(\beta_{ij}) = \begin{pmatrix} \cos\theta & \sin\theta & 0 \\ -\sin\theta & \cos\theta & 0 \\ 0 & 0 & 1 \end{pmatrix}.$$

Hence, by virtue of Eq. (3.6-3) and (3.7-3), we have

(3.7-5)
$$\sigma_x = \sigma_r \cos^2\theta + \sigma_\theta \sin^2\theta - \tau_{r\theta} \sin 2\theta,$$
$$\sigma_y = \sigma_r \sin^2\theta + \sigma_\theta \cos^2\theta + \tau_{r\theta} \sin 2\theta,$$
$$\sigma_z = \sigma_z,$$
$$\tau_{xy} = (\sigma_r - \sigma_\theta) \sin\theta \cos\theta + \tau_{r\theta}(\cos^2\theta - \sin^2\theta),$$
$$\tau_{zx} = \tau_{zr} \cos\theta - \tau_{z\theta} \sin\theta,$$
$$\tau_{zy} = \tau_{zr} \sin\theta + \tau_{z\theta} \cos\theta.$$

Spherical or other orthogonal curvilinear coordinates can be treated in a similar manner.

3.8 STRESS BOUNDARY CONDITIONS

Problems in mechanics usually appear this way: We know something about the forces or velocities or displacements on the surface of a body of solids or fluids and inquire into what happens inside the body. For example, the wind blows on a building whose foundation we know is firm. What are the stresses acting in the columns and beams? Are they safe? To resolve such questions we set down the known facts concerning the external world in the form of boundary conditions and then use the differential equations (field equations) to extend the information to the interior of the body. If a solution is found to satisfy all the field equations and boundary conditions, then complete information is obtained for the entire interior of the body.

On the surface of a body or at an interface between two bodies the traction (force per unit area) acting on the surface must be the same on both sides of the surface. This indeed is the basic concept of stress which defines the inter-action of one part of a body on another.

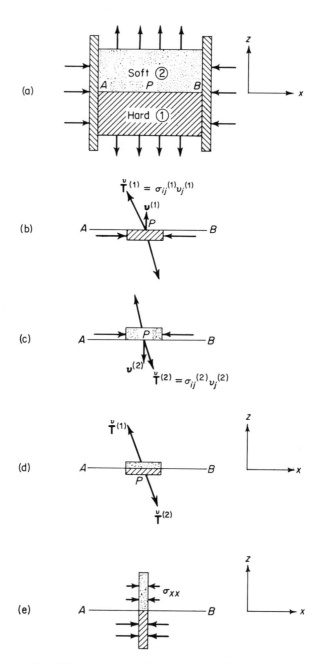

Fig. 3.12 Derivation of the stress boundary condition.

Consider a cube composed of a hard material joined to a soft material, as shown in Fig. 3.12. Let the block be compressed between two plane walls. Both the soft material and the hard material will be stressed. At a point P on the interface AB the situation may be illustrated by a sequence of free-body diagrams as shown in the figure. For the hard material, in the positive side of the interface at P there acts a surface traction $\overset{v}{\mathbf{T}}{}^{(1)}$ whose components are given by $\sigma_{ij}^{(1)}v_j^{(1)}$, where $\sigma_{ij}^{(1)}$ is the stress tensor in the hard material. For the soft material there must exist a similar traction $\overset{v}{\mathbf{T}}{}^{(2)}$, with components $\sigma_{ij}^{(2)}v_j^{(2)}$. The equilibrium of an infinitesimally thin pillbox, as shown in the free-body diagram 3.12(d), requires that

$$(3.8\text{-}1) \qquad\qquad \overset{v}{\mathbf{T}}{}^{(1)} = \overset{v}{\mathbf{T}}{}^{(2)}.$$

This is the condition, referred to above, of equality of surface traction on both sides of an interface. More explicitly, let the interface be the xy-plane ($x = x_1, y = x_2$), and let the z-axis (x_3) be normal to xy. Then, the vector equation (3.8-1) implies the three equations:

$$(3.8\text{-}2) \qquad \sigma_{zz}^{(1)} = \sigma_{zz}^{(2)}, \qquad \sigma_{xz}^{(1)} = \sigma_{xz}^{(2)}, \qquad \sigma_{yz}^{(1)} = \sigma_{yz}^{(2)},$$

which are the boundary conditions on the stresses in the media 1 and 2 at their interface.

Note that the interface conditions (3.8-2) indicate nothing about the stress components $\sigma_{xx}, \sigma_{yy}, \sigma_{xy}$. These components are not required to be continuous across the boundary. Indeed, if the elastic modulus of materials 1 and 2 are unequal and the compressive strain is uniform, then, in general,

$$(3.8\text{-}3) \qquad \sigma_{xx}^{(1)} \neq \sigma_{xx}^{(2)}, \qquad \sigma_{yy}^{(1)} \neq \sigma_{yy}^{(2)}, \qquad \sigma_{xy}^{(1)} \neq \sigma_{xy}^{(2)}.$$

That these discontinuities are not in conflict with any conditions of equilibrium can be seen in Fig. 3.12(e).

A special case of the above is one in which the medium 2 is so soft that its stresses are completely negligible compared with those in 1 (for example, air vs. steel). Then the surface is said to be *free* and the boundary conditions are

$$(3.8\text{-}4) \qquad \sigma_{zz} = 0, \qquad \sigma_{xz} = 0, \qquad \sigma_{yz} = 0.$$

On the other hand, if the traction in medium 2 is known, then it can be considered as the "external" load acting on medium 1. Thus the stress boundary conditions on a solid body usually take the form

(3.8-5) $\sigma_{nn} = p_1,$ $\sigma_{nt_1} = p_2,$ $\sigma_{nt_2} = p_3,$

where p_1, p_2, p_3 are specific functions of location and time and n, t_1, t_2 are a set of local orthogonal axes with n pointing in the direction of outer normal.

Although every surface is an interface between two spaces, it is a general practice to confine one's attention to one side of it and call the other side "external." For example, structures engineers speak of the wind load on a building as the "external load" applied to the building. Reciprocally, to the fluid dynamicist the building is merely a rigid border to the flow of air. The same interface presents two different kinds of boundary conditions to the two media. The basic justification for such a divergence of attitude is that the small elastic deformation of the structure is unimportant to the aerodynamicist who computes the aerodynamic pressure acting on the structure, whereas the elastic deformation is all-important to the structures analyst who determines the safety of the building. Hence, for the aerodynamicist the building is rigid, whereas for the elastician it is not. In other words, both boundary conditions are approximations.

PROBLEMS

3.1 Consider a long string. If you pull on it with a force T, it is clear that the same total tension T acts on every cross section of the string. If we consider the strength of the string, it is intuitively clear that the larger the cross section, the stronger it would be. Thus if we have several strings and wish to compare the strengths of their materials, the comparison should be based on the *stress* (which in this case is equal to the tension T divided by the cross-sectional area) rather than on the total force. It is not too much to hope that all strings of the same material will break at the same ultimate stress. In fact, the problem would be very interesting if an experiment were done and one discovered that all strings made of the same material did not break at the same stress. Could you conceive such a contingency? What if the strings were extremely small? For concreteness, consider nylon threads of diameters 1 cm, 0.1 cm, 10^{-2} cm, 10^{-3} cm, $\ldots$, 10^{-6} cm. When would you begin to feel a little uncertain that some other factors might enter the picture in defining the strength of the threads? What are the factors?

3.2 Take a piece of chalk and break it (a) by bending, (b) by twisting. The way the piece of chalk breaks will be different in these two cases. Why? Can we predict the mode of failure? The cleavage surface?

3.3 A gentle breeze blowing over an expanse of water generates ripples.

Describe the stress vector acting on the water surface. Write down the boundary conditions at the water surface.

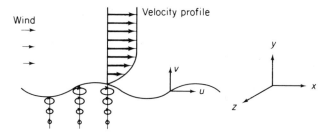

Fig. P3.3   Dynamic boundary conditions at the water-air interface.

3.4 In Fig. P3.4 is shown water in a reservoir. At a point P, let us consider surfaces A-A, B-B, etc. Draw stress vectors acting on these surfaces. Consider all possible surfaces passing through P. What is the locus of all the stress vectors?

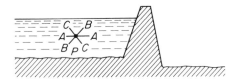

Fig. P3.4   Water in a reservoir.

Answer: A sphere.

3.5 Water in a reservoir is pouring over a dam (Fig. P3.5). Consider a point close to the top of the dam, say 10 cm above it. Again (as in Prob. 3.4), consider all surfaces passing through this point and describe the stress vectors acting on these surfaces. Is the locus of all the stress vectors a sphere?

Now consider a sequence of points closer and closer to the solid surface on the top of the dam, say at distances 1 cm, 10^{-1} cm, 10^{-2} cm, 10^{-3} cm, 10^{-4} cm. Would you expect the stress-vector locus to change as the distance becomes very small? Pay particular attention to the viscosity of water.

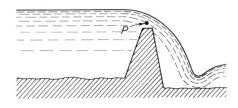

Fig. P3.5   Water pouring over a dam.

3.6 Label the stresses shown in Fig. P3.6.

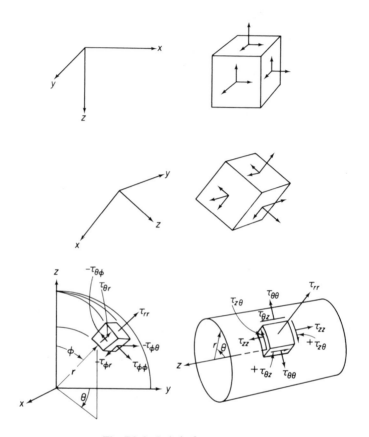

Fig. P3.6 Label of stresses.

3.7 The components of a stress tensor at a certain place in a body may be presented as a matrix:

$$
\begin{array}{c@{\quad}ccc}
 & x & y & z \\
x & 0 & 1 & 2 \\
y & 1 & 2 & 0 \\
z & 2 & 0 & 1
\end{array}.
$$

What is the stress vector acting on the outer side (the side away from the origin) of the following plane, which passes through the place in question?

$$x + 3y + z = 1.$$

What are the normal and tangential components of the stress vector on this plane?

Answer: $\overset{v}{T}_i = (5, 7, 3)/\sqrt{11}$; $T^{(n)} = \frac{29}{11}$, shear $= 0.771$.

Solution: The plane has a normal vector with direction cosines $(1, 3, 1)/\sqrt{11}$. Hence

$$\overset{v}{T}_1 = \frac{(0\cdot1) + (1\cdot3) + (2\cdot1)}{\sqrt{11}} = \frac{5}{\sqrt{11}}, \quad \overset{v}{T}_2 = \frac{7}{\sqrt{11}}, \quad \overset{v}{T}_3 = \frac{3}{\sqrt{11}}.$$

If we use $\mathbf{i}, \mathbf{j}, \mathbf{k}$ to denote unit vectors in the directions of the x-, y-, z-axes, respectively, we have $\overset{v}{\mathbf{T}} = (5\mathbf{i} + 7\mathbf{j} + 3\mathbf{k})/\sqrt{11}$. The normal component equals $\overset{v}{T}_i v_i = \frac{29}{11}$. The shear (tangential component) can be obtained by several methods:

(1) Let shear $= s$ and normal component $= n$. Then

$$s^2 + n^2 = |\overset{v}{T}_i|^2 = \frac{25 + 49 + 9}{11} = \frac{83}{11}.$$

$$s^2 = \frac{83}{11} - \left(\frac{29}{11}\right)^2;$$

hence

$$s = \frac{6\sqrt{2}}{11}.$$

(2) Vector of the normal component plus vector of the shear component equals $\overset{v}{\mathbf{T}}$ vector. The normal component lies in the direction of unit normal $(1\mathbf{i} + 3\mathbf{j} + 1\mathbf{k})29/(11\sqrt{11})$. Let the shear component vector be $x\mathbf{i} + y\mathbf{j} + z\mathbf{k}$; then $29/(11\sqrt{11}) + x = 5/\sqrt{11}$, implying $x = (55 - 29)/36.5 = 0.712$. Similarly $y = -0.274$, $z = 0.109$, and shear $= (x^2 + y^2 + z^2)^{1/2} = 0.771$.

3.8 With reference to the x-, y-, and z-coordinates, the state of stress at a certain point of a body is given by the following matrix

$$(\sigma_{ij}) = \begin{pmatrix} 200 & 400 & 300 \\ 400 & 0 & 0 \\ 300 & 0 & -100 \end{pmatrix} \text{ psi.}$$

Find the stress vector acting on a plane passing through the point and parallel to the plane $x + 2y + 2z - 6 = 0$.

Answer: $\overset{v}{\mathbf{T}} = 533\mathbf{i} + 133\mathbf{j} + 33\mathbf{k}$.

3.9 Does equilibrium exist for the following stress distribution in the absence of body force?

$$\sigma_x = 3x^2 + 4xy - 8y^2, \quad \tau_{xy} = -\tfrac{1}{2}x^2 - 6xy - 2y^2,$$
$$\sigma_y = 2x^2 + xy + 3y^2, \quad \sigma_z = \tau_{xz} = \tau_{yz} = 0.$$

Answer: Yes, according to Eq. (3.5-2).

3.10 The stress at a point is $\sigma_x = 5000$ psi, $\sigma_y = 5000$ psi, $\tau_{xy} = \sigma_z = \tau_{xz} = \tau_{yz} = 0$. Consider all planes passing through this point. On each plane there acts a stress vector which can be resolved into two components: one normal to the plane and one tangential to it. These components are called the *normal* and *shear stress*, respectively. Consider planes oriented in all directions. Show that the maximum shear stress in the material at the point is 2500 psi.

Note: It is instructive to solve this problem according to the first principle. However, the results to be discussed in Chapter 4 (especially p. 95) will simplify the solution greatly.

Solution: Let a coordinate system be so chosen that the point in question is located at the origin. A plane passing through this point may be represented by the equation

$$lx + my + nz = 0 \tag{1}$$

where (l, m, n) is the direction cosine of the normal to the plane. Hence $(v_1, v_2, v_3) = (l, m, n)$ and $l^2 + m^2 + n^2 = 1$. By permitting the normal vector to assume all possible directions, we obtain all the planes named in the problem. Now the stress vector acting on the plane [Eq. (1)] is $(\overset{v}{T_1}, \overset{v}{T_2}, \overset{v}{T_3})$ $= (5000l, 5000m, 0)$. The normal component of surface traction is the component of $(\overset{v}{T_i})$ in the direction of (v_i), i.e., the scalar product of these vectors: Normal stress $= 5000(l^2 + m^2)$. Hence (shear stress)^2 $= (\overset{v}{T_i})^2 - $ (normal stress)$^2 = (5000)^2(l^2 + m^2) - (5000)^2(l^2 + m^2)^2 = (5000)^2[l^2 + m^2 - (l^2 + m^2)^2]$. But $l^2 + m^2 + n^2 = 1$; hence

$$\text{(shear stress)}^2 = (5000)^2[1 - n^2 - (1 - n^2)^2] \tag{2}$$

To find the value of n which is less than 1 and which renders the shear stress a relative maximum, we set

$$0 = \frac{\partial}{\partial n} \text{(shear stress)}^2 = (5000)^2[-2n + 2(1 - n^2) \cdot 2n].$$

The solution is $n^2 = \frac{1}{2}$. Hence from Eq. (2) we obtain the maximum shear stress squared, $(5000)^2/4$, and the final result that the maximum shear stress is 2500 psi.

3.11 If the state of stress at a point (x_0, y_0, z_0) is

$$(\sigma_{ij}) = \begin{pmatrix} 100 & 0 & 0 \\ 0 & 50 & 0 \\ 0 & 0 & -100 \end{pmatrix} \text{psi},$$

find the stress vector and the magnitude of the normal and shearing stress acting on the plane $x - x_0 + y - y_0 + z - z_0 = 0$.

Answer: $\overset{v}{\mathbf{T}} = \frac{1}{\sqrt{3}}(100, 50, -100)$ psi, $\sigma^{(n)} = 16.7$ psi, $\tau = 81.7$ psi.

3.12 For the keyed shaft shown in Fig. P3.12, determine the maximum permissible value of the load P if the stress in the shear key is not to exceed 10,000 lb/sq in.

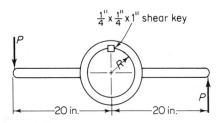

$\frac{1}{4}'' \times \frac{1}{4}'' \times 1''$ shear key

Fig. P3.12   Key on a shaft.

Solution: The key has a cross section $\frac{1}{4}$ in. $\times \frac{1}{4}$ in. and is 1 in. long. The effective area for shear stress resisting the torque is $\frac{1}{4}$ in.². The limiting shear stress in the shear key is 10,000 psi. The limiting shear force $= 10{,}000 \cdot \frac{1}{4} = 2500$ lb. The torque is therefore $2500 \cdot R$, where R is the radial distance of the key from the center of the shaft. The external torque is $P(20 + 20) = 40P$. Therefore, $40P = 2500R$, and we have $P = 62.5R$ lb.

3.13 Prove that if $\sigma_z = \sigma_{zx} = \sigma_{zy} = 0$, then under the coordinate transformation (2.4-3), p. 48, we have $\sigma_x + \sigma_y = \sigma_{x'} + \sigma_{y'}$; that is, in a planar stress distribution the sum of the two normal stresses is an invariant.

3.14 Two sheets of plywood are spliced together as shown in Fig. P3.14. If the allowable shear stress in the glue is 200 lb/sq in., what must be the minimum length L of the splice pieces if a 10,000 lb load is to be carried.

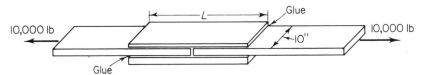

Fig. P3.14   Glued seam.

Answer: $L_{\min} = 5$ in.

3.15 Consider oceanography. What can you say about the stress vector acting on the bottom of the ocean? How much of the stress state on the surface of the ocean floor can be stated immediately (a) on the water side and (b) on the solid side? What components of stress cannot be determined without considering the entire earth?

3.16 The set of eight planes with direction numbers $(\pm 1, \pm 1, \pm 1)$, with one of the $\pm$ signs chosen in each case, e.g., $(1, 1, -1)$ corresponding to the plane $x + y - z = 0$, are called the *octahedral* planes. Let the state of stress be specified by τ_{ij} for which $\tau_{ij} = 0$ whenever $i \neq j$. Determine the stress vector and the shear stress acting on each of the octahedral planes.

3.17 Have you ever seen a tree branch broken by wind and observed how it split? What does it tell us about the strength characteristics of the wood?

3.18 Experiment by attempting to break various materials such as macaroni, celery, carrots, high carbon steel tools like drills and files, strips of aluminum and magnesium, or silicone Silly Putty. Discuss the strength characteristics of these materials.

3.19 A circular cylindrical rod is twisted. Describe the stress state in the rod. Use the notations shown in Fig. P3.6 (p. 84) or Fig. 3.11 (p. 78). Discuss, in particular, the stress components at a point on the outer, free surface of the rod.

3.20 A water tower which consists of a big raindrop-shaped tank supported on top of a column is hit by an earthquake. The tower shakes. The maximum lateral acceleration (in a direction perpendicular to the column) is estimated to be 0.2 times the gravitational acceleration. The maximum lateral inertia force induced by the earthquake is therefore equal to 20% of the weight of the tank and water acting in a horizontal direction. The maximum vertical acceleration is about the same magnitude. Discuss the stress state in the column.

3.21 Discuss the state of stress distribution in an airplane wing during flight and landing.

3.22 *Couette Flow.* The space between two concentric cylinders is filled with a fluid (Fig. P3.22). Let the inner cylinder be stationary while the outer cylinder is rotated at an angular speed ω radians/sec. If the torque measured on the inner cylinder is T, what is the torque acting on the outer cylinder? Why?

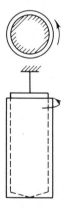

Fig. P3.22   Couette flow meter.

3.23 In designing a tie rod, it is decided that the maximum shear stress must not exceed 20,000 lb/sq in. (because of possible yielding). What is the maximum tension the rod can take? Use steel.

Answer: 40,000 psi.

3.24 Take a thin strip of steel of rectangular cross section (say, $\frac{1}{16}$ in. $\times$ 1 in. $\times$ 18 in.). Using a handbook, find the ultimate strength of the steel (of the order of 100,000 lb/sq in.). Let the strip be used to sustain a compressive load in the direction of the longest edge. On the basis of the ultimate strength alone, how many pounds of load should the strip be able to resist?

Now try to compress the strip with your hand. The strip buckles at a load far smaller than is expected. Explain this phenomenon of elastic buckling.

3.25 Roll a sheet of paper into a circular cylinder of radius about 1 or 2 ins. Such a tube can sustain a fairly sizable end compression.

Stand the tube on the table and compress it with the palm of your hand. The cylinder will fail by buckling. Describe the buckling pattern. How large is the buckling load compared with the strength of the paper in compression if buckling can be avoided?

Since the paper does not tear after buckling, nor does it stretch, the *metric* of the deformed surface is identical with that of the original one. Hence the transformation from the cylinder to the buckled surface is an *isometric* transformation.

It is known in differential geometry that if a surface can be transformed isometrically into another, their total curvature must be the same at corresponding points. Now the total curvature of a surface is the product of the principal curvatures. For a flat sheet of paper the total curvature is zero; so is that of the cylinder; so also must be the post-buckling surface. In this way we expect the post-buckling surface to be composed of areas with zero total curvature, namely, flat triangular portions that are assembled together into a diamond pattern. Compare this with the experimental findings.

Note: The subject in this problem is of great interest to aeronautical and astronautical engineering. Thin-walled structures are used extensively when light weight is mandatory. Elastic stability designs these structures.

3.26 A rope is hung from the ceiling. Let the density of the rope be 2 g/cm^3. Find the stress in the rope.

Solution: Let the x-axis be chosen in the direction of the rope. The only stress of concern is σ_x. We shall assume $\tau_{xy} = \tau_{xz} = 0$. Then the equation of equilibrium is

$$\frac{\partial \sigma_x}{\partial x} + \rho g = 0$$

where g is the gravitational acceleration. The solution is

$$\sigma_x = -\rho g x + \text{const.}$$

But $\sigma_x = 0$ when $x = L$, the length of the rope. Hence the constant is $\rho g L$. Thus $\sigma_x = \rho g(L - x)$. The maximum tension is at the ceiling where $\sigma_x = \rho g L$.

3.27 Consider a vertical column of an isothermal atmosphere that obeys the gas law $p/\rho = RT$ or $\rho = p/RT$, where ρ is the density of the gas, p is the pressure, R is the gas constant, and T is the absolute temperature. This gas is subjected to a gravitational acceleration g so that the body force is ρg per unit volume, pointing to the ground. If the pressure at the ground level $z = 0$ is p_0, determine the relation between the pressure and the height z above the ground.

Answer: $p = p_0 \exp[-(g/RT)z]$.

3.28 Discuss why the solution given in Prob. 3.27 is unrealistic for the earth's atmosphere. If the temperature T is a known function of the height z, what would the solution be?

3.29 Consider a two-dimensional state of stress in a thin plate in which $\tau_{zz} = \tau_{zx} = \tau_{zy} = 0$. The equations of equilibrium acting in the plate in the absence of body force are

$$\frac{\partial \sigma_x}{\partial x} + \frac{\partial \tau_{xy}}{\partial y} = 0, \qquad \frac{\partial \tau_{xy}}{\partial x} + \frac{\partial \sigma_y}{\partial y} = 0.$$

Show that these equations are satisfied identically if $\sigma_x, \sigma_y, \sigma_{xy}$ are derived from an arbitrary function $\Phi(x, y)$:

$$\sigma_x = \frac{\partial^2 \Phi}{\partial y^2}, \qquad \sigma_y = \frac{\partial^2 \Phi}{\partial x^2}, \qquad \tau_{xy} = -\frac{\partial^2 \Phi}{\partial x\, \partial y}.$$

Thus the equations of equilibrium can be satisfied by infinitely many solutions. To select the correct one from among these solutions is a problem to be resolved in subsequent chapters.

CHAPTER FOUR

Principal Stresses and Principal Axes

Principal stresses, stess invariants, stress deviations, and the maximum shear are important concepts. They tell us the state of stress in the simplest numerical way. They are directly related to the failure strength of materials. One has to evaluate them frequently; therefore, we devote a chapter to them.

4.1 INTRODUCTION

We have seen that nine components of stress, of which six are independent, are necessary to specify the state of material interaction at any given point in a body. These nine components of stress form a symmetric matrix

$$\sigma = \begin{pmatrix} \sigma_{11} & \sigma_{12} & \sigma_{13} \\ \sigma_{21} & \sigma_{22} & \sigma_{23} \\ \sigma_{31} & \sigma_{32} & \sigma_{33} \end{pmatrix}, \qquad (\sigma_{ij} = \sigma_{ji}),$$

the components of which transform as the components of a tensor under rotation of coordinates. Later we shall show that because the stress tensor is symmetric, a set of coordinates can be found with respect to which the matrix of stress components can be reduced to a diagonal matrix of the form

$$\boldsymbol{\sigma} = \begin{pmatrix} \sigma_1 & 0 & 0 \\ 0 & \sigma_2 & 0 \\ 0 & 0 & \sigma_3 \end{pmatrix}.$$

The particular set of coordinates axes with respect to which the stress matrix is diagonal are called the *principal axes*, and the corresponding stress components are called the *principal stresses*. The coordinate planes determined by the principal axes are called the *principal planes*. Physically, each of the principal stresses is a normal stress acting on a principal plane. On the principal plane the stress vector is normal to the plane, and there is no shear component.

To know the principal axes and principal stresses is obviously useful, because they help us visualize the state of stress at any point. In fact, the matter is so important that in solving problems in continuum mechanics we rarely stop before the final answer is reduced to principal values. We need to know, therefore, not only that the principal stresses exist and can be found *in principle* but also the practical means of finding them. We shall show that the symmetry of the stress tensor is the basic reason for the existence of principal axes. Other symmetric tensors, such as the strain tensor, by analogy and identical mathematical process, must also have principal axes and principal values. Indeed, the proof for the possibility of reducing a real-valued symmetric matrix into a principal one is not limited to three dimensions but can be extended to *n* dimensions, and even to infinite dimensions. We shall find that such an extension is of great importance when we consider mechanical vibrations of elastic bodies, or acoustics in general. In the vibration theory, the principal values correspond to the *vibration frequencies*, and the principal coordinates describe the *normal modes* of vibration. We shall not discuss these subjects now but merely point out that the subject we are going to study has much broader applications than to stress alone.

As an introduction, we shall consider the two-dimensional case in greater details. Then we shall proceed to the three-dimensional case in abridged notations. Finally, we shall use the principal stresses to discuss some geometric representations of the stress state, as well as to introduce some additional definitions.

4.2 PLANE STATE OF STRESS

Let us consider a simplified physical situation in which a thin membrane is stretched by forces acting on the edges of the membrane and lying in the plane of the membrane. An example is shown in Fig. 4.1. We shall leave the faces $z = h$ and $z = -h$ free (unstressed). In this case, we can safely say that

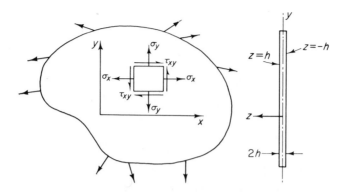

Fig. 4.1   An approximate plane state or stress.

since the stress components σ_{zz}, σ_{zy}, σ_{zx} are zero on the surfaces, they are approximately zero throughout the membrane because the membrane is very thin:

(4.2-1) $\sigma_{zz} = \sigma_{zx} = \sigma_{zy} = 0.$

The state of stress in which the equations above hold is called a *plane state of stress* in the xy-plane. Obviously, in plane stress we are concerned only with the stress components in the symmetric matrix

$$\begin{pmatrix} \sigma_x & \tau_{xy} & 0 \\ \tau_{xy} & \sigma_y & 0 \\ 0 & 0 & 0 \end{pmatrix}.$$

Here for clarity we write σ_x for σ_{xx}, σ_y for σ_{yy}, and τ_{xy} for σ_{xy}.

We shall now consider a rotation of coordinates from xy to $x'y'$ and apply the results of Sec. 3.6 to find the stress components in the new frame of reference:

$$\begin{pmatrix} \sigma_{x'} & \tau_{x'y'} & 0 \\ \tau_{x'y'} & \sigma_{y'} & 0 \\ 0 & 0 & 0 \end{pmatrix}.$$

In this case, the direction cosines between the two systems of rectangular Cartesian coordinates can be expressed in terms of a single angle θ. See Fig. 4.2. The matrix of direction cosines is

(4.2-2) $$\begin{pmatrix} \beta_{11} & \beta_{12} & \beta_{13} \\ \beta_{21} & \beta_{22} & \beta_{23} \\ \beta_{31} & \beta_{32} & \beta_{33} \end{pmatrix} = \begin{pmatrix} \cos\theta & \sin\theta & 0 \\ -\sin\theta & \cos\theta & 0 \\ 0 & 0 & 1 \end{pmatrix}.$$

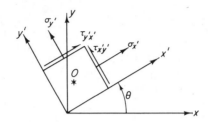

Fig. 4.2 Change of coordinates in plane state of stress.

Writing x, y and x', y' in place of x_1, x_2 and x'_1, x'_2; σ_x for τ_{11}; τ_{xy} for τ_{12}, etc., and identifying direction cosines β_{ij} according to Eq. (4.2-2), we obtain, on substituting into Eq. (3.6-3), the new components:

(4.2-3) $\sigma_{x'} = \sigma_x \cos^2 \theta + \sigma_y \sin^2 \theta + 2\tau_{xy} \sin \theta \cos \theta,$

(4.2-4) $\sigma_{y'} = \sigma_x \sin^2 \theta + \sigma_y \cos^2 \theta - 2\tau_{xy} \sin \theta \cos \theta,$

(4.2-5) $\tau_{x'y'} = (-\sigma_x + \sigma_y) \sin \theta \cos \theta + \tau_{xy}(\cos^2 \theta - \sin^2 \theta).$

Since

$$\sin^2 \theta = \tfrac{1}{2}(1 - \cos 2\theta), \qquad \cos^2 \theta = \tfrac{1}{2}(1 + \cos 2\theta),$$

we may write the above equations as

(4.2-6) $\sigma_{x'} = \dfrac{\sigma_x + \sigma_y}{2} + \dfrac{\sigma_x - \sigma_y}{2} \cos 2\theta + \tau_{xy} \sin 2\theta,$

(4.2-7) $\sigma_{y'} = \dfrac{\sigma_x + \sigma_y}{2} - \dfrac{\sigma_x - \sigma_y}{2} \cos 2\theta - \tau_{xy} \sin 2\theta,$

(4.2-8) $\tau_{x'y'} = -\dfrac{\sigma_x - \sigma_y}{2} \sin 2\theta + \tau_{xy} \cos 2\theta.$

From these equations we can deduce that

(4.2-9) $\sigma_{x'} + \sigma_{y'} = \sigma_x + \sigma_y,$

(4.2-10) $\dfrac{\partial \sigma_{x'}}{\partial \theta} = 2\tau_{x'y'}, \qquad \dfrac{\partial \sigma_{y'}}{\partial \theta} = -2\tau_{x'y'},$

(4.2-11) $\tau_{x'y'} = 0$ when $\tan 2\theta = \dfrac{2\tau_{xy}}{\sigma_x - \sigma_y}.$

The directions of the x'-, y'-axes corresponding to the particular values of θ given by Eq. (4.2-11) are called the *principal directions*; the axes x' and y' are then called the *principal axes*, and $\sigma_{x'}$, $\sigma_{y'}$ are called the *principal stresses*. If x', y' were principal axes, then $\tau_{x'y'} = 0$, and Eq. (4.2-10) show that

$\sigma_{x'}$ is either a maximum or a minimum with respect to all choices of θ. Similarly it is the case with $\sigma_{y'}$. On substituting θ from Eq. (4.2-11) into Eq. (4.2-6) and (4.2-7), we obtain the result

(4.2-12) ▲ $$\begin{matrix}\sigma_{max}\\\sigma_{min}\end{matrix} = \frac{\sigma_x + \sigma_y}{2} \pm \sqrt{\left(\frac{\sigma_x - \sigma_y}{2}\right)^2 + \tau_{xy}^2}.$$

On the other hand, on differentiating $\tau_{x'y'}$ from Eq. (4.2-8) with respect to θ and setting the derivative equal to zero, we can find an angle θ at which $\tau_{x'y'}$ reaches its extreme value. It can be shown that this angle is $\pm 45°$ from the principal directions given by Eq. (4.2-11) and that the maximum value of $\tau_{x'y'}$ is

(4.2-13) ▲ $$\tau_{max} = \frac{\sigma_{max} - \sigma_{min}}{2} = \sqrt{\left(\frac{\sigma_x - \sigma_y}{2}\right)^2 + \tau_{xy}^2}.$$

This is the maximum of the shear stress acting on all planes parallel to the z-axis. When planes inclined to the z-axis are also considered, some other planes may have a shear higher than this. See p. 102 and 108.

4.3 MOHR'S CIRCLE FOR PLANE STRESS

A geometric representation of the equations given in the previous section was given by Otto Mohr (*Zivilingenieur*, 1882, p. 113, and his *Technische Mechanik*, 2nd ed. 1914). An example is shown in Fig. 4.3. The normal and shear stresses acting on a surface are plotted on a *stress plane* in which the abscissa represents the normal stress and the ordinate, the shear. For the normal stress a tension is plotted as positive, and a compression, negative. For the shear a special rule is needed. We specify (for Mohr's circle construction only) that a shear is taken as positive on a face of an element when it yields a clockwise moment about the center point O of the element (see Fig. 4.2). A shear stress yielding a counterclockwise moment about the center O is taken to be negative. Thus, $\tau_{x'y'}$ in Fig. 4.2 is considered negative and $\tau_{y'x'}$ is considered positive. Following this special rule, we plot, in Fig. 4.3, the point A whose abscissa is σ_x and whose ordinate is $-\tau_{xy}$ and the point B with abscissa σ_y and ordinate τ_{yx}. Join the line AB, which intersects the σ axis at C. With C as a center, draw a circle through A and B. This is Mohr's circle.

To obtain the stresses acting on a surface whose normal makes an angle θ with the x-axis in the counterclockwise direction we draw a radius CP which makes an angle 2θ with the line CA, as shown in Fig. 4.3. Then the abscissa of P gives the normal stress on that surface; and the ordinate, the shear. The point located on the other end of the diameter PQ represents the stress acting on a surface whose normal makes an angle $\theta + (\pi/2)$ with the x-axis.

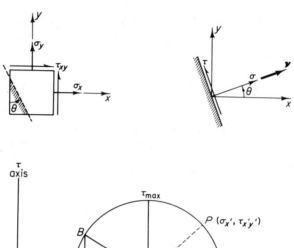

Fig. 4.3   Mohr's circle for plane stress.

To prove that this construction is valid, we note that Mohr's circle has a center located at C, where

$$(4.3\text{-}1) \qquad \overline{OC} = \frac{\sigma_x + \sigma_y}{2},$$

and a radius

$$(4.3\text{-}2) \qquad \overline{AC} = \overline{CP} = \sqrt{\left(\frac{\sigma_x - \sigma_y}{2}\right)^2 + \tau_{xy}^2}.$$

From Fig. 4.3 we see that the abscissa of P is

$$(4.3\text{-}3) \qquad \sigma_{x'} = \overline{OC} + \overline{CP}\cos(2\theta - 2\alpha)$$
$$= \overline{OC} + \overline{CP}(\cos 2\theta \cos 2\alpha + \sin 2\theta \sin 2\alpha).$$

But we see also from the diagram that

$$(4.3\text{-}4) \qquad \cos 2\alpha = \frac{\sigma_x - \sigma_y}{2\overline{CP}}, \qquad \sin 2\alpha = \frac{\tau_{xy}}{\overline{CP}}.$$

Substituting these results into Eq. (4.3-3), we get

(4.3-5) $\qquad \sigma_{x'} = \dfrac{\sigma_x + \sigma_y}{2} + \dfrac{\sigma_x - \sigma_y}{2}\cos 2\theta + \tau_{xy}\sin 2\theta,$

which is exactly the same as Eq. (4.2-6).

Similarly, we have the ordinate of P:

(4.3-6) $\quad \tau_{x'y'} = \overline{CP}\sin(2\theta - 2\alpha) = \overline{CP}(\sin 2\theta \cos 2\alpha - \cos 2\theta \sin 2\alpha)$

$\qquad\qquad = \dfrac{\sigma_x - \sigma_y}{2}\sin 2\theta - \tau_{xy}\cos 2\theta,$

which agrees with Eq. (4.2-8) in magnitude but differs in sign. The sign is fixed by the convention adopted here for Mohr's circle. A positive-valued $\tau_{x'y'}$ according to Eq. (4.2-8) would be a counterclockwise moment and would have been plotted with negative ordinate on Mohr's circle. Hence, everything agrees, and the validity of Mohr's circle is proved.

Mohr's circle gives a visual picture of how the stress varies at a point for plane stress. For such purposes, the construction is very helpful. It tells us how to locate the principal axes, and it makes the important formulas (4.2-12) and (4.2-13) graphic and easy to remember. It shows us that the planes on which the maximum shear occurs are 45° from the principal planes. It shows us that in the case of *pure shear*, $\sigma_x = -\sigma_y$, Mohr's circle is a circle with the origin as the center, and on the plane at 45° from the principal axes the shear reaches the maximum while the normal stress is zero. In the case of hydrostatic stress, $\sigma_x = \sigma_y$, $\tau_{xy} = 0$, Mohr's circle is a single point.

PROBLEM 4.1 Given the state of stress as shown in Fig. P4.1. Determine (a) the principal stresses, and (b) the maximum shearing stress. Show the results on properly oriented elements.

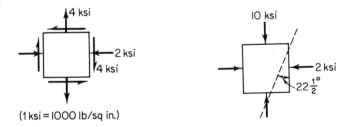

(1 ksi = 1000 lb/sq in.)

Fig. P4.1 A state of stress.   **Fig. P4.2** Stress on an inclined section.

PROBLEM 4.2 The stresses shown in Fig. P4.2 are principal stresses. Determine the stresses acting on the plane at an angle of $22\frac{1}{2}°$ with the vertical axis as shown.

4.4 PRINCIPAL STRESSES

Let us now generalize some of the results obtained in the previous sections, which were restricted to the plane state of stress. For the convenience of writing, we shall return to the index notation, with reference to rectangular Cartesian coordinates x_1, x_2, x_3.

In a general state of stress, the stress vector acting on a surface with normal $\mathbf{v}$ depends on the direction of $\mathbf{v}$. At a given point in a body, the angle between the stress vector and the normal $\mathbf{v}$ varies with the orientation of the surface. We shall show that we can always find a surface which is so oriented that the stress vector is exactly normal to it. In fact, we shall show that there are at least three mutually orthogonal surfaces that fulfill this requirement at any point in the body. Such a surface is called a *principal plane*; its normal, a *principal axis*; and the value of normal stress acting on the principal plane, a *principal stress*.

Let $\mathbf{v}$ be a unit vector in the direction of a principal axis and let σ be the corresponding principal stress. Then the stress vector acting on the surface normal to $\mathbf{v}$ has components σv_i. On the other hand, this same vector is given by the expression $\tau_{ji}v_j$, where τ_{ij} is the stress tensor. Hence, writing $v_i = \delta_{ji}v_j$, we have, on equating these two expressions and transposing to the same side,

$$(4.4\text{-}1) \qquad (\tau_{ji} - \sigma\delta_{ji})v_j = 0, \qquad (i = 1, 2, 3).$$

These three equations are to be solved for v_1, v_2, and v_3. Since $\mathbf{v}$ is a unit vector, we must find a set of nontrivial solutions for which $v_1^2 + v_2^2 + v_3^2 = 1$. Thus, Eq. (4.4-1) pose an eigenvalue problem. Since τ_{ij} as a matrix is real-valued and symmetric, we need only to recall a result in the theory of matrices to assert that *there exist three real-valued principal stresses and a set of orthonormal principal axes. Whether the principal stresses are all positive, all negative, or mixed depends on whether the quadratic form $\tau_{ij}x_ix_j$ is positive definite, negative definite, or uncertain, respectively.* However, because of the importance of these results, we shall derive them anew below.

The system of Eq. (4.4-1) has a set of nonvanishing solutions v_1, v_2, v_3 if and only if the determinant of the coefficients vanishes; i.e.,

$$(4.4\text{-}2) \qquad |\tau_{ij} - \sigma\delta_{ij}| = 0.$$

Equation (4.4-2) is a cubic equation in σ; its roots are the principal stresses. For each value of the principal stress, a unit normal vector $\mathbf{v}$ can be determined.

On expanding Eq. (4.4-2), we have

$$(4.4\text{-}3) \qquad |\tau_{ij} - \sigma\delta_{ij}| = \begin{vmatrix} \tau_{11} - \sigma & \tau_{12} & \tau_{13} \\ \tau_{21} & \tau_{22} - \sigma & \tau_{23} \\ \tau_{31} & \tau_{32} & \tau_{33} - \sigma \end{vmatrix}$$

$$= -\sigma^3 + I_1\sigma^2 - I_2\sigma + I_3 = 0,$$

where

$$(4.4\text{-}4) \qquad I_1 = \tau_{11} + \tau_{22} + \tau_{33},$$

$$(4.4\text{-}5) \qquad I_2 = \begin{vmatrix} \tau_{22} & \tau_{23} \\ \tau_{32} & \tau_{33} \end{vmatrix} + \begin{vmatrix} \tau_{33} & \tau_{31} \\ \tau_{13} & \tau_{11} \end{vmatrix} + \begin{vmatrix} \tau_{11} & \tau_{12} \\ \tau_{21} & \tau_{22} \end{vmatrix},$$

$$(4.4\text{-}6) \qquad I_3 = \begin{vmatrix} \tau_{11} & \tau_{12} & \tau_{13} \\ \tau_{21} & \tau_{22} & \tau_{23} \\ \tau_{31} & \tau_{32} & \tau_{33} \end{vmatrix}.$$

On the other hand, if σ_1, σ_2, σ_3 are the roots of Eq. (4.4-3), which can then be written as

$$(4.4\text{-}7) \qquad (\sigma - \sigma_1)(\sigma - \sigma_2)(\sigma - \sigma_3) = 0,$$

it can be seen that the following relations between the roots and the coefficients must hold:

$$(4.4\text{-}8) \qquad I_1 = \sigma_1 + \sigma_2 + \sigma_3,$$

$$(4.4\text{-}9) \qquad I_2 = \sigma_1\sigma_2 + \sigma_2\sigma_3 + \sigma_3\sigma_1,$$

$$(4.4\text{-}10) \qquad I_3 = \sigma_1\sigma_2\sigma_3.$$

Since the principal stresses characterize the physical state of stress at a point, they are independent of any coordinates of reference. Hence, Eq. (4.4-7) is independent of the orientation of the coordinates of reference. But (4.4-7) is exactly the same as (4.4-3). Therefore, Eq. (4.4-3) and the coefficients I_1, I_2, I_3 are invariant with respect to the rotation of coordinates. I_1, I_2, I_3 are called the *invariants* of the stress tensor with respect to rotation of coordinates.

We shall show now that for a symmetric stress tensor the three principal stresses are all real and the three principal planes are mutually orthogonal. These important properties can be established when the stress tensor is symmetric,

$$(4.4\text{-}11) \qquad \tau_{ij} = \tau_{ji}.$$

The proof is as follows. Let $\overset{1}{\mathbf{v}}, \overset{2}{\mathbf{v}}, \overset{3}{\mathbf{v}}$ be unit vectors in the direction of the principal axes, with components $\overset{1}{v_i}, \overset{2}{v_i}, \overset{3}{v_i} (i = 1, 2, 3)$ which are the solutions of Eq. (4.4-1) corresponding to the roots $\sigma_1, \sigma_2, \sigma_3$, respectively;

$$(\tau_{ij} - \sigma_1 \delta_{ij})\overset{1}{v_j} = 0,$$

(4.4-12) $$(\tau_{ij} - \sigma_2 \delta_{ij})\overset{2}{v_j} = 0,$$

$$(\tau_{ij} - \sigma_3 \delta_{ij})\overset{3}{v_j} = 0.$$

Multiplying the first equation by $\overset{2}{v_i}$ and the second by $\overset{1}{v_i}$, summing over i, and substracting the resulting equations, we obtain

(4.4-13) $$(\sigma_2 - \sigma_1)\overset{1}{v_i}\overset{2}{v_i} = 0$$

on account of the symmetry condition (4.4-11), which implies that

(4.4-14) $$\tau_{ij}\overset{1}{v_j}\overset{2}{v_i} = \tau_{ji}\overset{1}{v_j}\overset{2}{v_i} = \tau_{ij}\overset{2}{v_j}\overset{1}{v_i},$$

the last equality being obtained by interchanging the dummy indices i and j.

Now, if we assume tentatively that Eq. (4.4-3) has a complex root, then, since the coefficients in Eq. (4.4-3) are real-valued, a complex conjugate root must also exist, and the set of roots may be written as

$$\sigma_1 = \alpha + i\beta, \qquad \sigma_2 = \alpha - i\beta, \qquad \sigma_3$$

where α, β, σ_3 are real numbers and i stands for the imaginary number $\sqrt{-1}$. In this case, Eq. (4.4-12) show that $\overset{1}{v_j}$ and $\overset{2}{v_j}$ are conjugate to each other and can be written as

$$\overset{1}{v_j} \equiv a_j + ib_j, \qquad \overset{2}{v_j} \equiv a_j - ib_j.$$

Therefore,

$$\overset{1}{v_j}\overset{2}{v_j} = (a_j + ib_j)(a_j - ib_j)$$
$$= a_1^2 + a_2^2 + a_3^2 + b_1^2 + b_2^2 + b_3^2 \neq 0.$$

It follows from Eq. (4.4-13) that $\sigma_1 - \sigma_2 = 2i\beta = 0$ or $\beta = 0$. This contradicts the original assumption that the roots are complex. Thus, the assumption of the existence of complex roots is untenable, and the roots $\sigma_1, \sigma_2, \sigma_3$ are all real.

When $\sigma_1 \neq \sigma_2 \neq \sigma_3$, Eq. (4.4-13) and similar equations imply

(4.4-15) $$\overset{1}{v_i}\overset{2}{v_i} = 0, \qquad \overset{2}{v_i}\overset{3}{v_i} = 0, \qquad \overset{3}{v_i}\overset{1}{v_i} = 0;$$

i.e., the principal vectors are mutually orthogonal to each other. If $\sigma_1 = \sigma_2$ $\neq \sigma_3$, $\overset{3}{v_i}$ will be fixed, but we can determine an infinite number of pairs of vectors $\overset{1}{v_i}$ and $\overset{2}{v_i}$ orthogonal to $\overset{3}{v_i}$. If $\sigma_1 = \sigma_2 = \sigma_3$, then any set of orthogonal axes may be taken as the principal axes.

If the reference axes x_1, x_2, x_3 are chosen to coincide with the principal axes, then the matrix of stress components becomes

$$(4.4\text{-}16) \qquad (\tau_{ij}) = \begin{pmatrix} \sigma_1 & 0 & 0 \\ 0 & \sigma_2 & 0 \\ 0 & 0 & \sigma_3 \end{pmatrix}.$$

4.5 SHEARING STRESSES

We have seen that on an element of surface with a unit outer normal $\mathbf{v}$ (with components v_i) there acts a traction $\overset{v}{\mathbf{T}}$, ($\overset{v}{T_i} = \tau_{ji}v_j$). The component of $\overset{v}{\mathbf{T}}$ in the direction of $\mathbf{v}$ is the normal stress acting on the surface element. Let this normal stress be denoted by $\sigma_{(n)}$. Since the component of a vector in the direction of a unit vector is given by the scalar product of the two vectors, we obtain

$$(4.5\text{-}1) \qquad \sigma_{(n)} = \overset{v}{T_i}v_i = \tau_{ij}v_iv_j.$$

On the other hand, since the vector $\overset{v}{\mathbf{T}}$ can be decomposed into two orthogonal components $\sigma_{(n)}$ and τ, where τ denotes the shearing stress tangent to the surface (see Fig. 4.4), we see that the magnitude of the shearing stress on a surface having the normal $\mathbf{v}$ is given by the equation

$$(4.5\text{-}2) \qquad \tau^2 = |\overset{v}{T_i}|^2 - \sigma_{(n)}^2.$$

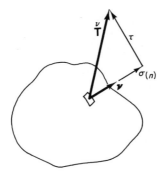

Fig. 4.4 Notations.

Let the principal axes be chosen as the coordinate axes, and let $\sigma_1, \sigma_2, \sigma_3$ be the principal stresses. Then

(4.5-3)
$$\overset{v}{T}_1 = \sigma_1 v_1, \qquad \overset{v}{T}_2 = \sigma_2 v_2, \qquad \overset{v}{T}_3 = \sigma_3 v_3,$$
$$|\overset{v}{T}_i|^2 = (\sigma_1 v_1)^2 + (\sigma_2 v_2)^2 + (\sigma_3 v_3)^2,$$

and, from Eq. (4.5-1),

(4.5-4)
$$\sigma_{(n)} = \sigma_1 v_1^2 + \sigma_2 v_2^2 + \sigma_3 v_3^2,$$

(4.5-5)
$$\sigma_{(n)}^2 = [\sigma_1 v_1^2 + \sigma_2 v_2^2 + \sigma_3 v_3^2]^2.$$

On substituting into Eq. (4.5-2) and noting that

(4.5-6)
$$(v_1)^2 - (v_1)^4 = (v_1)^2[1 - (v_1)^2] = (v_1)^2[(v_2)^2 + (v_3)^2],$$

we see that

(4.5-7)
$$\tau^2 = (v_1)^2(v_2)^2(\sigma_1 - \sigma_2)^2 + (v_2)^2(v_3)^2(\sigma_2 - \sigma_3)^2$$
$$+ (v_3)^2(v_1)^2(\sigma_3 - \sigma_1)^2.$$

For example, if $v_1 = v_2 = 1/\sqrt{2}$ and $v_3 = 0$, then $\tau = \pm\frac{1}{2}(\sigma_1 - \sigma_2)$.

PROBLEM 4.3 Show that $\tau_{max} = \frac{1}{2}(\sigma_{max} - \sigma_{min})$ and that the plane on which τ_{max} acts makes an angle of $45°$ with the direction of the largest and the smallest principal stresses.

Solution: The problem is to find the maximum or minimum of τ. Now τ^2 is given by Eq. (4.5-7). We must find the extremum of τ^2 as a function of v_1, v_2, v_3 under the restriction that $v_1^2 + v_2^2 + v_3^2 = 1$. Using the method of Lagrangian multiplier, we seek to minimize the function

$$f \equiv v_1^2 v_2^2(\sigma_1 - \sigma_2)^2 + v_2^2 v_3^2(\sigma_2 - \sigma_3)^2 + v_3^2 v_1^2(\sigma_3 - \sigma_1)^2$$
$$+ \lambda(v_1^2 + v_2^2 + v_3^2 - 1).$$

By the usual method, we compute the partial derivatives $\partial f/\partial v_i$, $\partial f/\partial \lambda$, equate them to zero, and solve for v_1, v_2, v_3, λ. This leads to the following equations:

$$\frac{\partial f}{\partial v_1} = 0: \quad 2v_1 v_2^2(\sigma_1 - \sigma_2)^2 + 2v_1 v_3^2(\sigma_3 - \sigma_1)^2 + 2\lambda v_1 = 0, \qquad (1)$$

$$\frac{\partial f}{\partial v_2} = 0: \quad 2v_2 v_1^2(\sigma_1 - \sigma_2)^2 + 2v_2 v_3^2(\sigma_2 - \sigma_3)^2 + 2\lambda v_2 = 0, \qquad (2)$$

$$\frac{\partial f}{\partial v_3} = 0: \quad 2v_3 v_2^2(\sigma_2 - \sigma_3)^2 + 2v_3 v_1^2(\sigma_3 - \sigma_1)^2 + 2\lambda v_3 = 0, \qquad (3)$$

$$\frac{\partial f}{\partial \lambda} = 0: \quad v_1^2 + v_2^2 + v_3^2 = 1. \qquad (4)$$

One of the solutions of Eq. (1) is obviously $v_1 = 0$. On setting $v_1 = 0$, Eq. (2) and (3) become

$$v_3^2(\sigma_2 - \sigma_3)^2 + \lambda = 0, \qquad v_2^2(\sigma_2 - \sigma_3)^2 + \lambda = 0.$$

These equations are consistent only if $v_2 = v_3$. On setting $v_2 = v_3$, Eq. (4) becomes $0 + v_2^2 + v_2^2 = 1$ or $v_2 = 1/\sqrt{2}$. Hence the first set of solutions is

$$v_1 = 0, \qquad v_2 = v_3 = \frac{1}{\sqrt{2}}, \qquad \lambda = -\frac{(\sigma_2 - \sigma_3)^2}{2}.$$

Substituting this back into f, or Eq. (4.5-7), we find the extremum of τ^2:

$$\tau_{ext}^2 = \frac{(\sigma_2 - \sigma_3)^2}{4} \quad \text{or} \quad \tau_{max} \quad \text{or} \quad \tau_{min} = \frac{\sigma_2 - \sigma_3}{2}.$$

Other sets of solutions of Eq. (1), (2), (3), (4) can be obtained by setting in turn $v_2 = 0$, $v_3 = 0$. We have then the relative maxima or minima

$$\frac{\sigma_2 - \sigma_3}{2}, \qquad \frac{\sigma_3 - \sigma_1}{2}, \qquad \frac{\sigma_1 - \sigma_2}{2}.$$

The largest of the three is the absolute maximum of τ. Hence the answer.

The direction of the normal to the plane on which the absolute maximum shear occurs is given by the appropriate v's. Whichever the solution is, we have

$$v_i = v_j = \frac{1}{\sqrt{2}} \qquad (i \neq j),$$

which implies a 45° inclination the x_i-, x_j-axes.

Note: The solution to this problem appears quite lengthy because we have not yet presented the Mohr circles for three-dimensional stress states. If the results of Sec. 4.8 were known, then the answer to the present problem becomes apparent immediately. However, the derivation of the results in Sec. 4.8 is fairly lengthy.

4.6* STRESS-DEVIATION TENSOR

The tensor

$$(4.6-1) \qquad \tau_{ij}' = \tau_{ij} - \sigma_0 \delta_{ij}$$

is called the *stress-deviation tensor*, in which δ_{ij} is the Kronecker delta and σ_0 is the mean *stress*:

$$(4.6-2) \qquad \sigma_0 = \tfrac{1}{3}(\sigma_1 + \sigma_2 + \sigma_3) = \tfrac{1}{3}(\tau_{11} + \tau_{22} + \tau_{33}) = \tfrac{1}{3}I_1,$$

where I_1 is the first invariant of Sec. 4.4. The separation of τ_{ij} into a hydrostatic part $\sigma_0\delta_{ij}$ and the deviation τ'_{ij} is very important in describing the plastic behavior of metals.

The first invariant of the stress-deviation tensor always vanishes:

$$(4.6\text{-}3) \qquad I'_1 = \tau'_{11} + \tau'_{22} + \tau'_{33} = 0.$$

To determine the principal stress deviations, the procedure of Sec. 4.4 may be followed. The determinental equation

$$(4.6\text{-}4) \qquad |\tau'_{ij} - \sigma'\delta_{ij}| = 0$$

may be expanded in the form

$$(4.6\text{-}5) \qquad \sigma'^3 - J_2\sigma' - J_3 = 0.$$

It is easy to verify the following equations relating J_2, J_3 to the invariant I_2, I_3 defined in Sec. 4.4,

$$(4.6\text{-}6) \qquad J_2 = 3\sigma_0^2 - I_2,$$

$$(4.6\text{-}7) \qquad J_3 = I_3 - I_2\sigma_0 + 2\sigma_0^3 = I_3 + J_2\sigma_0 - \sigma_0^3,$$

and the alternative expressions below on account of Eq. (4.6-3),

(4.6-8)
$$J_2 = -\tau'_{11}\tau'_{22} - \tau'_{22}\tau'_{33} - \tau'_{33}\tau'_{11} + (\tau_{12})^2 + (\tau_{23})^2 + (\tau_{31})^2$$
$$= \tfrac{1}{2}[(\tau'_{11})^2 + (\tau'_{22})^2 + (\tau'_{33})^2] + (\tau_{12})^2 + (\tau_{23})^2 + (\tau_{31})^2$$
$$= \tfrac{1}{6}[(\tau_{11} - \tau_{22})^2 + (\tau_{22} - \tau_{33})^2 + (\tau_{33} - \tau_{11})^2] + (\tau_{12})^2 + (\tau_{23})^2 + (\tau_{31})^2,$$

$$(4.6\text{-}9) \qquad J_2 = \tfrac{1}{2}\tau'_{ij}\tau'_{ij}.$$

To show this, we note first that since J_2, J_3 and I_2, I_3 are all invariants, it is sufficient to verify the relations (4.6-6) and (4.6-7) with a particular choice of frame of reference. We observe that the principal axes of the stress tensor and the stress-deviation tensor coincide. Choose x_1, x_2, x_3 in the direction of the principal axes. Then if $\sigma'_1, \sigma'_2, \sigma'_3$ are the principal stress deviations, we have

$$(4.6\text{-}10) \qquad \sigma'_1 = \sigma_1 - \sigma_0, \qquad \sigma'_2 = \sigma_2 - \sigma_0, \qquad \sigma'_3 = \sigma_3 - \sigma_0,$$

$$(4.6\text{-}11) \qquad J_2 = -(\sigma'_1\sigma'_2 + \sigma'_2\sigma'_3 + \sigma'_3\sigma'_1),$$

$$(4.6\text{-}12) \qquad J_3 = \sigma'_1\sigma'_2\sigma'_3.$$

Note the negative sign in (4.6-11) because of our choice of signs in (4.6-5). The reason for this choice will become evident if we observe from the last two lines of (4.6-8) that J_2 so defined is indeed positive definite. From (4.6-11), by direct substitution, we have

$$J_2 = -(\sigma_1 - \sigma_0)(\sigma_2 - \sigma_0) - (\sigma_2 - \sigma_0)(\sigma_3 - \sigma_0) - (\sigma_3 - \sigma_0)(\sigma_1 - \sigma_0)$$

$$= -(\sigma_1\sigma_2 + \sigma_2\sigma_3 + \sigma_3\sigma_1) + 2\sigma_0(\sigma_1 + \sigma_2 + \sigma_3) - 3\sigma_0^2$$

$$= -I_2 + 6\sigma_0^2 - 3\sigma_0^2 = 3\sigma_0^2 - I_2,$$

which verifies (4.6-6). A similar substitution of (4.6-10) into (4.6-12) verifies (4.6-7). Revert now to an arbitrary orientation of frame of reference. A direct identification of the coefficients in (4.6-5) with (4.6-4) yields, as in (4.4-5),

$$J_2 = - \begin{vmatrix} \tau'_{22} & \tau'_{23} \\ \tau'_{32} & \tau'_{33} \end{vmatrix} - \begin{vmatrix} \tau'_{33} & \tau'_{31} \\ \tau'_{13} & \tau'_{11} \end{vmatrix} - \begin{vmatrix} \tau'_{11} & \tau'_{12} \\ \tau'_{21} & \tau'_{22} \end{vmatrix}.$$

Expansion of the determinants yields the first line of (4.6-8). The primes over τ'_{12}, τ'_{23}, τ'_{31} can be omitted because they are equal to τ_{12}, τ_{23}, τ_{31}, respectively. The second line of (4.6-8) is obtained if we add the null quantity $\frac{1}{2}(\tau'_{11} + \tau'_{22} + \tau'_{33})^2$ to the first line and simplify the results. To obtain the third line of (4.6-8), we note first that

$$\tau_{11} - \tau_{22} = (\tau_{11} - \sigma_0) - (\tau_{22} - \sigma_0) = \tau'_{11} - \tau'_{22}.$$

Hence,

$$(\tau_{11} - \tau_{22})^2 + (\tau_{22} - \tau_{33})^2 + (\tau_{31} - \tau_{11})^2$$
$$= 2(\tau'^2_{11} + \tau'^2_{22} + \tau'^2_{33}) - 2(\tau'_{11}\tau'_{22} + \tau'_{22}\tau'_{33} + \tau'_{33}\tau'_{11}).$$

Adding to the right-hand side the null quantity $(\tau'_{11} + \tau'_{22} + \tau'_{33})^2$ reduces the sum to $3(\tau'^2_{11} + \tau'^2_{22} + \tau'^2_{33})$. The equality of the third line of (4.6-8) with the second line is then evident. The last equation (4.6-9) is nothing but a restatement of the second line of (4.6-8). Thus, every equation is verified.

Example of Application. Testing of Material in a Pressurized Chamber

If we test a simply supported steel beam [Fig. 4.5(a)] in the laboratory by a lateral load P at the center, the relationship between P and the deflection δ under the load will be a curve as shown in Fig. 4.5(b). The spot at which the P-δ curve deviates by a specified amount from a straight line through the origin is the *yield point*. If the beam is designed to support an engineering structure, it should not be loaded beyond the yield point because beyond this point the deflection increases rapidly and irreversibly. "Permanent set" occurs.

Now, let us ask this question: If we are going to build a beam to be used as an instrument in the Mariana trench in the Pacific Ocean, 35,800 ft under

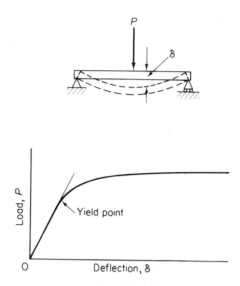

Fig. 4.5   Testing of a beam in a pressurized chamber.

the sea, what would the load-deflection curve be? Would the hydrostatic pressure of the ocean depth change the load-deflection curve of the beam?

Questions like this are of great interest to seismologists, geologists, engineers, and material scientists. Although nobody has performed such a test in the ocean depth, a simulated test was done by Percy Williams Bridgman (1882–1961) at Harvard. He built a test chamber in which high pressures that approach those in the ocean depth was achieved. The test results indicate that the P-δ curve is virtually unaffected by the hydrostatic pressure.

The yielding of the material, then, is unaffected by the hydrostatic pressure. Yielding is related to stress or strain, but only to that part of the stress or strain tensors that are independent of hydrostatic pressure. This leads to the consideration of stress-deviation tensor τ'_{ij} defined in Eq. (4.6-1), for which the hydrostatic part $\tau'_{\alpha\alpha}$ is zero. Yielding of material is related not to τ_{ij} but to τ'_{ij}.

4.7* LAMÉ'S STRESS ELLIPSOID

On any surface element with a unit outer normal vector $\mathbf{v}$, (v_i), there acts a traction vector $\overset{v}{\mathbf{T}}$, $(\overset{v}{T_i})$, with components given by

$$\overset{v}{T_i} = \tau_{ji}v_j.$$

Let the principal axes of the stress tensor be chosen as the coordinate axes x_1, x_2, x_3, and let the principal stresses be written as $\sigma_1, \sigma_2, \sigma_3$. Then,

$$\tau_{ij} = 0, \qquad (\text{if } i \neq j),$$

and

(4.7-1) $\overset{v}{T}_1 = \sigma_1 v_1, \qquad \overset{v}{T}_2 = \sigma_2 v_2, \qquad \overset{v}{T}_3 = \sigma_3 v_3.$

Since **v** is a unit vector, we have

(4.7-2) $(v_1)^2 + (v_2)^2 + (v_3)^2 = 1.$

On solving Eq. (4.7-1) for v_i and substituting into Eq. (4.7-2), we see that the components of $\overset{v}{T}_i$ satisfy the equation

(4.7-3) $\dfrac{(\overset{v}{T}_1)^2}{(\sigma_1)^2} + \dfrac{(\overset{v}{T}_2)^2}{(\sigma_2)^2} + \dfrac{(\overset{v}{T}_3)^2}{(\sigma_3)^2} = 1,$

which is the equation of an ellipsoid with reference to a system of rectangular coordinates with axes labeled $\overset{v}{T}_1, \overset{v}{T}_2, \overset{v}{T}_3$. This ellipsoid is the locus of the end points of vectors $\overset{v}{\mathbf{T}}$ issuing from a common center (see Fig. 4.6).

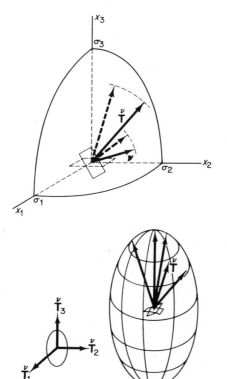

Fig. 4.6   Stress ellipsoid as the locus of the end of the vector $\overset{v}{\mathbf{T}}$ as **v** varies.

4.8* MOHR'S CIRCLES FOR THREE-DIMENSIONAL STRESS STATES

Let σ_1, σ_2, σ_3 be the principal stresses at a point. The components of the stress vector acting on any section can be obtained by the tensor transformation law, Eq. (3.6-3). Otto Mohr has shown an interesting result that if the normal stress $\sigma_{(n)}$ and the shearing stress τ acting on any section be plotted on a plane, with σ and τ as coordinates as shown in Fig. 4.7, they will necessarily fall in a closed domain represented by the shaded area bounded by the three circles with centers on the σ-axis.

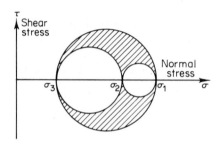

Fig. 4.7 Mohr's circles.

This result is very instructive in showing that indeed if $\sigma_1 \geqslant \sigma_2 \geqslant \sigma_3$, then σ_1 is the largest stress, and $(\sigma_1 - \sigma_3)/2$ is the largest shear for all possible surfaces. The plane on which the largest shear acts is inclined at 45° from the principal planes on which σ_1 and σ_3 act.

The meaning of the three bounding circles in Fig. 4.7 can be explained easily. Let x-, y-, z-axes be chosen in the directions of the principal axes. On a plane perpendicular to the x-axis there acts a normal stress, say σ_1, and no shear. On a plane normal to y there acts a normal stress, say σ_2, and no shear. Now consider all planes parallel to the z-axis. For these planes the normal and shear stresses acting on them are given exactly by Eq. (4.2-3) through (4.2-5) or Eq. (4.2-6) through (4.2-8). Hence, the Mohr-circle construction described in Sec. 4.3 applies, and the circle passing through σ_1, σ_2 represents the totality of all stress states on these planes. Similarly, the other two circles (one passing through σ_2, σ_3, and the other, through σ_3, σ_1) represent the totality of all stress states acting on all planes parallel to either the x-axis or the y-axis.

The proof for the general construction may proceed as follows. Consider a plane ABC whose unit normal $\mathbf{v}$ has direction cosines v_1, v_2, v_3 [$v_1 = \cos(\mathbf{v}, x)$, $v_2 = \cos(\mathbf{v}, y)$, etc.] with respect to the principal axes x, y, z (see Fig. 4.8). Then, since $\mathbf{v}$ is a unit vector,

$$(4.8\text{-}1) \qquad\qquad 1 = v_1^2 + v_2^2 + v_3^2.$$

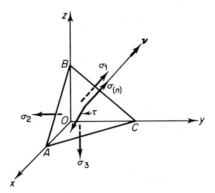

Fig. 4.8   Notations.

Further, the normal component $\sigma_{(n)}$ of the stress vector $\overset{v}{\mathbf{T}}$ acting on this plane is given by Eq. (4.5-4):

$$(4.8\text{-}2) \qquad\qquad \sigma_{(n)} = \sigma_1 v_1^2 + \sigma_2 v_2^2 + \sigma_3 v_3^2,$$

whereas the square of the magnitude of the stress vector $\overset{v}{\mathbf{T}}$ is given by Eq. (4.5-3):

$$(4.8\text{-}3) \qquad\qquad |\overset{v}{T_i}|^2 = \sigma_1^2 v_1^2 + \sigma_2^2 v_2^2 + \sigma_3^2 v_3^2.$$

We shall write, as in Sec. 4.5,

$$(4.8\text{-}4) \qquad\qquad |\overset{v}{T_i}|^2 = \sigma_{(n)}^2 + \tau^2,$$

so that τ is the shear stress (the tangential component of the stress vector) acting on the surface ABC. These equations can be solved for the direction cosines by the so-called Cramer's rule; i.e.,

$$v_1^2 = \frac{1}{D} \begin{vmatrix} 1 & 1 & 1 \\ \sigma_{(n)} & \sigma_2 & \sigma_3 \\ \sigma_{(n)}^2 + \tau^2 & \sigma_2^2 & \sigma_3^2 \end{vmatrix}, \qquad v_2^2 = \frac{1}{D} \begin{vmatrix} 1 & 1 & 1 \\ \sigma_1 & \sigma_{(n)} & \sigma_3 \\ \sigma_1^2 & \sigma_{(n)}^2 + \tau^2 & \sigma_3^2 \end{vmatrix},$$

(4.8-5)

$$v_3^2 = \frac{1}{D} \begin{vmatrix} 1 & 1 & 1 \\ \sigma_1 & \sigma_2 & \sigma_{(n)} \\ \sigma_1^2 & \sigma_2^3 & \sigma_{(n)}^2 + \tau^2 \end{vmatrix}, \qquad D = \begin{vmatrix} 1 & 1 & 1 \\ \sigma_1 & \sigma_2 & \sigma_3 \\ \sigma_1^2 & \sigma_2^2 & \sigma_3^2 \end{vmatrix}.$$

We can simplify these determinants as follows. Note first that D is a polynomial of $\sigma_1, \sigma_2, \sigma_3$. D vanishes whenever $\sigma_1 = \sigma_2, \sigma_2 = \sigma_3, \sigma_3 = \sigma_1$; hence D must contain factors $(\sigma_1 - \sigma_2), (\sigma_2 - \sigma_3), (\sigma_3 - \sigma_2)$. But D is a polynomial of degree 3 in σ_1, σ_2, σ_3; hence D must be representable as a constant times $(\sigma_1 - \sigma_2)(\sigma_2 - \sigma_3)(\sigma_3 - \sigma_1)$. We can easily identify the constant to be 1.

Alternatively, we may subtract the first column from the second and third columns in D and reduce the result directly to $D = (\sigma_1 - \sigma_2)(\sigma_2 - \sigma_3)$ $(\sigma_3 - \sigma_1)$.

Next, for the numerator of v_1^2, we subtract column 3 from the first two columns in the determinant to obtain

$$v_1^2 D = [\sigma_{(n)} - \sigma_3](\sigma_2^2 - \sigma_3^2) - (\sigma_2 - \sigma_3)[\sigma_{(n)}^2 + \tau^2 - \sigma_3^2]$$
$$= (\sigma_2 - \sigma_3)\{[\sigma_{(n)} - \sigma_3](\sigma_2 + \sigma_3) - [\sigma_{(n)}^2 + \tau^2 - \sigma_3^2]\}$$
$$= -(\sigma_2 - \sigma_3)\{\tau^2 + [\sigma_{(n)} - \sigma_2][\sigma_{(n)} - \sigma_3]\}.$$

Similarly, $v_2^2 D$, $v_3^2 D$ can be obtained by permuting the subscripts 1, 2, 3. Three cases must be discussed.

Case 1: All Principal Stresses Different, $\sigma_1 > \sigma_2 > \sigma_3$

In this case, $D \neq 0$, and we obtain

(4.8-6)
$$v_1^2 = \frac{\tau^2 + [\sigma_{(n)} - \sigma_2][\sigma_{(n)} - \sigma_3]}{(\sigma_1 - \sigma_2)(\sigma_1 - \sigma_3)},$$
$$v_2^2 = \frac{\tau^2 + [\sigma_{(n)} - \sigma_3][\sigma_{(n)} - \sigma_1]}{(\sigma_2 - \sigma_3)(\sigma_2 - \sigma_1)},$$
$$v_3^2 = \frac{\tau^2 + [\sigma_{(n)} - \sigma_1][\sigma_{(n)} - \sigma_2]}{(\sigma_3 - \sigma_1)(\sigma_3 - \sigma_2)}.$$

Consider a fixed value of v_3. We investigate the surface elements that envelop a right circular cone, the axes of which coincide with the z-axis. In other words, we consider all surfaces ABC which make a fixed angle with the z-axis. As it follows from the third equation of Eq. (4.8-6), the values of $\sigma_{(n)}$ and τ for these surface elements satisfy the equation

(4.8-7) $\tau^2 + [\sigma_{(n)} - \frac{1}{2}(\sigma_1 + \sigma_2)]^2 = \frac{1}{4}(\sigma_1 - \sigma_2)^2 + v_3^2(\sigma_1 - \sigma_3)(\sigma_2 - \sigma_3),$

which shows that on a plane with $\sigma_{(n)}$ and τ as coordinate axes the stress points corresponding to fixed values of v_3 form a circle with the center located at $(\sigma_1 + \sigma_2)/2$ on the σ-axis. This circle represents all the possible normal-stress-and-shear-stress combinations found on ABC by rotating ABC about the z-axis (v_3 fixed). The radius of this circle is given by the square root of the right-hand side of Eq. (4.8-7). As v_3 varies between 0 and 1, this radius varies between $\frac{1}{2}(\sigma_1 - \sigma_2)$ and $\frac{1}{2}(\sigma_1 + \sigma_2) - \sigma_3$ as will be shown presently. The family of circles given by Eq. (4.8-7) is bounded by these two radii as shown in Fig. 4.9.

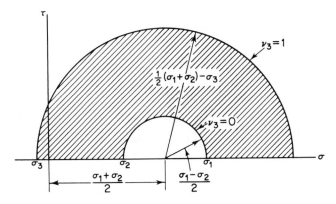

Fig. 4.9 Mohr's circles.

Note that when $\nu_3 = 1$, the right-hand side of Eq. (4.8-7) becomes

$$\tfrac{1}{4}(\sigma_1 - \sigma_2)^2 + (\sigma_1 - \sigma_3)(\sigma_2 - \sigma_3)$$
$$= \tfrac{1}{4}(\sigma_1 - \sigma_2)^2 + \sigma_1\sigma_2 - \sigma_3(\sigma_1 + \sigma_2) + \sigma_3^2$$
$$= \tfrac{1}{4}(\sigma_1 + \sigma_2)^2 - (\sigma_1 + \sigma_2)\sigma_3 + \sigma_3^2$$
$$= [\tfrac{1}{2}(\sigma_1 + \sigma_2) - \sigma_3]^2.$$

Hence the radius of the circle is $(\sigma_1 + \sigma_2)/2 - \sigma_3$.

Similar treatment of the other two equations of Eq. (4.8-6) yields two similar sets of circles with centers located on the σ-axis at $(\sigma_2 + \sigma_3)/2$, $(\sigma_3 + \sigma_1)/2$, respectively. The limiting radii are, respectively, $\tfrac{1}{2}|\sigma_2 - \sigma_3|$, $|\tfrac{1}{2}(\sigma_2 + \sigma_3) - \sigma_1|$, and $\tfrac{1}{2}|\sigma_3 - \sigma_1|$, $|\tfrac{1}{2}(\sigma_3 + \sigma_1) - \sigma_2|$. These are shown in Fig. 4.10 on p. 112.

Any point $(\sigma_{(n)}, \tau)$ must lie between the two limiting circles of each of the three families. Thus, the point can lie only within the shaded area shown in Fig. 4.7 (p. 108) and Fig. 4.10 (p. 112), since this is the only area common to the regions of the three families. This completes the proof.

Case 2: Two Principal Stresses Equal

Assume $\sigma_1 = \sigma_2 > \sigma_3$. Then the determinant $D = 0$ and Eq. (4.8-1), (4.8-2), and (4.8-3) have a solution only if the following equations hold:

$$\begin{vmatrix} 1 & 1 & 1 \\ \sigma_{(n)} & \sigma_1 & \sigma_3 \\ \sigma_{(n)}^2 + \tau^2 & \sigma_1^2 & \sigma_3^2 \end{vmatrix} = 0, \qquad \begin{vmatrix} 1 & 1 & 1 \\ \sigma_1 & \sigma_{(n)} & \sigma_3 \\ \sigma_1^2 & \sigma_{(n)}^2 + \tau^2 & \sigma_3^2 \end{vmatrix} = 0.$$

(4.8-8)

$$\begin{vmatrix} 1 & 1 & 1 \\ \sigma_1 & \sigma_1 & \sigma_{(n)} \\ \sigma_1^2 & \sigma_1^2 & \sigma_{(n)}^2 + \tau^2 \end{vmatrix} = 0.$$

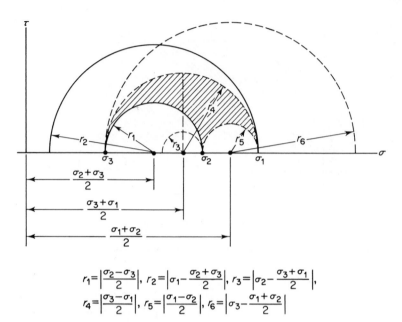

$$r_1 = \left| \frac{\sigma_2 - \sigma_3}{2} \right|, \quad r_2 = \left| \sigma_1 - \frac{\sigma_2 + \sigma_3}{2} \right|, \quad r_3 = \left| \sigma_2 - \frac{\sigma_3 + \sigma_1}{2} \right|,$$

$$r_4 = \left| \frac{\sigma_3 - \sigma_1}{2} \right|, \quad r_5 = \left| \frac{\sigma_1 - \sigma_2}{2} \right|, \quad r_6 = \left| \sigma_3 - \frac{\sigma_1 + \sigma_2}{2} \right|$$

Fig. 4.10 Mohr's circles for a three-dimensional state of stress defined by the three principal stresses $\sigma_1 > \sigma_2 > \sigma_3$.

The last is an identity. The first two are the same and give the result

$$-(\sigma_1 - \sigma_3)\{\tau^2 + [\sigma_{(n)} - \sigma_1][\sigma_{(n)} - \sigma_3]\} = 0;$$

i.e.,

$$\tau^2 + \sigma_{(n)}^2 - \sigma_{(n)}(\sigma_1 + \sigma_3) + \sigma_1 \sigma_3 = 0$$

or

(4.8-9) $$\tau^2 + [\sigma_{(n)} - \tfrac{1}{2}(\sigma_1 + \sigma_3)]^2 = \tfrac{1}{4}(\sigma_1 - \sigma_3)^2,$$

which is a circle. This is seen in Fig. 4.10 as the limiting case when $\sigma_2 \rightarrow \sigma_1$ so that the shaded area reduces to a semicircle joining σ_3 and σ_1. The case $\sigma_1 > \sigma_2 = \sigma_3$ similarly reduces to a circle.

Case 3: All Three Principal Stresses Equal, $\sigma_1 = \sigma_2 = \sigma_3$

In this case $\sigma_{(n)} = \sigma_1 = \sigma_2 = \sigma_3$ and $\tau = 0$ and all Mohr circles become a point.   Q.E.D.

PROBLEMS

4.4 Let $\tau_{xx} = 1000$ psi, $\tau_{yy} = -1000$ psi, $\tau_{zz} = 0$, $\tau_{xy} = 500$ psi, $\tau_{yz} = -200$ psi, $\tau_{zx} = 0$. What is total traction acting on a surface whose normal vector is

$$\mathbf{v} = 0.10\mathbf{i} + 0.30\mathbf{j} + \sqrt{0.90}\,\mathbf{k}?$$

What are the three components (in the x-, y-, z-directions) of the stress vector acting on the surface? What is the normal stress acting on the surface? What is the resultant shear stress acting on the surface?

Answer: $(\overset{v}{T_i}) = (250, -440, -60)$. Traction $= 509$ psi. Normal stress $= -164$ psi, shear $= 457$ psi.

4.5 George Stokes gave in 1850 the solution to the problem of a sphere (Fig. P4.5) moving in a (Newtonian) viscous fluid at a constant velocity U. On the surface of the sphere, the three components of the stress vector are

$$\overset{r}{T_x} = -\frac{x}{a}p_0 + \frac{3}{2}\,\mu\frac{U}{a}, \qquad \overset{r}{T_y} = -\frac{y}{a}p_0, \qquad \overset{r}{T_z} = -\frac{z}{a}p_0.$$

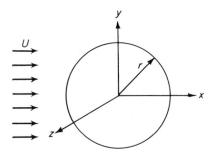

Fig. P4.5 A sphere in viscous fluid (Stokes's problem).

What is the resultant force acting on the sphere? Robert Millikan's Nobel prize-winning paper on the measurement of the charge of an electron using oil drops in a cloud chamber was based on this formula, which is not entirely satisfactory from a theoretical point of view. There is a large amount of literature concerned with the modifications and improvements of this formula.

Solution: The total surface force acting on the sphere is

$$F_x = \oint \overset{v}{T_x}\,dS, \qquad F_y = \oint \overset{v}{T_y}\,dS, \qquad F_z = \oint \overset{v}{T_z}\,dS.$$

By symmetry,

$$\oint x\,dS = \oint y\,dS = \oint z\,dS = 0.$$

Hence the only nonvanishing component of the resultant force is

$$F_x = \oint \frac{3}{2}\,\mu\frac{U}{a}\,dS = 4\pi a^2\,\frac{3}{2}\,\mu\frac{U}{a} = 6\pi\mu aU.$$

4.6 A thick-walled elastic circular cylinder is subjected to an internal pressure p_i. See Fig. P4.6. The stresses in the cylinder, at sections sufficiently far away from the ends, are

$$\sigma_{rr} = p_i \frac{a^2}{r^2} \frac{r^2 - b^2}{b^2 - a^2}, \qquad \sigma_{\theta\theta} = p_i \frac{a^2}{r^2} \frac{r^2 + b^2}{b^2 - a^2}, \qquad \sigma_{r\theta} = 0.$$

Find the radius r where the maximum principal stress occurs and the value of the absolute maximum principal stress.

This solution was due to Lamé. The nonuniformity of the stress distribution in the wall should be noted. The stress concentration in the inner wall is significant.

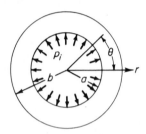

Fig. P4.6   Thick-walled cylinder subjected to internal pressure.

Solution: σ_{rr} and $\sigma_{\theta\theta}$ are principal stresses. The absolute maximum is to be sought with respect to r. The absolute maximum occurs when $r = a$ where

$$\sigma_{\theta\theta} = p_i \frac{a^2 + b^2}{b^2 - a^2}, \qquad \sigma_{rr} = -p_i.$$

4.7 *Stress concentration.* Describe the boundary conditions for a plate with a circular hole subjected to a static uniform tensile loading with the normal stress $\sigma_x = \text{const.} = p$ acting on the ends. See Fig. P4.7.

If this plate is made of a linear elastic material, it is known that the solution is

$$\sigma_r = \frac{p}{2}\left(1 - \frac{a^2}{r^2}\right)\left[1 + \left(1 - 3\frac{a^2}{r^2}\right)\cos 2\theta\right],$$

$$\sigma_\theta = \frac{p}{2}\left[1 + \frac{a^2}{r^2} - \left(1 + 3\frac{a^4}{r^4}\right)\cos 2\theta\right],$$

$$\tau_{r\theta} = -\frac{p}{2}\left(1 - \frac{a^2}{r^2}\right)\left(1 + 3\frac{a^2}{r^2}\right)\sin 2\theta.$$

(a) Check the stress boundary conditions to see if they are satisfied.

(b) Find the location of the point where the normal stress σ_θ is the maximum.

(c) Find the maximum shearing stress in the entire plate.

(d) Obtain the maximum principal stress in the plate.

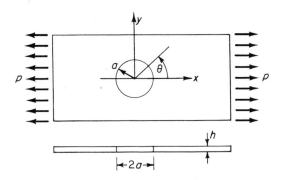

Fig. P4.7 Circular hole in a thin plate.

Note: You see that the maximum stress is increased around the hole. This is the phenomenon of stress concentration.

Answer: The horizontal edges and the circular hole are stress-free. On the hole, the boundary conditions are

$$\sigma_{rr} = 0, \qquad \tau_{r\theta} = 0, \qquad \text{when } r = a.$$

(b) σ_θ reaches the maximum $3p$ when $\theta = \pi/2$.

(c) Maximum shear equals $3p/2$ and occurs at $r = a$, $\theta = \pi/2$, acting in a plane inclined at 45° from the z-axis.

(d) Maximum principal stress is $3p$.

4.8 The problem of an elliptic hole in an infinite plate under a tensile stress S, with the minor axis of the ellipse parallel to the tension, yields the result that the tensile stress at the ends of the major axis of the hole is

$$\sigma = S\left(1 + 2\frac{a}{b}\right),$$

where $2a$ is the major axis of the ellipse, and $2b$ is the minor axis (Fig. P4.8). Show that the stress concentration becomes very large when a/b is large. A crack $(a/b \longrightarrow \infty)$ perpendicular to the direction of tension is therefore a severe stress raiser.

Explain the benefit that can be derived by drilling holes at the ends of the crack to stop its spreading.

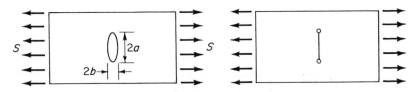

Fig. P4.8 Elliptical hole in plate. Stress relieving drilling.

4.9 Draw Mohr's circle for (a) pure uniaxial tension, (b) pure uniaxial compression, (c) pure biaxial tension, and (d) pure shear.

4.10 An earthquake is initiated at time $t = 0$ at an epicenter C. See Fig. P4.10. What are the stress conditions on the surface of the earth? What mathematical boundary conditions would you use to describe what is happening at the epicenter?

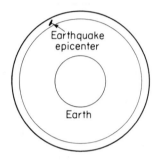

Fig. P4.10   Earthquake.

Fig. P4.11   Tension test specimen.

4.11 A specimen used for the static testing of material is shown in Fig. P4.11. When the specimen is pulled, what boundary conditions apply?

4.12 For normal impact of a hammer on a large flat surface of a semi-infinite elastic body (Fig. P4.12), what boundary conditions apply?

Solution: Initial condition: Deformation equals 0 everywhere. When the hammer strikes, the *boundary conditions* are

(a) On the flat surface but not under the hammer:

$$\overset{v}{T}_i = 0 \qquad (i = 1, 2, 3).$$

(b) Conditions at infinity in the semi-infinite body: Let u_i be the components of displacement caused by the deformation of the body, and let σ_{ij} be the stresses; then

$$(u_i) = 0, \qquad (\sigma_{ij}) = 0 \qquad (i, j = 1, 2, 3).$$

(c) On the surface of the table under the hammer, the traction and the displacement must be consistent. Hence if we denote the table and the hammer by (T) and (H), respectively, we must have, on the comon interface,

$$u_i^{(T)} = u_i^{(H)}, \qquad \overset{v}{T}_i^{(T)} = \overset{v}{T}_i^{(H)}.$$

The normal v is the normal to the interface, which may be described by the equation $z = f(x, y, t)$. Then $v_1 : v_2 : v_3 = \partial f/\partial x : \partial f/\partial y : -1$. However, we do not know the function $f(x, y, t)$, which can be determined rigorously only by solving the whole problem of stress distribution in the hammer and the table together.

In lieu of an exact solution, one may propose an approximate solution by formulating an approximate problem. In the present problem, we may assume, on the basis of our intuition that when the hammer strikes the table, the component of the stress vector normal to the table must be much larger than the tangential components. The latter must arise from friction. Hence if we ignore the latter components (in a case that might be called a "lubricated" hammer), then the boundary condition under the hammer may be written as

$$\overset{z}{T}_x = 0, \qquad \overset{z}{T}_y = 0, \qquad \overset{z}{T}_z = F(x, y)\, \delta(t)$$

where $\delta(t)$ is the Dirac unit-impulse function, which is zero when t is finite but tends to ∞ when $t \longrightarrow 0$ in such a way that the integral of $\delta(t)$ for t from $-\epsilon$ to $+\epsilon$ is exactly equal to 1, ϵ being a positive number. The function $F(x, y)$ cannot be determined exactly until the entire problem is solved. But if we are interested in an approximate solution, we may propose some functional form for $F(x, y)$. For example, we may set $F(x, y) = $ const. The solution based on such a formulation gives good results for stress distribution in the table at some distance from the hammer.

Note: Perhaps the language used in condition (c) is too strong. Perhaps there are uneven contacts between the hammer and the tabletop. There might be local failure, slippage, etc. One might imagine all kinds of unusual events happening. If one is serious about these possibilities, one must specify them precisely and then investigate their consequences, searching for experimental evidence as to whether some possibilities are plausible or not.

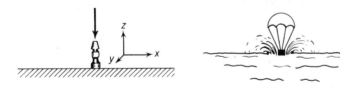

Fig. P4.12   Hammer.   **Fig. P4.13**   Package drop.

4.13 Suppose that the semi-infinite body of the preceding problem is a large expanse of water and the load is a package dropped from the airplane (Fig. P4.13). The water will surely splash. What boundary conditions are known in this case?

4.14 A palm tree supports its own weight.

(a) Assume that the tree top weighs 100 kg. The cross-sectional area at the top is 100 cm². The tree trunk has a specific gravity of 2. The tree is 10 m tall. If the tree trunk is a uniform cylinder, what is the stress in the tree trunk at a height x from the ground? Solve this problem by means of the equation of equilibrium $\tau_{ij,j} + X_i = 0$.

(b) The stress can be reduced by increasing the cross-sectional area of the

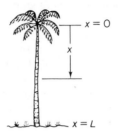

Fig. P4.14   A palm tree supporting its own weight.

trunk toward the base. Consider such a trunk of variable diameter. Compute the average stress by means of a free-body diagram.

(c) Solve part (b) by means of the exact equations of equilibrium. What additional consideration is needed?

4.15 What is the physical meaning of the sum $\tau_{\alpha\alpha}$?

Answer: It is the sum of the normal stresses in three orthogonal directions. If we consider a cube of water in static condition so that all its sides are subjected to a pressure p, and there is no shear stress acting on the surface, then $\tau_{\alpha\alpha} = -3p$, or $p = -\tau_{\alpha\alpha}/3$.

If a uniform tension of equal intensity acts on each side of a cube, then $\tau_{\alpha\alpha}/3$ represents the tension.

If the three stresses $\tau_{xx}, \tau_{yy}, \tau_{zz}$ are not equal, then $\tau_{\alpha\alpha}/3$ represents the mean normal stress.

4.16 The stress at a point in a body has the following components with respect to a set of rectangular Cartesian coordinates x_1, x_2, x_3:

$$(\sigma_{ij}) = \begin{pmatrix} 1 & 0 & -1 \\ 0 & -1 & 0 \\ -1 & 0 & 1 \end{pmatrix}.$$

Find the values of the invariants I_1, I_2, I_3 and the principal stresses.

Answer: $I_1 = 1, I_2 = -2, I_3 = 0.$ $(\sigma_1, \sigma_2, \sigma_3) = (0, 2, -1).$

4.17 Let τ_{ij} be a stress tensor. Evaluate the products (a) $e_{ijk}\tau_{jk}$ and (b) $e_{ijk}e_{ist}\tau_{kt}$.

4.18 A plate is stretched in the x-direction, compressed in the y-direction, and free in the z-direction. There is a flaw in a plane that is parallel to the z-axis and inclined at 45° to the x-axis. If the shear stress acting on the flaw exceeds a critical stress τ_{cr}, the plate will fail. Determine the critical combinations of σ_x and σ_y at which the plate fails.

4.19 Consider a rod that has a cross-sectional area of 1 cm².

(a) Assume that the material has the following strength characteristics, beyond which the rod breaks: maximum shear stress, 500 n/cm²; maximum tensile stress, 1000 n/cm²; maximum compressive stress, 10,000 n/cm². Let a tension P act on the rod. At what value of P will the rod break? What is the expected angle of inclination of the broken section?

(b) Answer the same if the strength characteristics are maximum shear, 500 n/cm²; maximum tension, 900 n/cm²; maximum compressions, 10,000 n/cm².

4.20 A circular cylindrical rod is stretched by an axial load, bent by a bending moment, and twisted by a torque, so that the stresses in a little element at a point on the surface of the cylinder are

$$\sigma_r = 0, \quad \tau_{rz} = \tau_{r\theta} = 0, \quad \sigma_z = 1 \text{ kn/cm}^2, \quad \tau_{z\theta} = 2 \text{ kn/cm}^2, \quad \sigma_\theta = 0.$$

What are the principal stresses at that point?

4.21 In the earth there is hydrostatic pressure due to the earth's weight, and there is a shear stress due to strain in the earth's crust. At a point in the earth, the hydrostatic pressure is 10 kn/cm², and the shear stress, evaluated with respect to a chosen frame of reference x_1, x_2, x_3, is $\tau_{12} = 5$ kn/cm², $\tau_{23} = \tau_{31} = 0$. Find the principal stresses and planes at that point.

Answer: $\sigma_1 = -5$ acts on a plane with a normal vector $v_1 = -v_2 = \sqrt{2}/2$, $v_3 = 0$, which is inclined at 45° to the negative x_1-axis. The principal axis associated with the principal stress $\sigma_2 = -10$ kn/cm² is the x_3-axis, and that associated with $\sigma_3 = -15$ is a vector inclined at 45° to the positive x_1-axis.

4.22 A driver of a moving car that weighs 1600 kg made a sudden panic stop by slamming on the brakes; this promptly locked the wheels. Assume a maximum coefficient of friction between the tire and the ground of $\frac{1}{4}$, and assume that each of the car wheels is attached to the hub by four bolts.

(a) Compute the shear force that must act in each bolt. The bolt has a diameter of 1 cm and its axis is 6 cm away from the axis of the wheel, which is 36 cm above the ground.

(b) The allowable shear stress of the bolt material is 150 Mn/m². Are the shear stresses in the bolts within the allowable limit? (Assume that the bolts are initially stressfree).

(c) The garage mechanic who put on the wheels for the car named above used a large wrench and tightened the nuts most vigorously, so that a tensile stress of 140 Mn/m² was imposed on the bolts. This tensile stress was the initial stress in the bolts. Now when the brakes were applied and a shear stress as computed above was induced, are the bolts still safe? To answer this question, compute the maximum shear stress in the bolts under the combined tension and shear; then compare it with the allowable shear stress.

Answer: (a) 1470 n. (b) shear stress due to braking = 18.71 *Mn/m²*. Shear stress due to car weight = 12.47 *Mn/m²*. Shear in most severe configuration = 31.18 *Mn/m²*. It is less than the allowable 150 *Mn/m²*; hence the bolts are safe. (c) By Mohr's circle construction, or by Eqs. (4.2-1.2), $\tau_{max} = 76.63$ *Mn/m²*, $\sigma_{max} = 146.63$ *Mn/m²*. The bolts are clearly safe with respect to shear. But better check the handbook with regard to tensile stress to see if σ_{max} is allowable.

CHAPTER FIVE

Analysis of Deformation

Forces applied to solids cause deformation, and forces applied to liquids cause flow. Often the major objective of an analysis is to find the deformation or flow. It is our objective in this chapter to analyze the deformation of solid bodies in such a way as to be relevant to the state of stress.

5.1 DEFORMATION

If we pull a rubber band, it stretches. If we compress a cylinder, it shortens. If we bend a rod, it bends. If we twist a shaft, it twists. Tensile stress causes tensile strain. Shear stress causes shear strain. This is common sense. To express these phenomena quantitatively, it is necessary to define strain measures.

Take a string of an initial length L_0. If it is stretched to a length L as shown in Fig. 5.1(a), it is natural to describe the change by dimensionless ratios such as L/L_0, $(L - L_0)/L_0$, $(L - L_0)/L$. Use of dimensionless ratios eliminates the absolute length from the consideration. We feel that it is these ratios that are related to the stress in the string, and not the lengths L_0 or L. This expectation can be verified in the laboratory. The ratio L/L_0 is called *stretch ratio* and is denoted by the symbol λ. The ratios

$$(5.1\text{-}1) \qquad \epsilon = \frac{L - L_0}{L_0}, \qquad \epsilon' = \frac{L - L_0}{L}$$

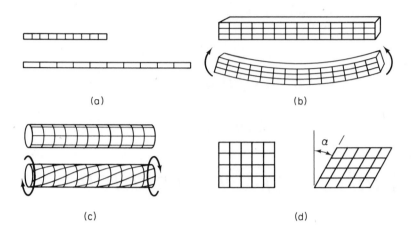

Fig. 5.1 Patterns of deformation.

are strain measures. Either of them can be used. Numerically, they are different. For example, if $L = 2$, $L_0 = 1$, we have $\epsilon = 1$, $\epsilon' = \frac{1}{2}$. We shall have reasons (to be named later) also to introduce the measures

(5.1-2)
$$e = \frac{L^2 - L_0^2}{2L^2}, \qquad \mathcal{E} = \frac{L^2 - L_0^2}{2L_0^2}.$$

If $L = 2$, $L_0 = 1$, we have $e = \frac{3}{8}$, $\mathcal{E} = \frac{3}{2}$. But if $L = 1.01$, $L_0 = 1.00$, then $e \doteq 0.01$, $\mathcal{E} \doteq 0.01$, $\epsilon \doteq 0.01$, and $\epsilon' \doteq 0.01$. Hence in infinitesimal elongations all the strain measures named above are equal. In finite elongations, however, they are different.

These strain measures can be used to describe more complex deformations. For example, if we bend a rectangular beam as shown in Fig. 5.1(b) by moments acting at the ends, the beam will deflect into a circular arc. The "fibers" on top will be shortened; those on the bottom will be elongated. These longitudinal strains are related to the bending moment acting on the beam.

To illustrate shear, consider a circular cylindrical shaft as shown in Fig. 5.1(c). When the shaft is twisted, the elements in the shaft are distorted in a manner shown in Fig. 5.1(d). In this case the angle α may be taken as a strain measure. It is more customary, however, to take $\tan \alpha$, or $\frac{1}{2} \tan \alpha$, as the shear strain; the reasons for this will be elucidated later.

The selection of proper strain measures is dictated basically by the stress-strain relationship (i.e., the constitutive equation of the material). For example, if we pull on a string, it elongates. The experimental results can be presented as a curve of the tensile stress σ plotted against the stretch ratio λ, or strain e. An empirical formula relating σ to e can be determined. The case of infinitesimal strain is simple because the different strain measures named above all coincide. It was found that, for most engineering materials subjected to an infinitesimal strain in uniaxial stretching, a relation like

(5.1-3) $\sigma = Ee$

is valid within certain range of stresses, where E is a constant called *Young's modulus*. Equation (5.1-3) is called *Hooke's law*. A material obeying (5.1-3) is said to be a *Hookean material*. Steel is a Hookean material if σ lies within certain bounds that are called *yield stresses*.

Corresponding to Eq. (5.1-3), the relationship for a Hookean material subjected to an infinitesimal shear strain is

(5.1-4) $\tau = G \tan \alpha$

where G is another constant called the *modulus of rigidity*. The range of validity of (5.1-4) is again bounded by *yield stresses*. The yield stresses in tension, in compression, and in shear are different in general.

Equations (5.1-3) and (5.1-4) are the simplest of constitutive equations. The more general cases will be discussed in Chapters 7, 8, and 9.

Deformations of most things in nature and in engineering are much more complex than those discussed above. We therefore need a general method of treatment. This is presented below.

Let us first consider the mathematical description of deformation. Let a body occupy a space S. Referred to a rectangular Cartesian frame of reference, every particle in the body has a set of coordinates. When the body is deformed, every particle takes up a new position, which is described by a new set of coordinates. For example, a particle P located orginally at a place with coordinates (a_1, a_2, a_3) is moved to the place Q with coordinates (x_1, x_2, x_3) when the body moves and deforms. Then the vector $\overrightarrow{PQ}$ is called the displacement vector of the particle. See Fig. 5.2. The components of the displacement vector are, clearly,

$$x_1 - a_1, \qquad x_2 - a_2, \qquad x_3 - a_3.$$

If the displacement is known for every particle in the body, we can construct the deformed body from the original. Hence, a deformation can be described by the displacement field. Let the variable (a_1, a_2, a_3) refer to any particle in the original configuration of the body, and let (x_1, x_2, x_3) be the coordinates of that particle when the body is deformed. Then the deformation of the body is known if x_1, x_2, x_3 are known functions of a_1, a_2, a_3:

(5.1-5) $x_i = x_i(a_1, a_2, a_3).$

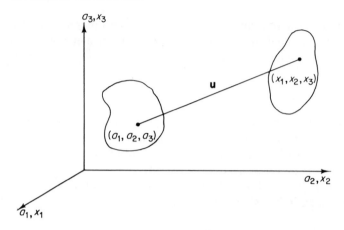

Fig. 5.2 Displacement vector.

This is a transformation (mapping) from a_1, a_2, a_3 to x_1, x_2, x_3. In continuum mechanics we assume that deformation is continuous. A neighborhood is transformed into a neighborhood. We also assume that the transformation is one-to-one; i.e., the functions in Eq. (5.1-5) are single-valued, continuous, and have unique inverse

$$(5.1\text{-}6) \qquad a_i = a_i(x_1, x_2, x_3)$$

for every point in the body.

The displacement vector **u** is then defined by its components

$$(5.1\text{-}7) \qquad u_i = x_i - a_i.$$

If a displacement vector is associated with every particle in the original position, we may write

$$(5.1\text{-}8) \qquad u_i(a_1, a_2, a_3) = x_i(a_1, a_2, a_3) - a_i.$$

If that displacement is associated with the particle in the deformed position, we write

$$(5.1\text{-}9) \qquad u_i(x_1, x_2, x_3) = x_i - a_i(x_1, x_2, x_3).$$

PROBLEM 5.1 In order that the transformation (5.1-5) be single-valued, continuous, and differentiable, what conditions must be satisfied by the functions $x_i(a_1, a_2, a_3)$?

Note: If the transformation is single-valued, continuous, and differentiable, then the functions $x_i(a_1, a_2, a_3)$ must be single-valued, continuous, and differentiable, and the Jacobian $|\partial x_i/\partial a_j|$ must not vanish in the space occupied by the body. The last statement is nontrivial. It will be useful to prove this statement by reviewing the "implicit function" theorem in calculus.

5.2 THE STRAIN

The idea that the stress in a body is related to the strain was first announced by Robert Hooke (1635–1703) in 1676 in the form of an anagram, *ceiiinossssttuv.* He explained it in 1678 as

$$Ut \ tensio \ sic \ vis,$$

or "The power of any springy body is in the same proportion with the extension." The meaning of this statement is clear to anyone who ever handled a spring or pulled a rubber band.

A rigid-body motion induces no stress. Thus, the displacements themselves are not directly related to the stress. To relate deformation with stress we must consider the stretching and distortion of the body. For this purpose, let us consider three neighboring points P, P', P'' in the body. See Fig. 5.3. If they are transformed to the points Q, Q', Q'' in the deformed configuration, the change in area and angles of the triangle is completely determined if we know the change in length of the sides. But the "location" of the triangle is undetermined by the change of the sides. Similarly, if the change of length between any two arbitrary points of the body is known, the new configuration of the body will be completely defined except for the location of the body in space. The description of the change in distance between any two points of the body is the key to the analysis of deformation.

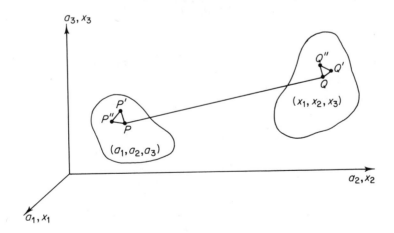

Fig. 5.3 Deformation of a body.

Consider an infinitesimal line element connecting the point $P(a_1, a_2, a_3)$ to a neighboring point $P'(a_1 + da_1, a_2 + da_2, a_3 + da_3)$. The square of the length ds_0 of PP' in the original configuration is given by

(5.2-1) $$ds_0^2 = da_1^2 + da_2^2 + da_3^2.$$

When P and P' are deformed to the points $Q(x_1, x_2, x_3)$ and $Q'(x_1 + dx_1, x_2 + dx_2, x_3 + dx_3)$, respectively, the square of the length ds of the new element QQ' is

(5.2-2) $$ds^2 = dx_1^2 + dx_2^2 + dx_3^2.$$

By Eq. (5.1-5) and (5.1-6) we have

(5.2-3) $$dx_i = \frac{\partial x_i}{\partial a_j} da_j, \qquad da_i = \frac{\partial a_i}{\partial x_j} dx_j.$$

Hence, on introducing the Kronecker delta, we may write

(5.2-4) $$ds_0^2 = \delta_{ij} da_i\, da_j = \delta_{ij} \frac{\partial a_i}{\partial x_l} \frac{\partial a_j}{\partial x_m} dx_l\, dx_m,$$

(5.2-5) $$ds^2 = \delta_{ij} dx_i\, dx_j = \delta_{ij} \frac{\partial x_i}{\partial a_l} \frac{\partial x_j}{\partial a_m} da_l\, da_m.$$

The difference between the squares of the length elements may be written, after several changes in the symbols for dummy indices, either as

(5.2-6) $$ds^2 - ds_0^2 = \left(\delta_{\alpha\beta} \frac{\partial x_\alpha}{\partial a_i} \frac{\partial x_\beta}{\partial a_j} - \delta_{ij}\right) da_i\, da_j,$$

or as

(5.2-7) $$ds^2 - ds_0^2 = \left(\delta_{ij} - \delta_{\alpha\beta} \frac{\partial a_\alpha}{\partial x_i} \frac{\partial a_\beta}{\partial x_j}\right) dx_i\, dx_j.$$

We define the *strain tensors*

(5.2-8) $$E_{ij} = \frac{1}{2}\left(\delta_{\alpha\beta} \frac{\partial x_\alpha}{\partial a_i} \frac{\partial x_\beta}{\partial a_j} - \delta_{ij}\right),$$

(5.2-9) $$e_{ij} = \frac{1}{2}\left(\delta_{ij} - \delta_{\alpha\beta} \frac{\partial a_\alpha}{\partial x_i} \frac{\partial a_\beta}{\partial x_j}\right),$$

so that

(5.2-10) $$ds^2 - ds_0^2 = 2E_{ij}\, da_i\, da_j,$$

(5.2-11) $$ds^2 - ds_0^2 = 2e_{ij}\, dx_i\, dx_j.$$

The strain tensor E_{ij} was introduced by Green and St.-Venant and is called Green's strain tensor. The strain tensor e_{ij} was introduced by Cauchy for infinitesimal strains and by Almansi and Hamel for finite strains and is known as Almansi's strain tensor. In analogy with terminology in hydrodynamics, E_{ij} is often referred to as Lagrangian and e_{ij} as Eulerian.

That E_{ij} and e_{ij} thus defined are tensors in the coordinate systems a_i and x_i, respectively, follows from the quotient rule when it is applied to Eq. (5.2-10) and (5.2-11). The tensors E_{ij} and e_{ij} are obviously *symmetric*; i.e.,

(5.2-12) $$E_{ij} = E_{ji}, \qquad e_{ij} = e_{ji}.$$

An immediate consequence of Eq. (5.2-10) and (5.2-11) is that $ds^2 - ds_0^2 = 0$ implies $E_{ij} = e_{ij} = 0$ and vice versa. But a deformation in which the length of every line element remains unchanged is a rigid-body motion. Hence, the *necessary and sufficient condition that a deformation of a body be a rigid-body motion is that all components of the strain tensor E_{ij} or e_{ij} be zero throughout the body.*

5.3 STRAIN COMPONENTS IN RECTANGULAR CARTESIAN COORDINATES

If we introduce the *displacement vector* **u** with components

(5.3-1) $$u_\alpha = x_\alpha - a_\alpha, \qquad (\alpha = 1, 2, 3),$$

then

(5.3-2) $$\frac{\partial x_\alpha}{\partial a_i} = \frac{\partial u_\alpha}{\partial a_i} + \delta_{\alpha i}, \qquad \frac{\partial a_\alpha}{\partial x_i} = \delta_{\alpha i} - \frac{\partial u_\alpha}{\partial x_i},$$

and the strain tensors reduce to the simple form

(5.3-3) $$E_{ij} = \frac{1}{2}\left[\delta_{\alpha\beta}\left(\frac{\partial u_\alpha}{\partial a_i} + \delta_{\alpha i}\right)\left(\frac{\partial u_\beta}{\partial a_j} + \delta_{\beta j}\right) - \delta_{ij}\right]$$
$$= \frac{1}{2}\left[\frac{\partial u_j}{\partial a_i} + \frac{\partial u_i}{\partial a_j} + \frac{\partial u_\alpha}{\partial a_i}\frac{\partial u_\alpha}{\partial a_j}\right]$$

and

(5.3-4) $$e_{ij} = \frac{1}{2}\left[\delta_{ij} - \delta_{\alpha\beta}\left(-\frac{\partial u_\alpha}{\partial x_i} + \delta_{\alpha i}\right)\left(-\frac{\partial u_\beta}{\partial x_j} + \delta_{\beta j}\right)\right]$$
$$= \frac{1}{2}\left[\frac{\partial u_j}{\partial x_i} + \frac{\partial u_i}{\partial x_j} - \frac{\partial u_\alpha}{\partial x_i}\frac{\partial u_\alpha}{\partial x_j}\right].$$

In unabridged notations (x, y, z for x_1, x_2, x_3; a, b, c for a_1, a_2, a_3; and u, v, w for u_1, u_2, u_3), we have the typical terms

$$E_{aa} = \frac{\partial u}{\partial a} + \frac{1}{2}\left[\left(\frac{\partial u}{\partial a}\right)^2 + \left(\frac{\partial v}{\partial a}\right)^2 + \left(\frac{\partial w}{\partial a}\right)^2\right],$$

$$e_{xx} = \frac{\partial u}{\partial x} - \frac{1}{2}\left[\left(\frac{\partial u}{\partial x}\right)^2 + \left(\frac{\partial v}{\partial x}\right)^2 + \left(\frac{\partial w}{\partial x}\right)^2\right],$$

(5.3-5)

$$E_{ab} = \frac{1}{2}\left[\frac{\partial u}{\partial b} + \frac{\partial v}{\partial a} + \left(\frac{\partial u}{\partial a}\frac{\partial u}{\partial b} + \frac{\partial v}{\partial a}\frac{\partial v}{\partial b} + \frac{\partial w}{\partial a}\frac{\partial w}{\partial b}\right)\right],$$

$$e_{xy} = \frac{1}{2}\left[\frac{\partial u}{\partial y} + \frac{\partial v}{\partial x} - \left(\frac{\partial u}{\partial x}\frac{\partial u}{\partial y} + \frac{\partial v}{\partial x}\frac{\partial v}{\partial y} + \frac{\partial w}{\partial x}\frac{\partial w}{\partial y}\right)\right].$$

Note that u, v, w are considered as functions of a, b, c, the position of points in the body in unstrained configuration, when the Lagrangian strain tensor is evaluated; whereas they are considered as functions of x, y, z, the position of points in the strained configuration, when the Eulerian strain tensor is evaluated.

If the components of displacement u_i are such that their first derivatives are so small that the squares and products of the partial derivatives of u_i are negligible, then e_{ij} reduces to Cauchy's *infinitesimal strain tensor*,

(5.3-6) $$e_{ij} = \frac{1}{2}\left[\frac{\partial u_j}{\partial x_i} + \frac{\partial u_i}{\partial x_j}\right].$$

In unabridged notation,

(5.3-7)

$$e_{xx} = \frac{\partial u}{\partial x}, \qquad e_{xy} = \frac{1}{2}\left(\frac{\partial u}{\partial y} + \frac{\partial v}{\partial x}\right) = e_{yx},$$

$$e_{yy} = \frac{\partial v}{\partial y}, \qquad e_{xz} = \frac{1}{2}\left(\frac{\partial u}{\partial z} + \frac{\partial w}{\partial x}\right) = e_{zx},$$

$$e_{zz} = \frac{\partial w}{\partial z}, \qquad e_{yz} = \frac{1}{2}\left(\frac{\partial v}{\partial z} + \frac{\partial w}{\partial y}\right) = e_{zy}.$$

In the infinitesimal displacement case, the distinction between the Lagrangian and Eulerian strain tensor disappears, since then it is immaterial whether the derivatives of the displacements are calculated at the position of a point before or after deformation.

WARNING. NOTATION FOR SHEAR STRAIN.

In most books and papers the strain components are defined as

$$e_x = \frac{\partial u}{\partial x}, \qquad \gamma_{xy} = 2e_{xy} = \frac{\partial u}{\partial y} + \frac{\partial v}{\partial x},$$

$$e_y = \frac{\partial v}{\partial y}, \qquad \gamma_{yz} = 2e_{yz} = \frac{\partial v}{\partial z} + \frac{\partial w}{\partial y},$$

$$e_z = \frac{\partial w}{\partial z}, \qquad \gamma_{zx} = 2e_{zx} = \frac{\partial u}{\partial z} + \frac{\partial w}{\partial x}.$$

In other words, the shear strains, denoted by $\gamma_{xy}, \gamma_{yz}, \gamma_{zx}$, are twice as large as our components e_{xy}, e_{yz}, e_{zx}, respectively. We shall not use this notation because the components e_x, γ_{xy}, etc., together do not form a tensor, and we lose a great deal of mathematical convenience. But beware of this difference when you read other books and papers!

5.4 GEOMETRIC INTERPRETATION OF INFINITESIMAL STRAIN COMPONENTS

Let x, y, z be a set of rectangular Cartesian coordinates. Consider a line element of length dx parallel to the x-axis $(dy = dz = 0)$. The change of the square of the length of this element due to deformation is

$$ds^2 - ds_0^2 = 2e_{xx}(dx)^2.$$

Hence,

$$ds - ds_0 = \frac{2e_{xx}(dx)^2}{ds + ds_0}.$$

But $ds = dx$ in this case, and ds_0 differs from ds only by a small quantity of the second order if we assume the displacements u, v, w, and the strain components e_{ij} to be infinitesimal. Hence,

$$(5.4\text{-}1) \qquad \frac{ds - ds_0}{ds} = e_{xx},$$

and it is seen that e_{xx} represents the *extension*, or change of length per unit length of a vector parallel to the x-axis. An application of the above discussion to a volume element is illustrated in Fig. 5.4, Case 1.

To see the meaning of the component e_{xy}, let us consider a small rectangle in the body with edges dx, dy. It is evident from Fig. 5.4, Cases 2, 3, and 4, that the sum $\partial u/\partial y + \partial v/\partial x$ represents the change of angle xOy which was originally a right angle. Thus,

$$(5.4\text{-}2) \qquad e_{xy} = \frac{1}{2}\left[\frac{\partial u}{\partial y} + \frac{\partial v}{\partial x}\right] = \frac{1}{2}\tan(\text{change of angle } xOy).$$

In engineering usage, the strain components e_{ij} $(i \neq j)$ doubled, i.e., $2e_{ij}$, are called the *shearing strains* or *detrusions*. The name is particularly suggestive in Case 3 of Fig. 5.4, which is called the case of *simple shear*.

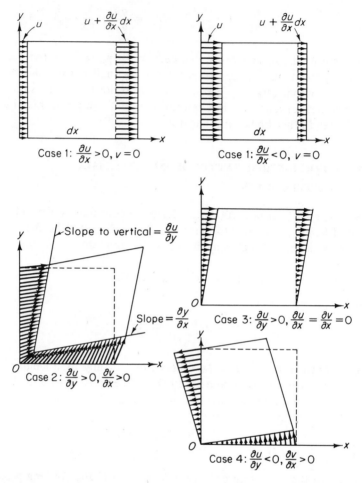

Fig. 5.4 Deformation gradients and interpretation of infinitesimal strain components.

5.5 INFINITESIMAL ROTATION

Consider an infinitesimal displacement field $u_i(x_1, x_2, x_3)$. From u_i, form the Cartesian tensor

(5.5-1) $$\omega_{ij} = \frac{1}{2}\left(\frac{\partial u_j}{\partial x_i} - \frac{\partial u_i}{\partial x_j}\right),$$

which is antisymmetric; i.e.,

(5.5-2) $$\omega_{ij} = -\omega_{ji}.$$

Hence, the tensor ω_{ij} has only three independent components ω_{12}, ω_{23}, and ω_{31}, because ω_{11}, ω_{22}, ω_{33} are zero. From such an antisymmetric tensor we can always build a *dual vector*

$$(5.5\text{-}3) \qquad \omega_k = \tfrac{1}{2}e_{kij}\omega_{ij},$$

where e_{kij} is the permutation symbol (Sec. 2.3). On the other hand, from the relation (5.5-3) and the *e-δ* identity, Eq. (2.3-11), it follows that $e_{ijk}\omega_k = \tfrac{1}{2}(\omega_{ij} - \omega_{ji})$ which, by Eq. (5.5-2), is ω_{ij}. Hence

$$(5.5\text{-}4) \qquad \omega_{ij} = e_{ijk}\omega_k.$$

Thus, ω_{ij} may be called the *dual* (antisymmetric) *tensor* of a vector ω_k. We shall call ω_k and ω_{ij}, respectively, the *rotation vector* and *rotation tensor* of the displacement field u_i.

A slight modification of the proof given at the end of Sec. 5.2 will convince us that *the vanishing of the symmetric strain tensor E_{ij} or e_{ij} is the necessary and sufficient condition for a neighborhood of a particle to be moved like a rigid body.* A rigid-body motion consists of a translation and a rotation. The translation is u_i. What is the rotation? We shall show that *in an infinitesimal displacement field for which the strain tensor vanishes at a point P the rotation of a neighborhood of P is given by the vector ω_i.* To show this, consider a point P' in the neighborhood of P. Let the coordinates of P and P' be x_i and $x_i + dx_i$, respectively. The relative displacement of P' with respect to P is

$$(5.5\text{-}5) \qquad du_i = \frac{\partial u_i}{\partial x_j}\,dx_j.$$

This can be written as

$$du_i = \frac{1}{2}\left(\frac{\partial u_i}{\partial x_j} + \frac{\partial u_j}{\partial x_i}\right)dx_j + \frac{1}{2}\left(\frac{\partial u_i}{\partial x_j} - \frac{\partial u_j}{\partial x_i}\right)dx_j.$$

The first quantity in parentheses is the infinitesimal strain tensor, which is zero by assumption. The second quantity in parentheses may be identified with Eq. (5.5-1). Hence,

$$(5.5\text{-}6) \qquad \begin{aligned} du_i &= -\omega_{ij}\,dx_j = \omega_{ji}\,dx_j \\ &= -e_{ijk}\omega_k\,dx_j \qquad \text{[by Eq. (5.5-4)]} \\ &= (\boldsymbol{\omega}\times d\mathbf{x})_i \qquad \text{(by definition).} \end{aligned}$$

Thus the relative displacement is the vector product of $\boldsymbol{\omega}$ and $d\mathbf{x}$. This is exactly what would have been produced by an infinitesimal rotation $|\boldsymbol{\omega}|$ about an axis through P in the direction of $\boldsymbol{\omega}$.

It should be noted that we have restricted ourselves to infinitesimal angular displacements. Angular measures for finite displacements are related to ω_{ij} in a more complicated way.

5.6* FINITE STRAIN COMPONENTS

When the strain components are not small, it is also easy to give simple geometric interpretations for the components of the strain tensors.

Consider a set of rectangular Cartesian coordinates with respect to which the strain components are defined as in Sec. 5.2. Let a line element before deformation be **da**, with components $da_1 = ds_0$, $da_2 = 0$, $da_3 = 0$. Let the extension E_1 of this element be defined by

$$(5.6\text{-}1) \qquad\qquad E_1 = \frac{ds - ds_0}{ds_0}$$

or

$$(5.6\text{-}2) \qquad\qquad ds = (1 + E_1)\, ds_0.$$

From Eq. (5.2-10) we have

$$(5.6\text{-}3) \qquad ds^2 - ds_0^2 = 2E_{ij}\, da_i\, da_j = 2E_{11}(da_1)^2.$$

Combining Eq. (5.6-2) and (5.6-3) we obtain

$$(5.6\text{-}4) \qquad\qquad (1 + E_1)^2 - 1 = 2E_{11}$$

which gives the meaning of E_{11} in terms of E_1. Conversely

$$(5.6\text{-}5) \qquad\qquad E_1 = \sqrt{1 + 2E_{11}} - 1.$$

This reduces to

$$(5.6\text{-}6) \qquad\qquad E_1 \doteq E_{11}$$

when E_{11} is small compared to 1.

To get the physical significance of the component E_{12}, let us consider two line elements $\mathbf{ds}_0$ and $\mathbf{d\bar{s}}_0$ which are at a right angle in the original state:

$$(5.6\text{-}7) \quad \begin{array}{llll} \mathbf{ds}_0: & da_1 = ds_0, & da_2 = 0, & da_3 = 0; \\ \mathbf{d\bar{s}}_0: & da_1 = 0, & da_2 = d\bar{s}_0, & da_3 = 0. \end{array}$$

After deformation these line elements become $\mathbf{ds}$, (dx_i) and $\mathbf{d\bar{s}}$, $(d\bar{x}_i)$. Forming the scalar product of the deformed elements, we obtain

$$ds \, d\bar{s} \cos \theta = dx_k \, d\bar{x}_k = \frac{\partial x_k}{\partial a_i} da_i \frac{\partial x_k}{\partial a_j} d\bar{a}_j$$

$$= \frac{\partial x_k}{\partial a_1} \frac{\partial x_k}{\partial a_2} ds_0 \, d\bar{s}_0.$$

But according to the definition (5.2-8), we have, since $\delta_{12} = 0$,

$$E_{12} = \frac{1}{2} \frac{\partial x_k}{\partial a_1} \frac{\partial x_k}{\partial a_2}.$$

Hence,

(5.6-8) $$ds \, d\bar{s} \cos \theta = 2E_{12} \, ds_0 \, d\bar{s}_0.$$

But, from Eq. (5.6-1) and (5.6-5), we have

$$ds = \sqrt{1 + 2E_{11}} \, ds_0, \qquad d\bar{s} = \sqrt{1 + 2E_{22}} \, d\bar{s}_0.$$

Hence, Eq. (5.6-8) yields

(5.6-9) $$\cos \theta = \frac{2E_{12}}{\sqrt{1 + 2E_{11}} \sqrt{1 + 2E_{22}}}.$$

The angle θ is the angle between the line elements $\mathbf{ds}$ and $\mathbf{d\bar{s}}$ after deformation. The change of angle between the two line elements, which in the original state are orthogonal, is $\alpha_{12} = \pi/2 - \theta$. From Eq. (5.6-9) we therefore obtain

(5.6-10) $$\sin \alpha_{12} = \frac{2E_{12}}{\sqrt{1 + 2E_{11}} \sqrt{1 + 2E_{22}}}.$$

These equations exhibit the relationship of E_{12} to the angles θ and α_{12}. The interpretation is not as simple as in the infinitesimal case because of the involvement off E_{11} and E_{22} in these equations.

A completely analogous interpretation can be made for the Eulerian strain components. Defining the extension e_1 per unit *deformed* length as

(5.6-11) $$e_1 = \frac{ds - ds_0}{ds},$$

we find

(5.6-12) $$e_1 = 1 - \sqrt{1 - 2e_{11}}.$$

Furthermore, if the deviation from a right angle between two elements in the original state which after deformation become orthogonal is denoted by β_{12}, we have

(5.6-13) $$\sin \beta_{12} = \frac{2e_{12}}{\sqrt{1 - 2e_{11}}\,\sqrt{1 - 2e_{22}}}.$$

In case of infinitesimal strains, Eqs. (5.6-10) and (5.6-13) reduce to the familiar results

(5.6-14) $e_1 \doteq e_{11}, \qquad E_1 \doteq E_{11}, \qquad \alpha_{12} \doteq 2E_{12}, \qquad \beta_{12} \doteq 2e_{12}.$

5.7 PRINCIPAL STRAINS. MOHR'S CIRCLE

Without much ado we can extend the results of Sec. 4.1 through 4.8 to the strain, because these properties are derived from the simple fact that the tensor concerned is symmetric. All we have to do is to change the word *stress* for *strain*. Thus:

(a) There exist three principal strains e_1, e_2, e_3 which are the roots of the determinantal equation

(5.7-1) $$|e_{ij} - e\,\delta_{ij}| = 0.$$

The roots of the cubic equation (5.7-1) are all real numbers.

(b) Associated with each principal strain, say e_1, there is a principal axis, with direction cosines $v_1^{(1)}, v_2^{(1)}, v_3^{(1)}$, which are the solutions of the equations

(5.7-2) $$(e_{ij} - e_1\,\delta_{ij})v_j^{(1)} = 0, \qquad (i = 1, 2, 3).$$

The three sets of solutions $(v_1^{(1)}, v_2^{(1)}, v_3^{(1)})$, $(v_1^{(2)}, v_2^{(2)}, v_3^{(2)})$, $(v_1^{(3)}, v_2^{(3)}, v_3^{(3)})$ are components of three unit vectors. If the roots e_1, e_2, e_3 of Eq. (5.7-1) are distinct $(e_1 \neq e_2 \neq e_3)$, then the three principal axes are orthogonal to one another. If two of the principal strains are the same, then Eq. (5.7-2) have infinitely many solutions, out of which an infinite number of pairs of orthogonal vectors can be selected and regarded as the principal axes. If all three roots are the same, then any set of three mutually orthogonal unit vectors may be regarded as principal.

(c) A plane perpendicular to a principal axis is called a principal plane.

(d) If the coordinate axes x_1, x_2, x_3 coincide with the principal axes, then the strain tensor assumes the canonical form

$$\begin{pmatrix} e_1 & 0 & 0 \\ 0 & e_2 & 0 \\ 0 & 0 & e_3 \end{pmatrix}.$$

(e) We can define a strain deviation tensor $e'_{ij} = e_{ij} - \frac{1}{3}e_{\alpha\alpha}\delta_{ij}$. Tensors e_{ij} and e'_{ij} have the following independent strain invariants:

$$(5.7\text{-}3) \quad \begin{aligned} I_1 &= e_{ij}\delta_{ij}. & J_1 &= e'_{ij}\delta_{ij} = 0, \\ I_2 &= \tfrac{1}{2}e_{ik}e_{ik}, & J_2 &= \tfrac{1}{2}e'_{ik}e'_{ik}, \\ I_3 &= \tfrac{1}{3}e_{ik}e_{km}e_{mi}. & J_3 &= \tfrac{1}{3}e'_{ik}e'_{km}e'_{mi}. \end{aligned}$$

(f) Mohr's circle may be used for the graphical analysis of strain. Lamé's ellipsoid is applicable to strain.

5.8* INFINITESIMAL STRAIN COMPONENTS
IN POLAR COORDINATES

As we indicated in Sec. 3.7, it is often desired to introduce curvilinear coordinates for reference. The strain components can be referred to a local rectangular frame of reference oriented in the direction of the curvilinear coordinates. For example, in polar coordinates r, θ, z the strain components may be designated $\epsilon_{rr}, \epsilon_{\theta\theta}, \epsilon_{zz}, \epsilon_{r\theta}, \epsilon_{rz}, \epsilon_{z\theta}$, and they are related to $\epsilon_{xx}, \epsilon_{yy}, \epsilon_{zz}, \epsilon_{xy}, \epsilon_{yz}, \epsilon_{zx}$ by the tensor transformation law, as in the cases of stresses. See Sec. 3.7.

However, if displacement vectors are resolved into components in the directions of the curvilinear coordinates, the strain-displacement relationship involves derivatives of the displacement components and therefore is influenced by the curvature of the coordinate system. The strain-displacement relations may appear quite different from the corresponding formulas in rectangular coordinates.

A truly general method for handling curvilinear coordinates is that of general tensor analysis. The reader is referred to more advanced treatises. An introduction is given in the author's *Foundations of Solid Mechanics.* Limiting ourselves in the present book to Cartesian tensors, we must treat each set of curvilinear coordinates in an ad hoc manner.

We shall illustrate two ad hoc approaches in the case of cylindrical polar coordinates: by transformation of coordinates and by detailed enumeration. The former will be discussed in this section; the latter, in Sec. 5.9.

In the first approach, we start from the relations between the polar coordinates r, θ, z and the rectangular coordinates x, y, z.

$$(5.8\text{-}1) \quad \begin{cases} x = r\cos\theta, \\ y = r\sin\theta, \end{cases} \quad \begin{cases} \theta = \tan^{-1}\dfrac{y}{x}, \\ r^2 = x^2 + y^2, \end{cases} \quad z = z.$$

(5.8-2)
$$\frac{\partial r}{\partial x} = \frac{x}{r} = \cos\theta, \qquad\qquad \frac{\partial r}{\partial y} = \frac{y}{r} = \sin\theta,$$

$$\frac{\partial \theta}{\partial x} = -\frac{y}{r^2} = -\frac{\sin\theta}{r}, \qquad \frac{\partial \theta}{\partial y} = \frac{x}{r^2} = \frac{\cos\theta}{r}.$$

It follows that any derivative with respect to x and y in the Cartesian equations may be transformed into derivatives with respect to r and θ by

(5.8-3)
$$\frac{\partial}{\partial x} = \frac{\partial r}{\partial x}\frac{\partial}{\partial r} + \frac{\partial \theta}{\partial x}\frac{\partial}{\partial \theta} = \cos\theta\,\frac{\partial}{\partial r} - \frac{\sin\theta}{r}\frac{\partial}{\partial \theta},$$

$$\frac{\partial}{\partial y} = \frac{\partial r}{\partial y}\frac{\partial}{\partial r} + \frac{\partial \theta}{\partial y}\frac{\partial}{\partial \theta} = \sin\theta\,\frac{\partial}{\partial r} + \frac{\cos\theta}{r}\frac{\partial}{\partial \theta}.$$

Now in polar coordinates we denote the components of the displacement vector $\mathbf{u}$ by u_r, u_θ, u_z as shown in Fig. 5.5. The components of the same vector resolved in the directions of rectangular coordinates are u_x, u_y, u_z. From Fig. 5.5 it is seen that these displacements are related by the equations

(5.8-4)
$$u_x = u_r \cos\theta - u_\theta \sin\theta,$$

$$u_y = u_r \sin\theta + u_\theta \cos\theta,$$

$$u_z = u_z.$$

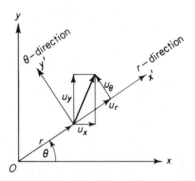

Fig. 5.5 Displacement vector in polar coordinates.

The strain components in polar coordinates are designated as

(5.8-6)

$$
\begin{array}{ccc}
\epsilon_{rr} & \epsilon_{r\theta} & \epsilon_{rz} \\
\epsilon_{\theta r} & \epsilon_{\theta\theta} & \epsilon_{\theta z} \\
\epsilon_{zr} & \epsilon_{z\theta} & \epsilon_{zz}
\end{array}
$$

They are really the strain components referred to a local frame of rectangular coordinates $x'y'z'$, with x' coinciding with the r-direction, y' coinciding with the θ-direction, and z' with z. The direction cosines between the two sets of coordinates are

	x	y	z
r or x'	$\cos\theta$	$\sin\theta$	0
θ or y'	$-\sin\theta$	$\cos\theta$	0
z or z'	0	0	1

(5.8-6)

The tensor transformation law holds, and we have

$$
\begin{aligned}
&\epsilon_{rr} = \epsilon_{xx}\cos^2\theta + \epsilon_{yy}\sin^2\theta + \epsilon_{xy}\sin 2\theta, \\
&\epsilon_{\theta\theta} = \epsilon_{xx}\sin^2\theta + \epsilon_{yy}\cos^2\theta - \epsilon_{xy}\sin 2\theta, \\
&\epsilon_{r\theta} = (\epsilon_{yy} - \epsilon_{xx})\cos\theta\sin\theta + \epsilon_{xy}(\cos^2\theta - \sin^2\theta), \\
&\epsilon_{zr} = \epsilon_{zx}\cos\theta + \epsilon_{zy}\sin\theta, \\
&\epsilon_{z\theta} = -\epsilon_{zx}\sin\theta + \epsilon_{zy}\cos\theta, \\
&\epsilon_{zz} = \epsilon_{zz}.
\end{aligned}
$$

(5.8-7)

Finally, we have

$$
\epsilon_{xx} = \frac{\partial u_x}{\partial x}, \qquad \epsilon_{yy} = \frac{\partial u_y}{\partial y}, \qquad \epsilon_{zz} = \frac{\partial u_z}{\partial z},
$$

(5.8-8)
$$
\epsilon_{xy} = \frac{1}{2}\left(\frac{\partial u_x}{\partial y} + \frac{\partial u_y}{\partial x}\right), \qquad \epsilon_{yz} = \frac{1}{2}\left(\frac{\partial u_y}{\partial z} + \frac{\partial u_z}{\partial y}\right),
$$

$$
\epsilon_{zx} = \frac{1}{2}\left(\frac{\partial u_z}{\partial x} + \frac{\partial u_x}{\partial z}\right).
$$

Now a substitution of (5.8-4) and (5.8-3) into (5.8-8) yields

$$
\epsilon_{xx} = \left(\cos\theta\,\frac{\partial}{\partial r} - \frac{\sin\theta}{r}\frac{\partial}{\partial\theta}\right)(u_r\cos\theta - u_\theta\sin\theta)
$$

$$
= \cos^2\theta\,\frac{\partial u_r}{\partial r} + \sin^2\theta\left(\frac{u_r}{r} + \frac{1}{r}\frac{\partial u_\theta}{\partial\theta}\right) - \cos\theta\sin\theta\left(\frac{\partial u_\theta}{\partial r} + \frac{\partial u_r}{r\,\partial\theta} - \frac{u_\theta}{r}\right),
$$

(5.8-9)
$$
\epsilon_{yy} = \sin^2\theta\,\frac{\partial u_r}{\partial r} + \cos^2\theta\left(\frac{u_r}{r} + \frac{\partial u_\theta}{r\,\partial\theta}\right) + \cos\theta\sin\theta\left(\frac{\partial u_\theta}{\partial r} + \frac{\partial u_r}{r\,\partial\theta} - \frac{u}{r}\right),
$$

$$
\epsilon_{xy} = \frac{\sin^2\theta}{2}\left(\frac{\partial u_r}{\partial r} - \frac{\partial u_\theta}{r\,\partial\theta} - \frac{u_r}{r}\right) + \frac{\cos^2\theta}{2}\left(\frac{\partial u_\theta}{\partial r} + \frac{\partial u_r}{r\,\partial\theta} - \frac{u_\theta}{r}\right).
$$

Substituting these and similar results into (5.8-7) and reducing, we obtain

(5.8-10)

$$\epsilon_{rr} = \frac{\partial u_r}{\partial r},$$

$$\epsilon_{\theta\theta} = \frac{u_r}{r} + \frac{1}{r}\frac{\partial u_\theta}{\partial \theta},$$

$$\epsilon_{r\theta} = \frac{1}{2}\left(\frac{1}{r}\frac{\partial u_r}{\partial \theta} + \frac{\partial u_\theta}{\partial r} - \frac{u_\theta}{r}\right),$$

$$\epsilon_{zr} = \frac{1}{2}\left(\frac{\partial u_r}{\partial z} + \frac{\partial u_z}{\partial r}\right),$$

$$\epsilon_{z\theta} = \frac{1}{2}\left(\frac{1}{r}\frac{\partial u_z}{\partial \theta} + \frac{\partial u_\theta}{\partial z}\right),$$

$$\epsilon_{zz} = \frac{\partial u_z}{\partial z}.$$

Thus, we see that the method transformation of coordinates is tedious but straightforward. Note that the structure of Eq. (5.8-10) and (5.8-8) are different. In the language of tensor analysis, the difference is caused by the differences in the fundamental metric tensors of the two coordinate systems.

The reader should be warned again that we have adopted the tensor notations for the strain, so that the shear strain components $\epsilon_{r\theta}$, ϵ_{rz}, $\epsilon_{z\theta}$ are one-half those ordinarily given as $\gamma_{r\theta}$, γ_{rz}, $\gamma_{z\theta}$ in most books.

5.9* DIRECT DERIVATION OF THE STRAIN-DISPLACEMENT
RELATIONS IN POLAR COORDINATES

The results of the preceding section can be derived directly from the geometric definition of the infinitesimal strain components. We recall that the normal strain components mean the ratio of change of length per unit length, whereas the shearing strain components mean one-half of the change of a right angle. For infinitesimal displacements, these changes can be seen directly from drawings such as those shown in Fig. 5.6.

Consider first the displacement in the r-direction, u_r. We see from Fig. 5.6(a) that

(5.9-1) $$\epsilon_{rr} = \frac{u_r + (\partial u_r/\partial r)\,dr - u_r}{dr} = \frac{\partial u_r}{\partial r}.$$

From the same figure we see also that a radial displacement of a circumferential element causes an elongation of that element and hence a strain in the θ-direction. The element ab, which was originally of length $r\,d\theta$, is displaced to $a'b'$ and becomes of length $(r + u_r)\,d\theta$. The tangential strain due to

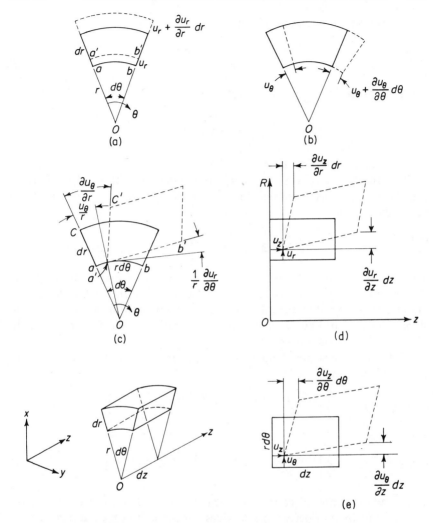

Fig. 5.6 Displacement in cylindrical polar coordinates. (From E. E. Sechler. Elasticity in Engineering, Courtesy Dover Publications).

this radial displacement is, therefore,

$$(5.9\text{-}2) \qquad \epsilon_{\theta\theta}^{(1)} = \frac{(r + u_r)\, d\theta - r\, d\theta}{r\, d\theta} = \frac{u_r}{r}.$$

On the other hand, as shown in Fig. 5.6(b), the tangential displacement u_θ gives rise to a tangential strain equal to

$$(5.9\text{-}3) \qquad \epsilon_{\theta\theta}^{(2)} = \frac{u_\theta + (\partial u_\theta/\partial\theta)\, d\theta - u_\theta}{r\, d\theta} = \frac{1}{r}\frac{\partial u_\theta}{\partial\theta}.$$

The total tangential strain $\epsilon_{\theta\theta}$ is the sum

$$(5.9\text{-}4) \qquad \epsilon_{\theta\theta} = \frac{u_r}{r} + \frac{1}{r}\frac{\partial u_\theta}{\partial \theta}.$$

The normal strain in the axial direction is

$$(5.9\text{-}5) \qquad \epsilon_{zz} = \frac{\partial u_z}{\partial z},$$

as in the case of rectangular coordinates.

The shearing strain $\epsilon_{r\theta}$ is equal to one-half of the change of angle $\angle c'a'b' - \angle cab$, as illustrated in Fig. 5.6(c). A direct examination of the figure shows that

$$(5.9\text{-}6) \qquad \epsilon_{r\theta} = \frac{1}{2}\left(\frac{1}{r}\frac{\partial u_r}{\partial \theta} + \frac{\partial u_\theta}{\partial r} - \frac{u_\theta}{r}\right).$$

The first term comes from the change in the radial displacement in the θ-direction; the second term comes from the change in the tangential displacement in the radial direction; and the last term appears since part of the slope change of the line $a'c'$ comes from the rotation of the element as a solid body about the axis through 0.

The remaining strain components $\epsilon_{z\theta}$, ϵ_{zr} can be derived with reference to Fig. 5.6(d) and (e).

$$(5.9\text{-}7) \qquad \epsilon_{z\theta} = \frac{1}{2}\left[\frac{(\partial u_z/\partial\theta)\,d\theta}{r\,d\theta} + \frac{(\partial u_\theta/\partial z)\,dz}{dz}\right] = \frac{1}{2}\left[\frac{1}{r}\frac{\partial u_z}{\partial\theta} + \frac{\partial u_\theta}{\partial z}\right]$$

and

$$(5.9\text{-}8) \qquad \epsilon_{zr} = \frac{1}{2}\left[\frac{(\partial u_r/\partial z)\,dz}{dz} + \frac{(\partial u_z/\partial r)\,dr}{dr}\right] = \frac{1}{2}\left[\frac{\partial u_r}{\partial z} + \frac{\partial u_z}{\partial r}\right].$$

These equations are, of course, the same as Eq. (5.8-10). Indeed, the direct geometric method of derivation provides a much clearer mental picture than the algebraic method of the preceding section.

5.10* OTHER STRAIN MEASURES

We must not think that the strain tensors defined above are the only ones suitable for the description of deformation. They are the most natural ones when we base our analysis of deformation on the change of the square of the distances between any two particles (Sec. 5.2). The *square* of distances is a convenient starting point because we have Pythagoras' theorem, which states that the square of the hypothenuse of a right triangle is equal to the sum of the squares of the legs. Using Pythagoras' theorem, we state that the square of the distance between two points x_i and $x_i + dx_i$, with coordinates referred to a rectangular Cartesian frame of reference, is

$$ds^2 = dx_1^2 + dx_2^2 + dx_3^2,$$

In Sec. 5.2 we based our analysis on this equation; the result was a natural definition of strain tensors.

Deformation does not have to be described this way. For example, we may insist on using the change of distance ds (instead of ds^2) as our starting point, or using the set of nine first derivatives of the displacement field:

(5.10-1)
$$\begin{vmatrix} \dfrac{\partial u}{\partial x} & \dfrac{\partial u}{\partial y} & \dfrac{\partial u}{\partial z} \\[2mm] \dfrac{\partial v}{\partial x} & \dfrac{\partial v}{\partial y} & \dfrac{\partial v}{\partial x} \\[2mm] \dfrac{\partial w}{\partial x} & \dfrac{\partial w}{\partial y} & \dfrac{\partial w}{\partial z} \end{vmatrix}.$$

Indeed, these derivatives, called "deformation gradients," are quite convenient. We may separate the matrix $(\partial u_i/\partial x_j)$ into a sum of a symmetric part and an antisymmetric part:

(5.10-2)
$$\begin{vmatrix} \dfrac{\partial u}{\partial x} & \dfrac{1}{2}\left(\dfrac{\partial u}{\partial y} + \dfrac{\partial v}{\partial x}\right) & \dfrac{1}{2}\left(\dfrac{\partial u}{\partial z} + \dfrac{\partial w}{\partial x}\right) \\[2mm] \dfrac{1}{2}\left(\dfrac{\partial v}{\partial x} + \dfrac{\partial u}{\partial y}\right) & \dfrac{\partial v}{\partial y} & \dfrac{1}{2}\left(\dfrac{\partial v}{\partial z} + \dfrac{\partial w}{\partial y}\right) \\[2mm] \dfrac{1}{2}\left(\dfrac{\partial w}{\partial x} + \dfrac{\partial u}{\partial z}\right) & \dfrac{1}{2}\left(\dfrac{\partial w}{\partial y} + \dfrac{\partial v}{\partial z}\right) & \dfrac{\partial w}{\partial z} \end{vmatrix}$$

$$+ \begin{vmatrix} 0 & \dfrac{1}{2}\left(\dfrac{\partial u}{\partial y} - \dfrac{\partial v}{\partial x}\right) & \dfrac{1}{2}\left(\dfrac{\partial u}{\partial z} - \dfrac{\partial w}{\partial x}\right) \\[2mm] -\dfrac{1}{2}\left(\dfrac{\partial u}{\partial y} - \dfrac{\partial v}{\partial x}\right) & 0 & \dfrac{1}{2}\left(\dfrac{\partial v}{\partial z} - \dfrac{\partial w}{\partial y}\right) \\[2mm] -\dfrac{1}{2}\left(\dfrac{\partial u}{\partial z} - \dfrac{\partial w}{\partial x}\right) & -\dfrac{1}{2}\left(\dfrac{\partial v}{\partial z} - \dfrac{\partial w}{\partial y}\right) & 0 \end{vmatrix}.$$

Then it is evident that the symmetric part of the deformation gradient matrix is the matrix of the infinitesimal strain as defined in Sec. 5.3.

Other well-known strain measures are Cauchy's strain tensors and Finger's strain tensors. When the mapping is given by Eq. (5.1-5) and (5.1-6), Cauchy's strain tensors are

$$C_{ij} = \frac{\partial a_k}{\partial x_i}\frac{\partial a_k}{\partial x_j}, \qquad \bar{C}_{ij} = \frac{\partial x_k}{\partial a_i}\frac{\partial x_k}{\partial a_j},$$

whereas Finger's strain tensors are

$$B_{ij} = \frac{\partial x_i}{\partial a_k}\frac{\partial x_j}{\partial a_k}, \qquad \bar{B}_{ij} = \frac{\partial a_i}{\partial x_k}\frac{\partial a_j}{\partial x_k}.$$

For these tensors, the absence of strain is not indicated by the vanishing of C_{ij} or B_{ij} but by $C_{ij} = \delta_{ij}$, $B_{ij} = \delta_{ij}$.

We shall not discuss these strain measures any further except to note that they may be convenient for some special purposes in advanced theories of continua.

PROBLEMS

5.2 (a) A state of deformation in which the displacement field u_i is a linear function of the coordinates x_i is called a "homogeneous deformation." What is the equation of a surface which will become a sphere $x^2 + y^2 + z^2 = r^2$ *after* a homogeneous deformation? [Use an equation of the type $f(x, y, z) = 0$, in which x, y, z are rectangular Cartesian coordinates.]

(b) Consider the following linear transformations of coordinates from (x, y, z) to (x', y', z'), both of which refer to the same Cartesian frame of reference. Fig. P5.2.

(1) *Pure shear:* $x' = kx, y' = k^{-1}y, z' = z.$
(2) *Simple shear:* $x' = x + 2sy, y' = y, z' = z.$

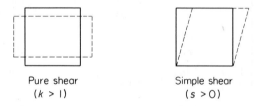

Pure shear   Simple shear
$(k > 1)$ $(s > 0)$

Fig. P5.2   Pure shear and simple shear.

We may regard (x, y, z) as the coordinates of a material particle before a deformation is imposed on a body and (x', y', z') as those after deformation. Show that a pure shear may be regarded as a simple shear referred to axes inclined at $\tan^{-1}(k^{-1})$ with Ox, Oy if $s = \frac{1}{2}(k - k^{-1})$. Equivalently, a simple shear may be regarded as a pure shear with $k = \sqrt{(s^2 + 1)} + s$ and the major axis of the strain ellipsoid inclined at $\frac{1}{4}\pi - \frac{1}{2}\tan^{-1} s = \tan^{-1}(k^{-1})$ with Ox.

(Drawings of the strain ellipses for these two cases can be found in J. C. Jaeger: *Elasticity, Fracture and Flow*. London: Methuen & Co., 1956, p. 32.)

Solution: We define a "homogeneous deformation" as one in which the displacement field u_i is a linear function of the coordinates so that a point x_i is moved to x_i' under the transformation

$$x_i' = x_i + u_i = x_i + u_i^{(0)} + a_{ik}x_k, \tag{1}$$

where $u_i^{(0)}$ and a_{ik} are constants. Under this transformation, a sphere $x'^2 + y'^2 + z'^2 = r^2$ corresponds to an ellipsoid

$$[u_i^{(0)} + x_i + a_{ik}x_k][u_i^{(0)} + x_i + a_{ik}x_k] = r^2. \tag{2}$$

Now, pure shear and simple shear are defined by the following equations and can be represented graphically for a square, as shown in Fig. P5.2.
Pure shear:

$$x' = kx, \qquad y' = y/k, \qquad z' = z. \tag{3}$$

Simple shear:

$$x' = x + 2sy, \qquad y' = y, \qquad z' = z. \tag{4}$$

The two transformations appear quite different in the figures. But in fact they are similar. The similarity is best shown by considering the strain ellipsoids.
Since $z' = z$, it is sufficient to consider transformation of curves in the x-, y-plane. By Eq. (3) a circle $x'^2 + y'^2 = 1$ is transformed into an ellipse:

$$k^2x^2 + \frac{y^2}{k^2} = 1; \tag{5}$$

Whereas by Eq. (4) the same circle is transformed into another ellipse:

$$x^2 + 4sxy + (1 + 4s^2)y^2 = 1. \tag{6}$$

Let us simplify Eq. (6) by a rotation of coordinates. By Eq. (2.4-2), if x, y is rotated to ξ, η through an angle θ, we have

$$x = \xi \cos\theta - \eta \sin\theta, \qquad y = \xi \sin\theta + \eta \cos\theta. \tag{7}$$

On substituting into Eq. (6) and simplifying, we obtain

$$\xi^2[\cos^2\theta + 4s\cos\theta\sin\theta + (1 + 4s^2)\sin^2\theta] + \eta^2[\sin^2\theta$$
$$- 4s\sin\theta\cos\theta + (1 + 4s^2)\cos^2\theta] + \xi\eta[-2\cos\theta\sin\theta$$
$$+ 4s(\cos^2\theta - \sin^2\theta) + 2\cos\theta\sin\theta(1 + 4s^2)] = 1. \tag{8}$$

The coefficient of $\xi\eta$ vanishes if $s = -\cot 2\theta$, or $\theta = -\frac{1}{2}\tan^{-1}(1/s)$. With this value of θ, the coefficient of ξ^2 in Eq. (8) becomes

$$\cos^2\theta - 2\cot 2\theta \sin 2\theta + (1 + 4\cot^2 2\theta)\sin^2\theta$$
$$= 1 - 2\cos 2\theta + \cos^2 2\theta/\cos^2\theta = \tan^2\theta.$$

Similarly, the coefficient of η^2 in Eq. (8) can be reduced to $\cot^2\theta$. Therefore Eq. (8) becomes

$$\tan^2\theta\, \xi^2 + \cot^2\theta\, \eta^2 = 1. \tag{9}$$

If we write $k = \tan \theta$, then Eq. (9) is reduced exactly to Eq. (5). Therefore these two strain ellipsoids are equal; one is rotated from the other by an angle θ. This verifies the equivalence of pure shear and simple shear.

To find the relation between k and s, we note that

$$\cot 2\theta = \frac{\cos 2\theta}{\sin 2\theta} = \frac{\cos^2 \theta - \sin^2 \theta}{2 \sin \theta \cos \theta} = \frac{1}{2}[\cot \theta - \tan \theta].$$

Therefore, since $s = -\cot 2\theta$ and $k = \tan \theta$, we have

$$-s = \frac{1}{2}\left[\frac{1}{k} - k\right], \quad \text{and} \quad k = s + \sqrt{(s^2 + 1)}. \tag{10}$$

5.3 A steel pipe of length 5 ft, diameter 6 in., and wall thickness $\frac{1}{16}$ in. is stretched 0.010 in. axially, expanded 0.001 in. in diameter, and twisted through 1°. Determine the strain components in the pipe.

Answer: $e_{zz} = e_{\theta\theta} = 1.66 \cdot 10^{-4}$, $e_{z\theta} = 4.38 \cdot 10^{-4}$.

5.4 For the truss as shown in Fig. P5.4, determine
    (a) The loads in the rods.
    (b) The stresses in the rods.
    (c) Assume a one-dimensional stress-strain relationship $e = \sigma/E$ for the rods, and assume that the Young's modulus for steel is $E = 3 \times 10^7$ psi. Determine the longitudinal strain e in the rods.
    (d) Determine the displacement vector at the point of loading B.

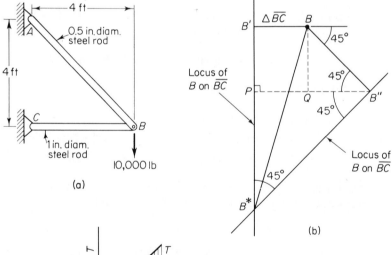

Fig. P5.4 A simple truss and a method of determining the displacement at the joint B.

Answer: (b) $\sigma_{AB} = 72{,}000$ lb/in.^{2}, $\sigma_{BC} = -12{,}800$ lb/in.^{2}
  (c) $e_{AB} = 2.4 \times 10^{-3}$ in./in., $\epsilon_{BC} = -4.25 \times 10^{-4}$ in./in.
  (d) 0.252 in.

Solution: The loads in the rods are determined by static equilibrium as in Chapter 1. We obtain a tension of $\sqrt{2} \cdot 10^4$ in AB and a compression of $-10{,}000$ lb in BC. The stresses are obtained by dividing the loads by the cross-sectional area of the members. A further division by Young's modulus gives the strains $e_{AB} = 2.405 \times 10^{-3}$ and $e_{BC} = -0.425 \times 10^{-3}$.

To determine the displacement at B, we note that the steel rods are pinended. As a consequence of shortening the rod BC, the point B moves to the left, but the rod B can swing around C, so that the locus of the possible location of B lies on an arc on a circle with C as center and $\overline{BC}$ as radius. For very small $\Delta\,\overline{BC}$ (as compared with $\overline{BC}$) this locus is a line segment perpendicular to $\overline{BC}$. Similarly, the bar AB extends $\Delta\,\overline{AB}$ in length, and the locus of B on AB lies on an arc $\perp AB$. The intersection of these arcs, B^*, is the final location of the displaced joint B.

To compute the displacement $\overline{BB^*}$, we see from the Fig. P5.4(b) that

$$\overline{BB^*} = \sqrt{\overline{BB'^2} + \overline{B'B^{*2}}} = \sqrt{\overline{BB'^2} + (\overline{B'P} + \overline{PB^*})^2}$$
$$= \sqrt{\overline{BB'^2} + (\overline{B'P} + \overline{PB''})^2} = \sqrt{\overline{BB'^2} + (\overline{B'P} + \overline{PQ} + \overline{QB''})^2}$$
$$= \sqrt{\overline{BB'^2} + (\overline{BQ} + \overline{BB'} + \overline{QB''})^2}.$$

Now

$$\overline{BB'} = |e_{BC}| \cdot 4 \cdot 12 = 0.425 \cdot 10^{-3} \cdot 48 = 2.07 \cdot 10^{-2} \text{ in.}$$
$$\overline{BQ} = \overline{BB''} \cos 45° = e_{AB}\overline{AB} \cos 45° = 0.1154 \text{ in.}$$

Similarly, $\overline{QB''} = \overline{BQ} = 0.1154$ in. Hence we obtain $\overline{BB^*} = 0.252$ in. by substitution.

Note: Alternative Method of Finding Displacement at B. The work done by the load is equal to the strain energy stored in the rods. When a rod is subject to a gradually increasing tension from zero to T, its length changes by amount $eL = TL/EA$ where L is the length of the rod, and A is its cross-sectional area. The strain energy stored in the rod is equal to $\frac{1}{2}(T^2L/AE)$. See Fig. P5.4(c). Now, when a load of 10,000 lb is gradually applied onto the bracket, the work done by it is equal to $\frac{1}{2} \cdot 10{,}000 \cdot \delta$, where δ is the displacement in the direction of the load, i.e., the vertical component of the displacement. The factor $\frac{1}{2}$ is necessary because the structure being linearly elastic, the force-deflection relationship is linear, so that the area under the curve which represents the work done, is $\frac{1}{2}$ of load $\times$ deflection. Hence, on equating the work done with the strain energy stored, we obtain

$$\frac{1}{2} \cdot 10{,}000 \cdot \delta = \frac{1}{2}\frac{T_{AB}^2 L_{AB}}{EA_{AB}} + \frac{1}{2}\frac{T_{BC}^2 L_{BC}}{EA_{BC}}.$$

On substituting numerical values into the above, we obtain $\delta = 0.250$ in. The total displacement of the joint B is $(\delta^2 + \Delta \overline{BC}^2)^{1/2} = 0.252$ in.

5.5 A rocket-launching tower is affected by thermal deflection caused by nonuniform heating of the rocket under the sun (Fig. P5.5). Assume that the rocket body is a circular cylinder, and estimate the horizontal displacement of the tip A if the following assumptions hold:

(a) The linear thermal coefficient of expansion is $\alpha = 10^{-5}$ in./in./°F.

(b) The maximum temperature in the rocket body on the side facing the sun is 20°F hotter than the minimum temperature on the shady side.

(c) The temperature distribution is uniform along the length (longitudinal axis) but varies linearly along the x-axis.

(d) As a consequence of (c), a plane section of the rocket remains plane in thermal expansion.

(e) The rocket is unloaded and is free to deform.

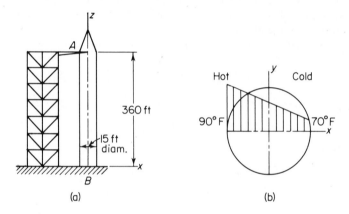

Fig. P5.5 Thermal deflection of a launching tower.

Hint: Compute thermal strain; then integrate to obtain the deflection.

Answer: Thermal strain difference from two sides $= \alpha T = 20 \times 10^{-5}$ ft/ft. Tip deflection $= 10.36$ in.

5.6 Derive an expression for the change in volume of an element of unit volume subjected to small strains e_{ij}. Show that the invariant $I_1 = e_{11} + e_{22} + e_{33}$ represents the change of volume per unit volume.

Solution: According to the general principle discussed in Sec. 5.7, we can find a set of rectangular Cartesian coordinates with respect to which the strain tensor assumes the form $e_k \delta_{ik}$ (k not summed), where e_1, e_2, e_3 are the principal strains. Let us consider a strained body and choose a unit cube whose edges are oriented along the principal axes of strain. Each edge, originally of length 1, becomes $1 + e_i$ after deformation. The new volume is, therefore,

$$(1 + e_1)(1 + e_2)(1 + e_3) = 1 + e_1 + e_2 + e_3 + \text{higher order terms.}$$

Hence, on ignoring the higher order terms, we have that the change of volume per unit volume is $e_1 + e_2 + e_3$.

We know from Eq. (5.7-3) that $I_1 = e_{ij}\delta_{ij}$ is an invariant. It is equal to $e_1 + e_2 + e_3$ with references to the principal axes. Hence $I_1 = e_1 + e_2 + e_3$ with reference to any Cartesian frame of reference. Hence $I_1 = e_{ij}\delta_{ij}$ means the change of volume per unit volume.

If the material is incompressible, its volume does not change; then $I_1 = e_{ii} = 0$.

5.7 Given a stress field σ_{ij}, with components referred to a system of coordinates x_1, x_2, x_3.

(a) What is the definition of principal stresses?

(b) What is the definition of principal axes?

(c) Describe briefly how the principal directions (i.e., directions of the principal axes) can be determined in principle.

(d) Consider a strain tensor e_{ij} referred to the same axes. How do you determine the principal strains and the corresponding principal directions?

(e) If the stress and strain tensors are related by the relation

$$\sigma_{ij} = \lambda e_{kk}\delta_{ij} + 2\mu e_{ij},$$

where λ, μ are constants, prove that the principal axes of stress coincide with the principal axes of strain.

5.8 In a study of earthquakes Lord Rayleigh investigated a solution of the linearized equations of elasticity in the form

$$u = Ae^{-by} \exp [ik(x - ct)],$$
$$v = Be^{-by} \exp [ik(x - ct)],$$
$$w = 0.$$

If the plane xz represents the ground while y represents the depth into the earth, and u, v, w are the displacements of the particles of the earth, then the solution above represents a wave propagating in the x-direction, with a speed c and an amplitude that decreases exponentially from the ground surface. The wave is assumed to be generated inside the earth. The ground surface is free; i.e., the stress vector acting on the ground surface is zero. After checking the equations of motion and the boundary conditions, he found the constants A, B, b, c, and obtained the solution

$$u = A(e^{-0.8475ky} - 0.5773e^{-0.3933ky}) \cos k (x - C_R t),$$
$$v = A(-0.8475e^{-0.8475ky} + 1.4679e^{-0.3933ky}) \sin k (x - C_R t),$$
$$w = 0.$$

The constant C_R is the so-called Rayleigh wave speed, which is equal to 0.9194 times the shear wave speed if Poisson's ratio is $\frac{1}{4}$. This solution satisfies the conditions of a wave propagating in a semi-infinite elastic solid with a free surface $y = 0$. The particles move in the xy-plane, with amplitude decreasing as the distance from the free surface increases (see Fig. P5.8). It represents one of the most prominent waves that can be seen on a seismograph when there is an earthquake.

(a) Sketch the waveform.

(b) Sketch the path of motion of particles on the free surface, $y = 0$, at several values of x. Do the same for several particles at different values of $y > 0$.

(c) Show that the motion of the particles is retrograde.

(d) Determine the places where the maximum principal strain occurs at any given instant, and the value of this strain.

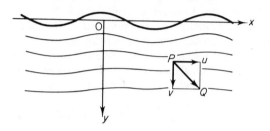

Fig. P5.8   Rayleigh surface wave.

Solution:

(d) Since $w = 0$, only the strain components e_{xx}, e_{yy}, e_{xy} are not identically zero. The exponential function e^{-by}, with $b > 0$, shows that the largest values of u, v, w and their derivatives will occur at $y = 0$. On this plane and at $t = 0$, we have

$$e_{xx} = \frac{\partial u}{\partial x} = -Ak(1 - 0.5773) \sin kx,$$

$$e_{yy} = \frac{\partial v}{\partial y} = Ak\,[(0.8475)^2 - 1.4679 \times 0.3933] \sin kx,$$

$$e_{xy} = \frac{Ak}{2}[(-0.8475 + 0.5773 \times 0.3933) + (-0.8475 + 1.4679)] \cos kx = 0.$$

Hence the maximum principal strains are

$$e_{xx} = \pm 0.4227Ak, \qquad e_{yy} = \pm 0.14094Ak.$$

5.9 Consider a square plate of unit size deformed as shown in Fig. P5.9. Find the strain components.

Solution: The deformation can be described by the following equations:

$$x_1 = a_1 + \frac{1}{\sqrt{3}}a_2, \qquad x_2 = a_2, \qquad x_3 = a_3,$$

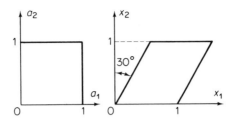

Fig. P5.9 Deformation of a square plate.

or

$$a_1 = x_1 - \frac{1}{\sqrt{3}} x_2, \qquad a_2 = x_2, \qquad a_3 = x_3.$$

Hence

$$ds^2 - ds_0^2$$

$$= \left\{ \left[\left(\frac{\partial x_1}{\partial a_1} \right)^2 - 1 \right] da_1^2 + 2 \frac{\partial x_1}{\partial a_1} \frac{\partial x_1}{\partial a_2} da_1\, da_2 + \left[\left(\frac{\partial x_1}{\partial a_2} \right)^2 + \left(\frac{\partial x_2}{\partial a_2} \right)^2 - 1 \right] da_2^2 \right\}$$

$$= \left\{ \left[1 - \left(\frac{\partial a_1}{\partial x_1} \right)^2 \right] dx_1^2 - 2 \frac{\partial a_1}{\partial x_1} \frac{\partial a_1}{\partial x_2} dx_1\, dx_2 + \left[1 - 1 - \left(\frac{\partial a_1}{\partial x_2} \right)^2 \right] dx_2^2 \right\}$$

$$= \frac{2}{\sqrt{3}} da_1\, da_2 + \left(\frac{1}{3} \right) da_2^2 = \frac{2}{\sqrt{3}} dx_1\, dx_2 - \frac{1}{3} dx_2^2.$$

But by (5.2-10) this is $2\,(E_{12} + E_{21})\, da_1\, da_2 + 2\, E_{22} da_2^2$. Hence

$$E_{12} = \frac{1}{2\sqrt{3}}, \qquad E_{22} = \frac{1}{6}. \qquad e_{12} = \frac{1}{2\sqrt{3}}, \qquad e_{22} = -\frac{1}{6};$$

whereas all other components of strain are zero.

5.10 Consider the square plate again, but this time shear to the right only a very small amount, so that

$$x_1 = a_1 + 0.01a_2, \qquad a_1 = x_1 - 0.01x_2, \qquad x_2 = a_2, \qquad x_3 = a_3.$$

Then

$$ds^2 - ds_0^2 = 0.01\, da_1\, da_2 + (0.01)^2\, da_2^2 = 0.01\, dx_1\, dx_2 - (0.01)^2\, dx_2^2.$$

Hence

$$E_{12} = 0.0025, \qquad E_{22} = 5 \times 10^{-5}, \qquad e_{12} = 0.0025, \qquad e_{22} = -5 \times 10^{-5}.$$

In this case, the E_{ij} and e_{ij} measures are approximately the same.

5.11 A square plate is deformed uniformily from configuration a to configuration b as shown in the three cases in Fig. P5.11. Determine the strain components $E_{11}, E_{22}, E_{12}; e_{11}, e_{22}, e_{12}$.

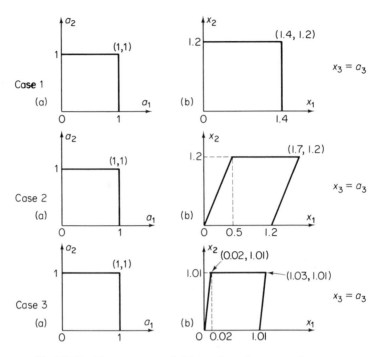

Fig. P5.11   Three patterns of deformation of a square plate.

Answer: The transformation that leads from configuation a to configuration b in Case 1 is $x_1 = 1.4a_1$, $x_2 = 1.2a_2$, $x_3 = a_3$. That in Case 2 is $x_1 = 1.2a_1 + 0.5a_2$, $x_2 = 1.2a_2$, $x_3 = a_3$. In Case 3, we have $x_1 = 1.01a_1 + 0.02a_2$, $x_2 = 1.01a_2$, $x_3 = a_3$. From these the strain components are obtained from Eq. (5.3-5). Case 3 qualifies for "infinitesimal" strains as given by Eq. (5.3-7).

5.12 A unit square $OABC$ is distorted to $OA'B'C'$ in three ways as shown in Fig. P5.12. In each of the Cases a, b, and c, write down the displacement field u_1, u_2 of every point in the square as functions of the location (a_1, a_2) of the point in the original position. Then determine the strains E_{ij}, e_{ij}. Assume

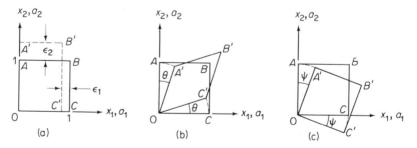

Fig. P5.12   Deformation of OABC to OA'B'C'.

$u_3 = 0$ and that u_1, u_2 are independent of x_3 or a_3. In Cases b and c, assume the lengths of OA, OA', OC, OC' are all 1. Also, obtain the simplified expressions of the strains e_{ij} if ϵ_1, ϵ_2, θ, ψ were infinitesimal.

5.13 A unit square $OABC$ is first subjected to a stretching as shown in Fig. P5.12(a), then to a distorsion as shown in Fig. P5.12(b), and finally to a rotation as shown in Fig. P5.12(c). After the three steps in succession, what are the values of the strains E_{ij}, e_{ij}? Answer this problem first for finite values of ϵ_1, ϵ_2, θ, ψ and then for infinitesimal values of ϵ_1, ϵ_2, θ, ψ.

5.14 Find the strain components E_{ij} and e_{ij} when one of the wedges in Fig. P5.14 is transformed into the other. The first wedge has an apex angle of $30°$; the other is $90°$. Radii are the same.

Fig. P5.14   Wedges changing angle.

5.15 Let $ABCD$ be a unit square in the x-, y-plane (Fig. P5.15). $ABCD$ is a part of a large deformable body subjected to a small strain that is uniform in the entire body and is given by

$$\begin{pmatrix} 1 & 2 & 3 \\ 2 & 1 & 0 \\ 3 & 0 & 2 \end{pmatrix} \times 10^{-3}.$$

What is the change of length of the lines AC and AE?

Answer: AC changes by 0.00423, AE changes by 0.00290.

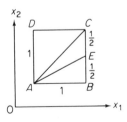

Fig. P5.15   Change of length of line segments in a plate of known strain.

5.16 A square membrane $-1 \leq x \leq 1$, $-1 \leq y \leq 1$ is stretched in such a manner that the displacement is described by

$$u = a(x^2 + y^2),$$

$$v = bxy,$$

$$w = 0.$$

What are the strain components at (x, y)? What is the principal strain at the origin $(0, 0)$. Assume the constants a, b to be infinitesimal.

5.17 A pin-jointed truss is shown in Fig. P5.17, where L is the length of the vertical and horizontal members. The cross-sectional area of all the members is the same, namely A. The material of all the members is the same, with Young's modulus E. The truss is loaded at the center by a load P. What would be the vertical deflection of the point under the load?

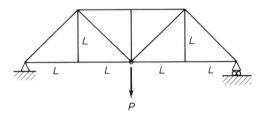

Fig. P5.17   Calculation of the vertical deflection of a joint of a truss.

Answer: Solve the problem by the strain-energy method illustrated in Prob. 5.4. The deflection is $5.828PL/AE$.

5.18 Compare the vertical deflection of the joint under the load W in the three trusses shown in Fig. P1.8, p. 26, and designed according to the principles stated in Prob. 1.8.

5.19 The following may happen in a number of situations, such as the flow of water, forming of metals, and in cell membranes. The material is incompressible. The displacement component w in the z-direction vanishes. The displacements u, v are infinitesimal and are functions of x, y. If, in a certain domain, we know that

$$u = (1 - y^2)(a + bx + cx^2),$$

where a, b, c are constants, compute the displacement v in the y-direction. *Hint:* Use the facts shown in Prob. 5.6.

5.20 A steel beam with a rectangular cross section of $\frac{1}{32} \times 1$ in.^{2} and a length of 10 in. is bent by couples acting at the ends into a circular arc of 60°. Determine the strain distribution in the beam. The midsurface of the beam may be assumed unstretched according to the ordinary beam theory (see Chapters 7 and 12) if the strain remains small.

Answer: Compute the radius of curvature R of the midsurface. Line elements parallel to the midsurface are elongated if they are outside the midsurface. Those inside the midsurface are shortened. The strain of the outer fiber of the bent beam is $(64R)^{-1}$.

5.21 A circular cylindrical shaft of 1 cm radius and of uniform isotropic material is twisted by an angle of 1°/cm axial length. See Fig. 12.1 on p. 316. Determine the strain components in the shaft.

CHAPTER SIX

Velocity Fields and Compatibility Conditions

We shall consider the velocity field and define the strain-rate tensor. Then we shall study the question of compatibility of the strain components or the strain-rate components.

6.1 VELOCITY FIELDS

For the study of fluid flow we are generally concerned with the velocity field, i.e., with the velocity of every particle in the body of the fluid. We refer the location of each fluid particle to a frame of reference O-xyz; then the field of flow is described by the velocity vector field $\mathbf{v}(x, y, z)$ which defines the velocity at every point (x, y, z). In terms of components, the velocity field is expressed by the functions

$$u(x, y, z), \qquad v(x, y, z), \qquad w(x, y, z),$$

or, if index notations are used, by $v_i(x_1, x_2, x_3)$.

For a continuous flow, we consider the continuous and differentiable functions $v_i(x_1, x_2, x_3)$. There are occasions, however, in which we must study the relationship of velocities at neighboring points. Let the particles P and P' be located instantaneously at x_i and $x_i + dx_i$, respectively. The difference in velocities at these two points is

$$(6.1\text{-}1) \qquad\qquad dv_i = \frac{\partial v_i}{\partial x_j} dx_j,$$

where the partial derivatives $\partial v_i / \partial x_j$ are evaluated at the particle P. Now

$$(6.1\text{-}2) \qquad \frac{\partial v_i}{\partial x_j} = \frac{1}{2}\left(\frac{\partial v_i}{\partial x_j} + \frac{\partial v_j}{\partial x_i}\right) - \frac{1}{2}\left(\frac{\partial v_j}{\partial x_i} - \frac{\partial v_i}{\partial x_j}\right).$$

Let us define the *rate-of-deformation* tensor V_{ij} and the *spin tensor* Ω_{ij} as

$$(6.1\text{-}3) \qquad\qquad V_{ij} \equiv \frac{1}{2}\left(\frac{\partial v_i}{\partial x_j} + \frac{\partial v_j}{\partial x_i}\right),$$

$$(6.1\text{-}4) \qquad\qquad \Omega_{ij} \equiv \frac{1}{2}\left(\frac{\partial v_j}{\partial x_i} - \frac{\partial v_i}{\partial x_j}\right).$$

It is evident that V_{ij} is symmetric and Ω_{ij} is antisymmetric; i.e.,

$$(6.1\text{-}5) \qquad\qquad V_{ij} = V_{ji}, \qquad \Omega_{ij} = -\Omega_{ji}.$$

Hence, the Ω_{ij} tensor has only three independent elements and there exists a vector $\mathbf{\Omega}$ *dual* to Ω_{ij}:

$$(6.1\text{-}6) \qquad\qquad \Omega_k \equiv e_{kij}\Omega_{ij}; \quad \text{i.e.,} \quad \mathbf{\Omega} = \tfrac{1}{2}\,\text{curl}\,\mathbf{v},$$

where e_{kij} is the permutation tensor (p. 45). The vector $\mathbf{\Omega}$ is called the *vorticity* vector.

These equations are similar to Eq. (5.5-3) and (5.5-5). Their geometric interpretations are also similar. Therefore, the analysis of the velocity field is very much like the analysis of an infinitesimal deformation field. Indeed, if we multiply v_i by an infinitesimal interval of time dt, the result is an infinitesimal displacement $u_i = v_i\, dt$. Hence, whatever we learned about the infinitesimal strain field can be immediately extended correspondingly to the *rate of change* of strain, with the word *velocity* replacing the word *displacement*.

6.2 THE SO-CALLED COMPATIBILITY CONDITION

Suppose we were given a set of two partial differential equations for one unknown function $u(x, y)$, such as

$$(6.2\text{-}1) \qquad\qquad \frac{\partial u}{\partial x} = x + 3y, \qquad \frac{\partial u}{\partial y} = x^2.$$

We know that these equations cannot be solved: We have too many equations that are mutually inconsistent. The inconsistency can be clarified if we compute the second derivative $\partial^2 u/\partial x\,\partial y$ from the two equations (6.2-1): The first yields 3, the second yields $2x$. They are unequal.

Therefore, when partial differential equations are given, the question of integrability arises. The differential equations

$$(6.2\text{-}2) \qquad \frac{\partial u}{\partial x} = f(x, y), \qquad \frac{\partial u}{\partial y} = g(x, y)$$

cannot be integrated unless the condition

$$(6.2\text{-}3) \qquad \frac{\partial f}{\partial y} = \frac{\partial g}{\partial x}$$

is satisfied. This condition, Eq. (6.2-3), is called the condition of integrability or the equation of compatibility.

Now consider a plane state of strain such as may exist in the solid propellant grain of a rocket. Suppose that a man made a laboratory model and obtained by various types of strain gauges, photoelastic equipment, laser holography combined with Moire pattern analysis, etc., a set of strain data which may be presented as

$$(6.2\text{-}4) \qquad \begin{aligned} e_{xx} = f(x, y), \qquad e_{yy} = g(x, y), \qquad e_{xy} = h(x, y), \\ e_{zz} = e_{zx} = e_{zy} = 0. \end{aligned}$$

The question arises whether his data are self-consistent. Could the consistency be checked? And if they are consistent, can we compute the displacements $u(x, y)$ and $v(x, y)$ from his data?

If the strain is small, the last question can be formulated as a mathematical question of integrating the differential equations

$$(6.2\text{-}5) \qquad \begin{aligned} \frac{\partial u}{\partial x} &= f(x, y), & (= e_{xx}) \\ \frac{\partial v}{\partial y} &= g(x, y), & (= e_{yy}) \\ \frac{\partial u}{\partial y} + \frac{\partial v}{\partial x} &= 2h(x, y). & (= 2e_{xy}) \end{aligned}$$

Now if we differentiate the first equation above with respect to y twice, the second with respect to x twice, and the third with respect to x and y once each, we obtain

$$(6.2\text{-}6) \qquad \frac{\partial^3 u}{\partial x \, \partial y^2} = \frac{\partial^2 f}{\partial y^2}, \qquad \frac{\partial^3 v}{\partial x^2 \, \partial y} = \frac{\partial^2 g}{\partial x^2},$$

$$(6.2\text{-}7) \qquad \frac{\partial^3 u}{\partial x \, \partial y^2} + \frac{\partial^3 v}{\partial x^2 \, \partial y} = 2 \frac{\partial^2 h}{\partial x \, \partial y}.$$

Substituting Eq. (6.2-6) into Eq. (6.2-7), we have

$$(6.2\text{-}8) \qquad \frac{\partial^2 f}{\partial y^2} + \frac{\partial^2 g}{\partial x^2} = 2 \frac{\partial^2 h}{\partial x \, \partial y}.$$

The experimental data must satisfy this equation. If not, the data are not consistent and there must have been errors.

Expressing the results above in terms of strain components, we have

$$(6.2\text{-}9) \qquad \frac{\partial^2 e_{xx}}{\partial y^2} + \frac{\partial^2 e_{yy}}{\partial x^2} = 2 \frac{\partial^2 e_{xy}}{\partial x \, \partial y},$$

which is called the *equation of compatibility.*

A similar descussion may be directed to a two-dimensional velocity field of a fluid. The components of the rate-of-strain tensor may be measured, for example, by the method of optical birefringence, if the fluid is birefringent. Or a set of strain rates might have been obtained theoretically. To check the consistency, we must have

$$(6.2\text{-}10) \qquad \frac{\partial^2 V_{xx}}{\partial y^2} + \frac{\partial^2 V_{yy}}{\partial x^2} = 2 \frac{\partial^2 V_{xy}}{\partial x \, \partial y},$$

where V_{ij} are the components of the rate-of-strain tensor. (See Sec. 6.1.) In fluid mechanics, however, this equation is referred to as the *condition of integrability*. Thus compatibility and integrability mean the same thing.

6.3* COMPATIBILITY OF STRAIN COMPONENTS
IN THREE DIMENSIONS

Extending the question discussed in the previous section to three dimensions, how do we integrate the differential equations

$$(6.3\text{-}1) \qquad e_{ij} = \frac{1}{2} \left[\frac{\partial u_j}{\partial x_i} + \frac{\partial u_i}{\partial x_j} \right]$$

to determine u_i?

Inasmuch as there are six equations for three unknown functions u_i, the system of Eq. (6.3-1) will have a single-valued solution only if the functions e_{ij} satisfy the conditions of compatibility.

Since strain components only determine the relative positions of points in the body, and since any rigid-body motion corresponds to zero strain, we expect that the solution u_i can be determined only up to an arbitrary rigid-body motion. But, if e_{ij} were specified arbitrarily, we could expect cases similar to those shown in Fig. 6.1. Here a continuous triangle (portion of material in a body) is given. If we deform it by following an arbitrarily specified strain field starting form the point A, we might end at the point C and D either with a gap between them or with overlapping of material. For a single-valued continouus solution to exist (up to a rigid-body motion), the ends C and D must meet perfectly in the strained configuration. This cannot be guaranteed unless the specified strain field along the edges of the triangle obeys the compatibility conditions.

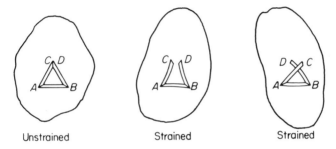

Unstrained   Strained   Strained

Fig. 6.1 Illustrations for the requirement of compatibility.

By differentiation of Eq. (6.3-1), we have

(6.3-2) $e_{ij,kl} = \tfrac{1}{2}(u_{i,jkl} + u_{j,ikl})$,

where the indices k and l following a comma indicates successive partial differentiations with respect to x_k and x_l. Interchanging subscripts, we have

$$e_{kl,ij} = \tfrac{1}{2}(u_{k,lij} + u_{l,kij}),$$
$$e_{jl,ik} = \tfrac{1}{2}(u_{j,lik} + u_{l,jik}),$$
$$e_{ik,jl} = \tfrac{1}{2}(u_{i,kjl} + u_{k,ijl}).$$

From these we verify at once that

(6.3-3) $e_{ij,kl} + e_{kl,ij} - e_{ik,jl} - e_{jl,ik} = 0$.

This is the *equation of compatibility* of St. Venant.

Of the 81 equations represented by Eq. (6.3-3), only 6 are essential. The rest are either identities or repetitions on account of the symmetry of e_{ij} and of kl in $e_{ij,kl}$. The 6 equations written in unabridged notations are

(6.3-4)

$$\frac{\partial^2 e_{xx}}{\partial y \, \partial z} = \frac{\partial}{\partial x}\left(-\frac{\partial e_{yz}}{\partial x} + \frac{\partial e_{zx}}{\partial y} + \frac{\partial e_{xy}}{\partial z}\right),$$

$$\frac{\partial^2 e_{yy}}{\partial z \, \partial x} = \frac{\partial}{\partial y}\left(-\frac{\partial e_{zx}}{\partial y} + \frac{\partial e_{xy}}{\partial z} + \frac{\partial e_{yz}}{\partial x}\right),$$

$$\frac{\partial^2 e_{zz}}{\partial x \, \partial y} = \frac{\partial}{\partial z}\left(-\frac{\partial e_{xy}}{\partial z} + \frac{\partial e_{yz}}{\partial x} + \frac{\partial e_{zx}}{\partial y}\right),$$

$$2\frac{\partial^2 e_{xy}}{\partial x \, \partial y} = \frac{\partial^2 e_{xx}}{\partial y^2} + \frac{\partial^2 e_{yy}}{\partial x^2},$$

$$2\frac{\partial^2 e_{yz}}{\partial y \, \partial z} = \frac{\partial^2 e_{yy}}{\partial z^2} + \frac{\partial^2 e_{zz}}{\partial y^2},$$

$$2\frac{\partial^2 e_{zx}}{\partial z \, \partial x} = \frac{\partial^2 e_{zz}}{\partial x^2} + \frac{\partial^2 e_{xx}}{\partial z^2}.$$

These conditions are derived for infinitesimal strains referred to rectangular Cartesian coordinates. If the strains are finite, the conditions are more complex.*

Let us now return to the question posed at the beginning of this section and inquire whether conditions (6.3-3) or (6.3-4) are sufficient to assure the existence of a single-valued continuous solution of the differential Eq. (6.3-1) up to a rigid-body motion. The answer is affirmative if the region is simply connected. Various proofs are available.† However, for multiconnected regions additional conditions of sufficiency are required. See the author's *Foundations of Solid Mechanics* for details.

PROBLEMS

6.1 Consider the motion of a body of fluid with velocity components u and v derived from a potential Φ:

$$u = \frac{\partial \Phi}{\partial x}, \qquad v = \frac{\partial \Phi}{\partial y},$$

while the component w is identically zero. Sketch the velocity field for the following potentials:

---

*These can be found in Green and Zerna, *Theoretical Elasticity*, London: Oxford University Press. 1954, p. 62.

†One of the simplest was given by E. Cesaro in 1906 and can be found in Fung, *Foundations of Solid Mechanics*, pp. 101–108.

(a) $\Phi = \dfrac{1}{4\pi} \log (x^2 + y^2) = \dfrac{1}{2\pi} \log r,$ $(r^2 = x^2 + y^2)$

(b) $\Phi = x$

(c) $\Phi = Ar^n \cos n\theta,$ $\left(\theta = \tan^{-1} \dfrac{y}{x}\right)$

(d) $\Phi = \dfrac{\cos \theta}{r}$

Note: A flow field whose velocity components are derived from a potential function $\Phi(x, y, z)$ is called a potential flow. In the examples named in this problem we have several cases in which Φ is expressed in terms of the polar coordinates r, θ. If we notice that the velocity vector (u, v) is exactly the gradient of the scalar function $\Phi(x, y)$ (see Chapter 2), we see from vector analysis that the velocity components in the polar coordinates are

$$u_r = \frac{\partial \Phi (r, \theta)}{\partial r}, \qquad u_\theta = \frac{1}{r} \frac{\partial \Phi (r, \theta)}{\partial \theta} \tag{1}$$

where u_r, u_θ are velocity components in the radial and tangential directions, respectively. See Fig. P6.1.

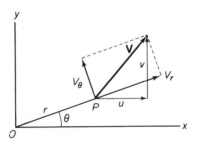

Fig. P6.1 Velocity components in polar coordinates.

These relations can be derived formally as follows: Since

$$r^2 = x^2 + y^2, \qquad \theta = \tan^{-1} \frac{y}{x},$$
$$x = r \cos \theta, \qquad y = r \sin \theta. \tag{2}$$

$$\frac{\partial x}{\partial r} = \cos \theta, \qquad \frac{\partial y}{\partial r} = \sin \theta,$$
$$\frac{\partial x}{\partial \theta} = -r \sin \theta, \qquad \frac{\partial y}{\partial \theta} = r \cos \theta, \tag{3}$$

we have

$$\frac{\partial \Phi}{\partial r} = \frac{\partial \Phi}{\partial x} \frac{\partial x}{\partial r} + \frac{\partial \Phi}{\partial y} \frac{\partial y}{\partial r} = u \cos \theta + v \sin \theta. \tag{4}$$

$$\frac{1}{r} \frac{\partial \Phi}{\partial \theta} = -\frac{\partial \Phi}{\partial x} \sin \theta + \frac{\partial \Phi}{\partial y} \cos \theta = -u \sin \theta + v \cos \theta. \tag{5}$$

But it is seen from Fig. P6.1 that

$$u_r = u \cos \theta + v \sin \theta, \qquad u_\theta = -u \sin \theta + v \cos \theta. \tag{6}$$

Hence Eq. (1) follows from (4) and (5).

6.2 The motion of an incompressible fluid in two dimensions may be derived from a stream function ψ as follows:

$$u = -\frac{\partial \psi}{\partial y}, \qquad v = \frac{\partial \psi}{\partial x}, \qquad w = 0.$$

Sketch the lines $\psi = $ const. for the following functions and compare with the results of the preceding problem.

(a) $\psi = c\theta$

(b) $\psi = y$

(c) $\psi = Ar^n \sin n\theta$

(d) $\psi = -\dfrac{\sin \theta}{r}$

6.3 For the flows described by the potentials listed above,
(a) Show that the vorticity vanishes in every case.
(b) Derive expressions for the rate-of-strain tensor.

Solution: The vorticity Ω, given by Eq. (6.1-6), has the components given in Eq. (6.1-4). In a two-dimensional flow, there is only one component of vorticity $\Omega_{12} = (\partial v/\partial x - \partial u/\partial y)/2$ that is not identically zero. If $u = \partial \varphi/\partial x$, $v = \partial \varphi/\partial y$, then $\Omega_{12} \equiv 0$. Hence all potential flow is irrotational. If

$$u = -\frac{\partial \psi}{\partial y}, \qquad v = \frac{\partial \psi}{\partial x}, \tag{1}$$

then Ω_{12} is

$$\text{vorticity} = \frac{1}{2}\left(\frac{\partial^2 \psi}{\partial x^2} + \frac{\partial^2 \psi}{\partial y^2}\right). \tag{2}$$

In polar coordinates, we have

$$\text{vorticity} = \frac{1}{2}\left(\frac{\partial^2 \psi}{\partial r^2} + \frac{1}{r}\frac{\partial \psi}{\partial r} + \frac{1}{r^2}\frac{\partial^2 \psi}{\partial \theta^2}\right). \tag{3}$$

That all the cases in Prob. 6.2 are irrotational can be verified by direct substitution.

For part b, the components of the rate-of-strain tensor in polar coordinates can be derived by transformation of coordinates as shown in Sec. 5.8, p. 135, or by direct derivation as shown in Sec. 5.9, p. 138. By a slight change of notations, we obtain

$$V_{rr} = \frac{\partial u_r}{\partial r}, \qquad V_{\theta\theta} = \frac{u_r}{r} + \frac{1}{r}\frac{\partial u_\theta}{\partial \theta},$$

$$V_{r\theta} = \frac{1}{2}\left(\frac{1}{r}\frac{\partial u_r}{\partial \theta} + \frac{\partial u_\theta}{\partial r} - \frac{u_\theta}{r}\right).$$

With these equations the problem is easily solved.

6.4 Suppose we were given the following displacement field defined in a unit circle,

$$u = ax^2 + bxy + c,$$
$$v = by^2 + cx + mz,$$
$$w = mz^3.$$

Is there any question of compatibility?

6.5 Suppose the displacement field in a unit circle is

$$u = ar \log \theta,$$
$$v = ar^2 + c \sin \theta,$$
$$w = 0.$$

Is it compatible?

6.6 In a two-dimensional, plane-strain field, the displacements are described by $u(x, y)$, $v(x, y)$, whereas that along the z-axis, w, is identically zero, x, y, z being a set of rectangular Cartesian coordinates.

  (a) Express the strain components e_{xx}, e_{xy}, e_{yy} in terms of u, v.

  (b) Derive the equation of compatibility for the strain system e_{xx}, e_{xy}, e_{yy}.

  (c) Is the following strain system a possible one?

$$e_{xx} = k(x^2 - y^2), \qquad e_{yy} = kxy, \qquad e_{xy} = k'xy,$$

where k, k' are constants. All other strain components are zero.

Answer: Possible if $k' = -k$.

FURTHER READING

FUNG, Y. C., *Foundations of Solid Mechanics*, Englewood Cliffs, N.J.: Prentice-Hall (1965), pp. 99–109.

GREEN, A. E., AND W. ZERNA, *Theoretical Elasticity*, London: Oxford University Press (1954), p. 62.

CHAPTER SEVEN

Constitutive Equations

The three most commonly used constitutive equations are presented. They are mathematical abstractions and are given here in the barest outline to exhibit their similarities and differences. They can be simplified greatly if the material is isotropic. Since the concept of isotropy is very important and is usually passed over too lightly by beginners, we shall devote Chapter 8 to it. The properties of real material are discussed in Chapter 9. An application of the results of the preceding chapters to simple beams is given at the end of this chapter.

7.1 SPECIFICATION OF THE PROPERTIES OF MATERIALS

The properties of materials are specified by constitutive equations. A wide variety of materials exist. Thus, we are not surprised that there are a great many constitutive equations describing an almost infinite variety of materials. What should be surprising, therefore, is the fact that three simple, idealized, stress-strain relationships give a good description of the mechanical properties of many materials around us: namely, the nonviscous fluid, the Newtonian viscous fluid, and the perfectly elastic solid. We shall describe these idealized relations in this chapter but hasten to add that real materials differ more or less from these idealized laws. When the differences are great, we speak of real gases, non-Newtonian viscous fluids, viscoelastic solids, plasticity, etc., whose descriptions, are, of course, much more complicated. Only the mathematical formulas will be presented here. Numerical constants, as well as deviations from these idealized laws, will be considered in Chapter 9.

An equation which describes a property of a material is called a *constitutive equation* of that material. A stress-strain relationship describes the mechanical property of a material and is therefore a constitutive equation. Our main objective in this chapter is to discuss the stress-strain relationship. There are other constitutive equations, such as those describing the heat transfer characteristics, electric resistance, mass transport, etc., but they are not our immediate concern.

7.2 THE NONVISCOUS FLUID

A nonviscous fluid is one for which the stress tensor is isotropic; i.e., it is of the form

$$(7.2\text{-}1) \quad \blacktriangle \qquad \sigma_{ij} = -p\delta_{ij},$$

where δ_{ij} is the Kronecker delta and p is a scalar called *pressure*. In matrix form, the components of stress in a nonviscous fluid may be displayed as

$$(7.2\text{-}2) \qquad (\sigma_{ij}) = \begin{pmatrix} -p & 0 & 0 \\ 0 & -p & 0 \\ 0 & 0 & -p \end{pmatrix}.$$

The pressure p in an *ideal gas* is related to the density ρ and temperature T by the equation of state

$$(7.2\text{-}3) \qquad \frac{p}{\rho} = RT,$$

where R is the gas constant. For a real gas or a liquid it is often possible to obtain an equation of state

$$(7.2\text{-}4) \qquad f(p, \rho, T) = 0.$$

See Chapter 9, especially Eq. (9.1-3), (9.5-1), and (9.5-3).

An anomaly exists in the case of an *incompressible fluid*, for which the equation of state is merely

$$(7.2\text{-}5) \qquad \rho = \text{const.}$$

Thus, the pressure p is left as an arbitrary variable for an incompressible fluid. It is determined solely by the equations of motion and the boundary conditions. For example, an incompressible fluid in the cylinder of a hydraulic press can assume any pressure depending on the force applied to the piston.

Since hydrodynamics is concerned mostly with incompressible fluids, we shall see that pressure is controlled by boundary conditions, whereas the variation of pressure (the pressure gradient) is calculated from the equations of motion.

Air and water can be treated as nonviscous in many problems. For example, in the problems of tides around the earth, waves in the ocean, flight of an airplane, flow in a jet, combustion in an automobile engine, etc., excellent results can be obtained by ignoring the viscosity of the media and treating it as a nonviscous fluid. On the other hand, there are important problems in which the viscosity of the media, though small, must not be neglected. Such are the problems of determining the drag force acting on an airplane, whether a flow is turbulent or laminar, the heating of a reentry spacecraft, the cooling of an automobile engine, etc.

7.3 NEWTONIAN FLUID

A Newtonian fluid is a viscous fluid for which the shear stress is linearly proportional to the rate of deformation. For a Newtonian fluid the stress-strain relationship is specified by the equation

$$(7.3\text{-}1) \quad \blacktriangle \qquad \sigma_{ij} = -p\delta_{ij} + \mathfrak{D}_{ijkl}V_{kl}.$$

where σ_{ij} is the stress tensor, V_{kl} is the rate-of-deformation tensor, $\mathfrak{D}_{ijkl}$ is a tensor of viscosity coefficients of the fluid, and p is the *static pressure*. The term $-p\delta_{ij}$ represents the state of stress possible in a fluid at rest (when $V_{kl} = 0$). The static pressure p is assumed to depend on the density and temperature of the fluid according to an equation of state. For Newtonian fluids we assume that the elements of the tensor $\mathfrak{D}_{ijkl}$ may depend on the temperature but not on the stress or the rate of deformation. The tensor $\mathfrak{D}_{ijkl}$, of rank 4, has $3^4 = 81$ elements. Not all these constants are independent. A study of the theoretically possible number of independent elements can be made by examining the symmetry properties of the tensors σ_{ij}, V_{kl} and the symmetry that may exist in the atomic constitution of the fluid. We shall not pursue it here because we know of no fluid that has been examined in such detail as to have all the constants in the tensor $\mathfrak{D}_{ijkl}$ determined. Most fluids appear to be isotropic, for which the structure of $\mathfrak{D}_{ijkl}$ is greatly simplified, as will be seen below. Those readers who are interested in the general structure of $\mathfrak{D}_{ijkl}$ should read Sec. 7.4 and the references referred to therein, because the tensor of elastic constants C_{ijkl} has a similar structure.

If the fluid is *isotropic*, i.e., if the tensor $\mathfrak{D}_{ijkl}$ has the same array of components in any system of rectangular Cartesian coordinates, then $\mathfrak{D}_{ijkl}$ can be expressed in terms of two independent constants λ and μ (see Sec. 8.4):

(7.3-2) $$\mathfrak{D}_{ijkl} = \lambda \delta_{ij} \delta_{kl} + \mu(\delta_{ik}\delta_{jl} + \delta_{il}\delta_{jk}),$$

and we obtain

(7.3-3) ▲ $$\sigma_{ij} = -p\delta_{ij} + \lambda V_{kk}\delta_{ij} + 2\mu V_{ij}.$$

A contraction of Eq. (7.3-3) gives

(7.3-4) $$\sigma_{kk} = -3p + (3\lambda + 2\mu)V_{kk}.$$

If it is assumed that the mean normal stress $\frac{1}{3}\sigma_{kk}$ is independent of the rate of dilation V_{kk}, then we must set

(7.3-5) $$3\lambda + 2\mu = 0;$$

thus, the constitutive equation becomes

(7.3-6) ▲ $$\sigma_{ij} = -p\delta_{ij} + 2\mu V_{ij} - \tfrac{2}{3}\mu V_{kk}\delta_{ij}.$$

This formulation is due to George G. Stokes and a fluid that obeys Eq. (7.3-6) is called a *Stokes fluid*, for which one material constant μ, the coefficient of viscosity, suffices to define its property. Some data on the coefficients of viscosity of fluids are given in Chapter 9.

If a fluid is *incompressible*, then $V_{kk} = 0$, and we have the constitutive equation for an *incompressible* viscous fluid:

(7.3-7) ▲ $$\sigma_{ij} = -p\delta_{ij} + 2\mu V_{ij}.$$

If $\mu = 0$, we obtain the constitutive equation of the *nonviscous fluid*:

(7.3-8) $$\sigma_{ij} = -p\delta_{ij}.$$

The presence of the static pressure term p marks a fundamental difference between fluid mechanics and elasticity. To accommodate this new variable, it is often assumed that an *equation of state* exists which relates the pressure p, the density ρ, and the absolute temperature T,

(7.3-9) $$f(p, \rho, T) = 0.$$

For example, for an *ideal gas*, Eq. (7.2-3) applies; for a real gas, Eq. (9.1-3) may be used; for water and seawater, Eq. (9.5-1) and (9.5-3) apply. An *incompressible* fluid specified by Eq. (7.2-5) is again a special case, for which the pressure p is a variable to be determined by the equations of motion and boundary conditions.

Fluids obeying Eq. (7.3-1) or Eq. (7.3-3), whose viscosity effects are represented by terms that are linear in the components of the rate of deformation, are called *Newtonian fluids*. Fluids that behave otherwise are said to be *non-Newtonian*. For example, a fluid whose coefficient of viscosity depends on the basic invariants of V_{ij} is non-Newtonian. See Sec. 9.11 for further discussions.

7.4 HOOKEAN ELASTIC SOLID

A Hookean elastic solid is a solid that obeys Hooke's law, which states that the stress tensor is linearly proportional to the strain tensor; i.e.,

(7.4-1) ▲ $$\sigma_{ij} = C_{ijkl}e_{kl},$$

where σ_{ij} is the stress tensor, e_{kl} is the strain tensor, and C_{ijkl} is a tensor of *elastic constants, or moduli*, which are independent of stress or strain. The tensorial quality of the constants C_{ijkl} follows the quotient rule (Sec. 2.9).

As a tensor of rank 4, C_{ijkl} has $3^4 = 81$ elements; but inasmuch as $\sigma_{ij} = \sigma_{ji}$, we must have

(7.4-2) $$C_{ijkl} = C_{jikl}.$$

Furthermore, since $e_{kl} = e_{lk}$, and in Eq. (7.4-1) the indices k and l are dummies for contraction, we can always symmetrize C_{ijkl} with respect to k and l without altering the sum. Thus, we can always write Eq. (7.4-1) as

(7.4-3) $$\sigma_{ij} = \tfrac{1}{2}(C_{ijkl} + C_{ijlk})e_{kl} = C'_{ijkl}e_{lk},$$

with the property

(7.4-4) $$C'_{ijkl} = C'_{ijlk}.$$

If such a symmetrization has been done, then C_{ijkl}, under the conditions (7.4-2) and (7.4-4), has a maximum of 36 independent constants.

That the total number of elastic constants cannot be more than 36 can be seen if we recall that because $\sigma_{ij} = \sigma_{ji}$ and $e_{ij} = e_{ji}$, there are only six independent elements in the stress tensor σ_{ij} and six in the strain e_{ij}. Hence, if each element of σ_{ij} is linearly related to all elements of e_{ij}, or vice versa, there will be six equations with 6 constants each, hence, 36 constants in total.

For most elastic solids the number of independent elastic constants are far smaller than 36. The reduction is caused by the existence of material symmetry. The reader is referred to the excellent discussions on this subject in the classical books on the theory of elasticity by Love, and Green and Adkins, listed on p. 185.

The greatest reduction in the number of elastic constants is obtained when the material is *isotropic,* i.e., when the elastic properties are identical in all directions. More precisely, isotropy for a material is defined by the requirement that the array of numbers C_{ijkl} has exactly the same numerical values no matter how the coordinate system is oriented. Because of the importance of the concept of isotropy, we shall discuss it in greater detail in Chapter 8. It will be shown that for an isotropic material exactly *two* independent elastic constants characterize the material. Hooke's law for an isotropic elastic solid reads

$$(7.4\text{-}5) \quad \blacktriangle \qquad \sigma_{ij} = \lambda e_{\alpha\alpha}\delta_{ij} + 2\mu e_{ij}.$$

The constants λ and μ are called the *Lamé constants.* In engineering literature the second Lamé constant μ is practically always written as G and identified as the *shear modulus.*

In Chapter 9, we shall present some data on the elastic constants of common materials and some other forms of stress-strain law.

It will be useful to write out Eq. (7.4-5) *in extenso.* With x, y, z as rectangular Cartesian coordinates, we have Hooke's law for an isotropic elastic solid:

$$(7.4\text{-}6) \quad \blacktriangle \qquad \begin{aligned} \sigma_{xx} &= \lambda(e_{xx} + e_{yy} + e_{zz}) + 2Ge_{xx} \\ \sigma_{yy} &= \lambda(e_{xx} + e_{yy} + e_{zz}) + 2Ge_{yy} \\ \sigma_{zz} &= \lambda(e_{xx} + e_{yy} + e_{zz}) + 2Ge_{zz} \\ \sigma_{xy} &= 2Ge_{xy}, \qquad \sigma_{yz} = 2Ge_{yz}, \qquad \sigma_{zx} = 2Ge_{zx}. \end{aligned}$$

These equations can be solved for e_{ij}. But customarily the inverted form is written as

$$(7.4\text{-}7) \quad \blacktriangle \qquad \begin{aligned} e_{xx} &= \frac{1}{E}[\sigma_{xx} - \nu(\sigma_{yy} + \sigma_{zz})], & e_{xy} &= \frac{1+\nu}{E}\sigma_{xy} = \frac{1}{2G}\sigma_{xy}, \\ e_{yy} &= \frac{1}{E}[\sigma_{yy} - \nu(\sigma_{zz} + \sigma_{xx})], & e_{yz} &= \frac{1+\nu}{E}\sigma_{yz} = \frac{1}{2G}\sigma_{yz}, \\ e_{zz} &= \frac{1}{E}[\sigma_{zz} - \nu(\sigma_{xx} + \sigma_{yy})], & e_{zx} &= \frac{1+\nu}{E}\sigma_{zx} = \frac{1}{2G}\sigma_{zx}. \end{aligned}$$

The constants E, ν, and G are related to the Lamé contants λ and G (or μ). See Eq. (9.6-9) on p. 217. E is called the *Young's modulus,* ν is called the *Poisson's ratio,* G is called the *modulus of elasticity in shear,* or *shear modulus.* In the one-dimensional case, in which σ_{xx} is the only nonvanishing component of stress, we have used the simplified version of the equations above in Chapter 5, Eq. (5.1-3) and (5.1-4).

It is very easy to remember Eq. (7.4-7). We recall the one-dimensional case, Eq. (5.1-3). Apply it to the simple block as illustrated in Fig. 1.5, p. 17. When the block is compressed in the z-direction, it shortens by a strain:

$$(7.4\text{-}8) \qquad\qquad e_{zz} = \frac{1}{E}\,\sigma_{zz}.$$

In the meantime the lateral sides of the block will bulge out somewhat. For a linear material the bulging strain is proportional to σ_{zz} and is in a sense opposite to the stress: A compression induces lateral bulging; a tension induces lateral shrinking. Hence we write

$$(7.4\text{-}9) \qquad\qquad e_{xx} = -\frac{\nu}{E}\sigma_{zz}, \qquad e_{yy} = -\frac{\nu}{E}\sigma_{zz}.$$

This is the case in which σ_{zz} is the only nonvanishing stress. If the block is subjected also to σ_{xx}, σ_{yy}, as is illustrated in Fig. 3.2, p. 66, and if the material is isotropic and linear (so that causes and effects are linearly superposable), then the influence of σ_{xx} on e_{yy}, e_{zz} and σ_{yy} on e_{xx}, e_{zz} must be the same as the influence of σ_{zz} on e_{xx}, e_{yy}. Hence (7.4-8) becomes

$$e_{zz} = \frac{1}{E}\,\sigma_{zz} - \frac{\nu}{E}\sigma_{xx} - \frac{\nu}{E}\sigma_{yy},$$

which is one of the equations of (7.4-7) and similarly for other equations in (7.4-7). For the shear stress and shear strain, each component produces its own effect.

7.5 EFFECT OF TEMPERATURE

In the preceding sections the stress-strain or strain-rate relations are determined at a given temperature. The viscosity of a fluid, however, varies with temperature (think of the motor oil in your car) as does the elastic modulus of a solid. In other words, $\mathfrak{D}_{ijkl}$ in Eq. (7.3-1), C_{ijkl} in Eq. (7.4-1) are functions of temperature and are coefficients determined under an isothermal experiment (with temperature kept uniform and constant).

It follows that these laws can be used only for isothermal processes. This is a severe limitation. For example, in the case of propagation of sound waves in fluids or in solids, the strain takes place so fast that there is no time to reach thermal equilibrium. In this case a better approximation is to identify the process as adiabatic. We have two alternatives: Either the material constants are redetermined under adiabatic conditions or the laws are modified to account for the change of temperature.

For an elastic body, Hooke's law can be modified into the *Duhamel-Neumann form* to account for the temperature change. Let the elastic constants C_{ijkl} be measured at a uniform constant temprerature T_0. Then if the temperature changes to T, we put

(7.5-1) ▲ $$\sigma_{ij} = C_{ijkl}e_{kl} - \beta_{ij}(T - T_0),$$

in which β_{ij} is a symmetric tensor, measured at zero strains. For an isotropic material, the second order tensor β_{ij} must also be isotropic. It follows that β_{ij} must be of the form $\beta\delta_{ij}$ (see Sec. 8.2). Hence, for an isotropic Hookean solid,

(7.5-2) $$\sigma_{ij} = \lambda e_{kk}\delta_{ij} + 2Ge_{ij} - \beta(T - T_0)\delta_{ij}.$$

Here λ and G are Lamé constants measured at constant tempreature. The constant β is proportional to the linear coefficient of expansion.

Further details can be found in Fung: *Foundations of Solid Mechanics*, Chapter 12, esp. p. 355.

7.6 MATERIALS WITH MORE COMPLEX MECHANICAL BEHAVIOR

As we said before, the nonviscous fluids, the Newtonian fluids, and the Hooken elastic solids are abstractions. No real material is known to behave exactly as one of these, although in limited ranges of temperature, stress, and strain some materials may follow one of these laws accurately. They are the simplest laws we can devise to relate the stress and strain, or strain rate. They are not expected to be exhaustive.

Almost any real material has a more complex behavior than these simple laws describe. For fluids, household paints and varnish are non-Newtonian. Wet clay and mud are non-Newtonian. Most colloidal solutions are non-Newtonian. For solids, most structural materials are, fortunately, Hookean in the useful range of stresses and strains; but beyond certain limits Hooke's law no longer applies. For example, virtually every known solid material can be broken (fractured) one way or another, under sufficiently large stresses or strains; but to break is to disobey Hooke's law.

Nevertheless, the vast literature on continuum mechanics is centered around these idealized materials, and the results have been remarkably useful. We shall discuss more complex behavior of liquids and solids in Chapter 9, but we shall leave the mathematical treatment of the non-Newtonian, nonlinearly elastic or inelastic solids to specialized treatises.

7.7 SIMPLE BEAM THEORY

In the following, we shall apply what we have learned so far to a topic that is one of the most useful in engineering mechanics: the classical, simple beam theory.

Beams are elongated solid bodies that are used by engineers to support lateral loads. Bridge girders, airplane wings, and beams of a house are examples of beams.

Consider a prismatic beam of a uniform isotropic Hookean material with a rectangular cross section as shown in Fig. 7.1(a) subjected to a pair of bending moments of magnitude M acting on its ends. If the cross section of the beam is symmetric with respect to the plane containing the end moments, the beam will deflect into a circular arc in the same plane, as shown in Fig. 7.1(b). The deflection curve must be a circular arc because of symmetry since every cross section is subjected to the same stress and strain. Let us assume that the deflection is small (compared with the length of the beam). We choose a rectangular frame of reference $x\,y\,z$, with the x-axis pointing in the direction of the longitudinal axis of the beam, y perpendicular to x but in the plane of bending, and z normal to x and y. See Fig. 7.1(a). The origin of the coordinates is to be chosen at the centroid of one cross section for reasons that will become clear below.

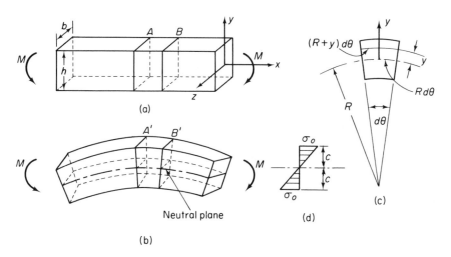

Fig. 7.1 Bending of a prismatic beam.

The deflection of the beam can be described by the deflection of the centroidal surface (the plane $y = 0$ when the beam is in the undeflected configuration) and any displacements relative to this surface. Consider two neighboring plane cross-sections A and B that are perpendicular to the plane

$y = 0$ when the beam is unloaded. When the beam is bent into a circular arc, the two planes A, B are deformed into planes A' and B', which remain normal to the arc. See Fig. 7.1(c). That A' and B' are planes is because of symmetry. That they are perpendicular to the centroidal arc is also because of symmetry. Let the radius of curvature of the centroidal arc be R. When the cross sections A' and B' are bent to a relative angle of $d\theta$, the centroidal arc length is $R\,d\theta$; whereas a line at a distance y above the centroidal line will have a length $(R + y)\,d\theta$. The change of length is $y\,d\theta$. A division by its original length $R\,d\theta$ yields the strain

$$(7.7\text{-}1) \qquad\qquad e_{xx} = \frac{y}{R}.$$

In response to the strain e_{xx} there will be stress σ_{xx}. We now make the assumption that σ_{xx} is the only nonvanishing component of stress, whereas $\sigma_{yy} = \sigma_{zz} = \tau_{xy} = \tau_{yz} = \tau_{zx} = 0$. Then according to Hooke's law (7.4-7) we have

$$(7.7\text{-}2) \qquad\qquad \sigma_{xx} = Ee_{xx} = E\frac{y}{R}.$$

Since only pure bending moments act on the beam, the resultant axial force must vanish:

$$(7.7\text{-}3) \qquad\qquad \int_A \sigma_{xx}\,dA = 0,$$

where A is the cross section, dA is an element of area in the cross section, and the integration extends over the entire cross section. Substituting (7.7-2) into (7.7-3) yields

$$(7.7\text{-}4) \qquad\qquad \int_A y\,dA = 0,$$

which says that the origin ($y = 0$) must be the *centroid* of the cross section. This explains our original choice of the centroid as origin. The centroidal plane $y = 0$ is unstressed during bending [according to Eq. (7.7-2)]. Material particles on this plane are not strained in the axial direction. This plane is therefore called the *neutral surface* of the beam.

The resultant moment of the bending stress σ_{xx} about the z-axis must be equal to the external moment M. A force $\sigma_{xx}\,dA$ acting on an element of area dA in a cross section has a moment arm y; hence the bending moment is

$$(7.7\text{-}5) \qquad\qquad M = \int_A y\sigma_{xx}\,dA.$$

A substitution of Eq. (7.7-2) into the above yields

(7.7-6)
$$M = \frac{E}{R} \int_A y^2 \, dA.$$

We now define the last integral as the *area moment of inertia of the cross section* and denote it by I.

(7.7-7)
$$I = \int_A y^2 \, dA.$$

Then the equations above may be written as

(7.7-8) ▲
$$\frac{M}{EI} = \frac{1}{R}, \qquad \sigma_{xx} = \sigma_0 \frac{y}{c}, \qquad \sigma_0 = \frac{Mc}{I},$$

where c is the largest distance from the neutral surface to the edge of the cross section. See Fig. 7.1(d). The stress σ_0 is the largest bending stress in the beam. It is called the *outer fiber stress* because it is associated with the outer edge of the beam cross section. I is a property of the cross-sectional geometry. See Prob. 12.2, p. 324. For a rectangular cross section with depth h and width b, as shown in Fig. 7.1(a), we have $c = h/2$ and $I = \frac{1}{12}bh^3$.

These formulas give us the stress and strain in a prismatic beam when it is subjected to pure bending. Can we use them for a prismatic beam subjected to a general loading such as the one shown in Fig. 7.2? Or to a beam with

Fig. 7.2 A beam subjected to a distributed loading.

variable cross sections? The answer is that although the solution is then no longer exactly correct, it is found empirically to be surprisingly good. The basic reason is that the shear stresses which must exist in the general case cause a deflection which usually is negligible compared with that due to the bending moment. Therefore, in general, the hypothesis that plane sections remain plane is very good, and Eq. (7.7-1) through (7.7-8) can be considered to be *locally* true along the beam.

Based on such an empirical observation, we can analyze the deflection of a beam under a lateral load. For example, consider the beams illustrated in Fig. 7.3. Let the beam deflection curve (deflection of the neutral surface) be $y(x)$. When $y(x)$ is small (much less than the length of the beam), its curvature can be approximated by d^2y/dx^2, and the use of Eq. (7.7-8) leads to

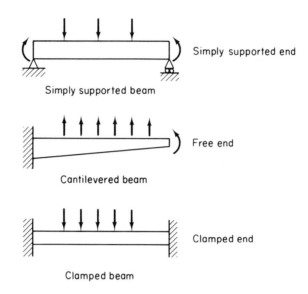

Fig. 7.3   End conditions of beams.

the basic equation

$$(7.7\text{-}9) \quad \blacktriangle \qquad \frac{d^2y}{dx^2} = \frac{1}{EI} M(x).$$

The beam deflection $y(x)$ can be obtained by solving this equation with appropriate boundary conditions, which are:

Simply supported end (deflection and moment vanish):

$$(7.7\text{-}10) \qquad \qquad y = 0, \qquad \frac{d^2y}{dx^2} = 0.$$

Clamped end (deflection and slope vanish):

$$(7.7\text{-}11) \qquad \qquad y = 0, \qquad \frac{dy}{dx} = 0.$$

Free end (moment and shear specified):

$$(7.7\text{-}12) \qquad EI \frac{d^2y}{dx^2} = M, \qquad EI \frac{d^3y}{dx^3} = S.$$

See Fig. 7.3. All these are pretty evident except the last one, for whose explanation we should recall Prob. 1.21, p. 32, which shows that the bending moment M, the transverse shear S, and the lateral load per unit length w are related by the equations

(7.7-13) ▲ $$\frac{dM}{dx} = S, \qquad \frac{dS}{dx} = w.$$

But since $M = EI\, d^2y/dx^2$, we must have $S = EI\, d^3y/dx^3$ as named above.

If the curvature is small (so that the analysis above can be valid) but the slope is finite, then we should use the exact expression for $1/R$, which leads to the following equation in place of (7.7-9):

(7.7-14) $$\frac{d^2y}{dx^2}\left[1 + \left(\frac{dy}{dx}\right)^2\right]^{-3/2} = \frac{M(x)}{EI}.$$

As an example, consider small deflection of a cantilever beam clamped at the left end, as shown in Fig. 7.4, and subjected to a constant bending moment. The right-hand side of Eq. (7.7-9) is constant in this case, and Eq. (7.7-9) can be integrated to obtain

$$y(x) = \frac{M}{EI}\frac{x^2}{2} + Ax + B,$$

where A and B are arbitrary constants. The boundary conditions $y = dy/dx = 0$ at $x = 0$ then yield $A = B = 0$, so that the solution is

$$y = \frac{M}{EI}\frac{x^2}{2}.$$

Fig. 7.4 Bending of a cantilever beam.

In this special case, the boundary conditions on the free end are also satisfied because $M = $ const.

Can we, however, satisfy all the boundary conditions in general? After all, our beam has two ends with two conditions each so that we have four boundary conditions, whereas our differential equation (7.7-9) is only of the second order. Are we going to have a sufficient number of arbitrary constants to satisfy all boundary conditions? The answer, as it stands, is no. A further reflection tells us, however, that for a general loading the differential equation must be obtained by combining Eq. (7.7-13) with Eq. (7.7-9).

Thus the general equation must be

(7.7-15)   ▲ $$w = \frac{dS}{dx} = \frac{d^2M}{dx^2} = \frac{d^2}{dx^2}\left(EI\frac{d^2y}{dx^2}\right),$$

which is a fourth-order differential equation, able to handle four boundary conditions. In the case of a uniform beam, we have

(7.7-16)   ▲ $$EI\frac{d^4y}{dx^4} = w(x).$$

Equation (7.7-15) is an approximate equation, exact only in pure bending of a prismatic beam, but is used often to describe beam deflection in the general case even for beams of variable cross section. In general, for a slender beam it yields close approximations. Significant deviation occurs only when the beam is not slender or for sandwich constructions with very soft core material in which shear deflection becomes significant.

PROBLEMS

7.1 A homogeneous, isotropic Stokes fluid in a domain $-\infty < x, y, z < \infty$ has the velocity components $u = ay$, $v = w = 0$ (see Fig. 9.4, p. 207). Show that the stresses in the fluid are

$$\sigma_{xy} = \mu a, \qquad \sigma_{yz} = \sigma_{zx} = 0, \qquad \sigma_{xx} = \sigma_{yy} = \sigma_{zz} = -p.$$

Design and sketch an instrument to measure the coefficient of viscosity μ.

7.2 Show that Hooke's law (7.4-5) may be rewritten as Eq. (9.6-10) on p. 218, if E, ν, and $G = \mu$ are related to the Lamé constants according to Eq. (9.6-9) on p. 217.

7.3 The simple beam theory presented in Sec. 7.7 gives us the stress distribution in pure bending, as well as the strain component e_{xx}. Find the strains e_{yy}, e_{zz}, and discuss the deformation of the beam in the y-, z-direction.

7.4 For a deformation pattern sketched in Fig. 5.4, Case 3, p. 130, what are the stresses in a material if it is isotropic and obeys Hooke's law? Design an experiment to determine the shear modulus G of the material.

7.5 Suppose you were given a large piece of solid material and were told to determine whether the material was isotropic or not. How would you do it? Design a *complete* program of experiments.

7.6 In plant and animal physiology there is a generally accepted hypothesis called Starling's law. It states that the rate of transfer of water across a cell membrane is proportional to the sum of the difference in hydrostatic pressure on the two sides of the membrane and the difference of the negative of the osmotic pressures. Thus,

$$\dot{m} = k(p_1 - p_2 - \pi_1 + \pi_2),$$

where m is the rate of movement of water (g/sec/cm²); p_1 and π_1 are the hydrostatic and osmotic pressures on one side of the membrane; p_2, π_2, those on the other side. What are the physical dimensions of the constant k? The permeability constant k is a material constant.

Suppose that you are not satisfied with this simple law—your experimental data seem to deviate from it—and you would like to propose a more complex law, for example, one that includes the stress deviation tensors on both sides of the membrane. Propose an equation which is correct dimensionally and tensorially and does not contradict any of the axioms of general physics. How many material constants are there in your proposal?

How would you verify your proposal? In other words, what experiments must be done so that you may say that your proposal is completely verified?

Discussion: The physical dimensions of $\dot{m}, p, \pi$, and k are $[MT^{-1}L^{-2}]$, $[MT^{-2}L^{-1}]$, $[MT^{-2}L^{-1}]$, and $[TL^{-1}]$, respectively. Hence the units of k are seconds per centimeter.

Any generalization of Starling's hypothesis must be guided by experimental results and logical consistency. Merely speculating on the basis of logic, we can offer many suggestions. For example, a different mechanism may work for hydrostatic pressure as against osmotic pressure; thus,

$$\dot{m} = k_1(p_1 - p_2) + k_2(\pi_2 - \pi_1).$$

Again, as proposed in the problem, the stress-deviation tensor τ_{ij} might be involved. Since $\dot{m}$ is a scalar, any involvement of τ_{ij} must be in the form of scalar invariants, such as $\tau'_{ij}\tau'_{ij}$, $\tau'_{ij}V'_{ij}$ (fluid), $\tau'_{ij}e'_{ij}$ (gel), $c_{ij}\tau_{ij}$, where V'_{ij} is the rate of the deformation deviation tensor, e'_{ij} is the strain deviation tensor, and c_{ij} is a tensorial set of constants. Hence we might have the following hypothetical relationships:

$$\dot{m} = k\,(\Delta p - \Delta\pi) + c\,\Delta\tau'_{ij}\tau'_{ij}.$$
$$\dot{m} = k\,(\Delta p - \Delta\pi) + c\,\Delta\tau'_{ji}V'_{ij}.$$

or

$$\dot{m} = k\,(\Delta p - \Delta\pi) + c\,\Delta\tau'_{ij}e'_{ij}.$$
$$\dot{m} = k\,(\Delta p - \Delta\pi) + \Delta c_{ij}\tau'_{ij}.$$

Δ means the difference of the quantities on the two sides of the membrane. What experiments must be designed to check these or other proposals?

7.7 By considering a thin elastic plate that obeys isotropic Hooke's law and is loaded by $\sigma_{xx} = -\sigma_{yy} = a$ constant, while $\sigma_{xy} = \sigma_{zz} = \sigma_{zx} = \sigma_{zy} = 0$ (Fig. P7.7), and a rectangular element in the plate with edges at 45° with respect to x- and y-axes, show that $G = E/[2(1 + v)]$.

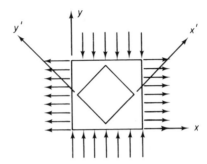

Fig. P7.7   A way to identify the relationship between E, v, and G.

7.8 Design an apparatus in which a liquid may be tested to see if it obeys the Newtonian fluid equation, and if it does, determine the viscosity constants.

Give special attention to isotropic fluids. Then discuss the general anisotropic case.

7.9 Do Prob. 7.8 for gases.

7.10 Describe some non-Newtonian fluids and some non-Hookean Solids.

7.11 A circular cylindrical shell of isotropic Hookean material is subjected to an internal pressure p_i and an external pressure p_o. Let R_i and R_o denote the initial inner and outer radius, respectively, when $p_i = p_o = 0$; whereas r_i, r_o are the corresponding radii in the loaded condition. Using the approximate solution given in Example 6, p. 21, for stress distribution, determine the relationship between r_i, r_o and R_i, R_o.

7.12 Consider the Coette flowmeter shown in Fig. P3.22, p. 88. Derive the velocity distribution in the channel and the torque-angular velocity relationship.

Solution: The constant torque that is transmitted through the fluid is $T = 2\pi r^2 L \tau$, where τ is the shear stress at radius r and L is the length of the viscometer. Let the coefficient of viscosity be μ; then $\tau = \mu \, dv/dr$. Hence $dv/dr = T/(2\pi L \mu r^2)$. An integration gives $v = [T/(2\pi L \mu)][-r^{-1} + C]$. To determine the integration constant C, let the boundary condition be that the inner cylinder is stationary, so that $v = 0$ when $r = R_1$. Then $C = R_1^{-1}$. Hence, on setting $r = R_2$ where $v = \omega R_2$, we obtain

$$\mu = \frac{T}{2\pi L R_2 \omega}\left[\frac{1}{R_1} - \frac{1}{R_2}\right].$$

7.13 A thin-walled cylinder of isotropic Hookean material has a radius a and a wall thickness h, $(h \ll a)$. The cylinder is stretched in the axial direction with a tensile force F and twisted by a torque T. Assume that the stresses are practically uniformly distributed in the wall of the cylinder. Determine the stresses and strains in the wall.

7.14 In a thin plate of isotropic Hookean material in which σ_{xx}, σ_{yy}, τ_{xy} are functions of x and y, and $\sigma_{zz} = \tau_{zx} = \tau_{zy} = 0$, the equations of equilibrium can be satisfied automatically if $\sigma_{xx}, \tau_{yy}, \tau_{xv}$ are derived from an arbitrary function $\Phi(x, y)$ in such a way that

$$\sigma_{xx} = \frac{\partial^2 \Phi}{\partial y^2}, \qquad \sigma_{yy} = \frac{\partial^2 \Phi}{\partial x^2}, \qquad \tau_{xy} = -\frac{\partial^2 \Phi}{\partial x\, \partial y}. \qquad (1)$$

See Prob. 3.30. What are the strains corresponding to the stresses given in Eq. (1)? Are the strains compatible? On substituting Eq. (1) into the equation of compatibility, show that the equation that governs the function Φ is

$$\frac{\partial^4 \Phi}{\partial x^4} + 2 \frac{\partial^4 \Phi}{\partial x^2\, \partial y^2} + \frac{\partial^4 \Phi}{\partial y^4} = 0$$

or $\nabla^4 \Phi = 0$ for short. A function $\Phi(x, y)$ that satisfies the "biharmonic" equation $\nabla^4 \Phi = 0$ and yields stresses according to the formulas above is called an *Airy stress function*.

7.15 Show that the polynomial

$$\Phi = \frac{a_2}{2} x^2 + b_2 xy + \frac{c_2}{2} y^2$$

is an Airy stress function. Examine the boundary conditions satisfied by this function on the edges of a rectangular plate: $x = \pm L, y = \pm C$; thus, identify the problem to which Φ is a solution.

7.16 Do the same for

$$\Phi_3 = \frac{a_3}{3 \cdot 2} x^3 + \frac{b_3}{2} x^2 y + \frac{c_3}{2} xy^2 + \frac{d_3}{3 \cdot 2} y^3.$$

7.17 Consider a stress field

$$\sigma_{xx} = y^2, \qquad \sigma_{yy} = x^2, \qquad \sigma_{xz} = \sigma_{yx} = \sigma_{zy} = \sigma_{zz} = 0.$$

Does it satisfy the equation of equilibrium? What boundary conditions does it satisfy on edges $x = \pm 1$ and $y = \pm 1$? Does it satisfy the compatibility condition? Is it a correct solution of some problem?

7.18 A thin plate of isotropic Hookean material is subjected to a uniform load σ_0 on the edges $y = \pm 1$ and no load on the edges $x = \pm 1$. The stresses $\sigma_{zz} = \tau_{zx} = \tau_{zy} = 0$. There is no body force. Find a solution $\sigma_{xx}, \tau_{xy}, \sigma_{yy}$ that is compatible and satisfies the equations of equilibrium and the boundary conditions.

7.19 A thin elastic plate of isotropic Hookean material is loaded on the boundary as shown in Fig. P7.19. There is no body force. It is suggested that the stress distribution is

$$\sigma_{xx} = \frac{q}{2I}(l^2 - x^2)y + \frac{q}{2I}\left(\frac{2}{3}y^3 - \frac{2}{5}c^2 y\right)$$

$$\sigma_{yy} = -\frac{q}{2I}\left(\frac{1}{3}y^3 - c^2 y + \frac{2}{3}c^3\right)$$

$$\sigma_{xy} = -\frac{q}{2I}(c^2 - y^2)x$$

$$\sigma_{zz} = \sigma_{xz} = \sigma_{zy} = 0,$$

where q is the load per unit area and $I = 2c^3/3$ is a constant. Determine whether the solution is correct or not.

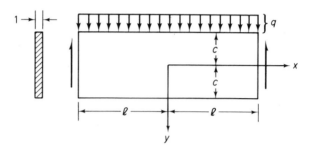

Fig. P7.19   A thin elastic plate.

7.20 As an example of an elementary three-dimensional problem of elasticity, formulate and solve the boundary-value problem associated with the stretching of a prismatic bar hanging at its top and loaded by its own weight (Fig. P7.20). Determine the stress and strain in the bar. Determine the elastic displacements.

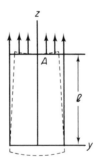

Fig. P7.20   A prismatic bar hanging at the top.

Hint: Try $\sigma_{zz} = \rho g z$, $\sigma_{xx} = \sigma_{yy} = \tau_{xy} = \tau_{yz} = \tau_{zx} = 0$ where ρ is the density and g is the gravitational acceleration.

Ref. Timoshenko and Goodier, *Theory of Elasticity*, New York: McGraw Hill (1951), Sec. 86. p. 246.

7.21 Draw diagrams showing the bending moment at every station for the beams in Fig. P7.21 loaded by their own weight.

Fig. P7.21   Beams loaded by uniformly distributed load.

7.22 Galileo, in his *Two New Sciences*, posed the following problem: A column of marble rested on two supports, in the manner of a simply supported beam. The citizens of Rome were worried about the safety of the column and sought to increase the support. They inserted a third support in the middle of the span, as in Fig. P7.22 upon which the column broke. Why?

Galileo answered in terms of the bending moments in the beam. Please explain the details.

Fig. P7.22   Galileo's problem of a fallen column.

7.23 If a hinge is added to the left end of a cantilever beam which is loaded by a constant force *P*, as shown in Fig. P7.23, can you draw the moment diagram?

This problem is said to be *statically indeterminate*. How would you solve the problem in principle?

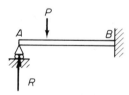

Fig. P7.23   Redundant supports.

Solution: A beam clamped at *B* and simply supported at *A* is called *statically indeterminate* because the reaction at *A* cannot be computed by statics alone. To solve the problem, we must consider the elasticity of the beam.

A method is as follows. Take the support at *A* away. Then the beam becomes a cantilever beam. We can find the deflection at *A* due to the load *P*. Let this be $\delta_A^{(P)}$, which is proportional to *P*.

Next, consider the same cantilever beam loaded by a force *R* at the tip. This produces a deflection $\delta_A^{(R)}$ at the end *A*. In reality, the end *A* does not move. Hence $\delta_A^{(R)} + \delta_A^{(P)} = 0$. From this equation, we can compute *R*. With *R* known, we can then complete the moment diagram.

7.24 A beam (Fig. P7.24) rests on three hinges which, unlike Galileo's rocks, are so rigidly attached to the foundation that both push and pull can be sus-

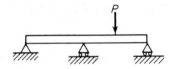

Fig. P7.24   Redundant supports.

tained. Sketch a method with which the bending moment distribution in the beam can be calculated.

Solution: First, withdraw one of the supports, so that the problem becomes statically determinate. Compute the deflection at the location of the withdrawn support due to the load *P*.

Next, apply a force *R* at the location of the withdrawn support, and compute the displacement at this point.

The condition that the net displacement at all the supports must vanish provides an equation to compute the reaction *R*. Then all the forces are known and the moment diagram can be completed.

7.25 A wooden cantilever beam of 2 in. × 6 in. cross section projects 10 ft from a wall. See Fig. P12.6, p. 327. Find the allowable concentrated load *F* applied at the free end of the beam if the wood weighs 120 lb/ft³, and if the permissible maximum bending stress is 2000 lb/in.² What is the difference between placing the 2 in. × 6 in. vertically and horizontally (i.e., compare the case in which *F* acts on the 2 in. side with the case in which *F* acts on the 6 in. side)?

7.26 A simply supported beam of span *L* has a uniform cross section. The allowable stress of the beam material is σ_{all}. A man of weight *W* wants to

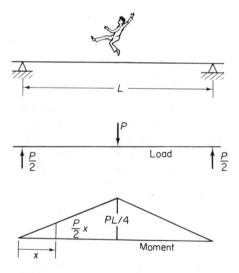

Fig. P7.26   Trying to break a plank.

break the beam by jumping on its center. How high does he have to jump? Ignore the effect of the weight of the beam. See Fig. P7.26.

Solution: Since the inertia force of the beam is ignored, only a concentrated load of the man acts on the beam. Let the load be P. The moment diagram is a triangle, as shown in Fig. P7.26. The maximum moment $PL/4$ produces the maximum bending stress. When the stress is σ_{all}, the maximum load P_{max} must be

$$P_{max} = \frac{4I\sigma_{all}}{Lc}, \tag{1}$$

where I is the cross-sectional moment of inertia and c is the distance of the outer fiber to the neutral axis. The beam deflection curve under this load $y(x)$ is governed by the equation

$$EI\frac{d^2y}{dx^2} = \frac{Px}{2}. \tag{2}$$

Integrate once and use the boundary condition $dy/dx = 0$ at $x = L/2$ (by symmetry). Then integrate once more and use the boundary condition $y = 0$ at $x = 0$ to obtain

$$y = \frac{P}{12EI}\left(x^3 - \frac{3L^2x}{4}\right). \tag{3}$$

Hence the maximum deflection at midspan ($x = L/2$) is

$$y_{max} = -\frac{P_{max}L^3}{48EI}. \tag{4}$$

Since the beam is linearly elastic, the total strain energy in the beam must be equal to

$$\frac{1}{2}P_{max} \cdot y_{max} = \frac{1}{2}\frac{P_{max}^2 L^3}{48EI} = \left(\frac{4I\sigma_{all}}{Lc}\right)^2 \frac{L^3}{96EI}.$$

Now we must compute how high the man has to jump in order to produce the breaking condition. If he jumps a height h above the beam, then in a free fall of height h plus the beam deflection y_{max} the work done (or the potential energy lost) is $Wh + Wy_{max}$. This must be equal to the strain energy $\frac{1}{2}P_{max}y_{max}$ at the instant of maximum deflection. Hence

$$Wh = \frac{1}{2}P_{max}y_{max} - Wy_{max} = (\frac{1}{2}P_{max} - W)y_{max}$$

where P_{max} and y_{max} are given by Eq. (1) and (4), respectively.

7.27 Eq. (7.7-16), derived for static equilibrium, can be extended to dynamic conditions with the application of D'Alembert's principle (p.7). We apply the negative of mass × acceleration as an inertia force on the structure, then the structure may be considered as in static equilibrium. Now, let $y(x, t)$ be the deflection of the neutral surface of a beam. If m is the mass of the beam per unit length, then the inertia force is

$$- m\frac{\partial^2 y}{\partial t^2}$$

per unit length of the beam. Hence, from Eq. (7.7-16), we obtain the equation of motion

$$EI\frac{\partial^4 y}{\partial x^4} = - m\frac{\partial^2 y}{\partial t^2}. \tag{1}$$

Let us use this equation to analyze the free vibration of a beam of uniform cross section and length L, simply supported at both ends. Find the vibration modes and frequencies by trying a solution of the type

$$y(x, t) = Ae^{i\omega t}f(x), \tag{2}$$

where A is a constant, and ω is a circular frequency, and i is the imaginary number $\sqrt{-1}$.

Solution: On substituting Eq. (2) into (1), we obtain

$$EI\frac{d^4 f}{dx^4} - m\omega^2 f = 0. \tag{3}$$

The boundary conditions are:

$$f = \frac{d^2 f}{dx^2} = 0 \qquad\qquad \text{when } x = 0 \text{ and } L. \tag{4}$$

A possible solution that satisfies Eqs. (4) is

$$f = \sin\frac{n\pi x}{L}, \qquad (n = 1, 2, 3, \ldots.) \tag{5}$$

A substitution into (3) shows that (3) is satisfied if we choose ω to be

$$\omega_n = \left(\frac{n\pi}{L}\right)^2 \sqrt{\frac{EI}{m}}. \qquad (n = 1, 2, 3, \ldots.) \tag{6}$$

Thus there are infinite number of vibration frequencies. The deflection mode corresponding to ω_n is the nth mode

$$y_n(x, t) = A_n e^{i\omega_n t} \sin\frac{n\pi x}{L} \tag{7}$$

7.28 A strong wind blows on a palm tree, Fig. P7.28. The wind load on the trunk is $w = kD$ per unit length of the trunk, where D is the local diameter of the trunk and k is a constant. How should the diameter vary with the height so that the tree is uniformly strong from the top to bottom with respect to bending in wind? Note that the area moment of inertia of the cross section of the tree trunk is proportional to D^4 and the outer fiber stress due to bending is proportional to $Mc/I \sim MD^{-3}$, where M is the bending moment. Ignore the bending moment contributed by the leaves.

Hint: Let x and ξ be measured downward from the treetop. The bending moment at x is

$$M(x) = \int_0^x (x - \xi)kD(\xi)\, d\xi.$$

The maximum bending stress at x is proportional to $M(x)/D^3(x)$. The problem is to determine $D(x)$ so that $M(x)/D^3(x)$ is constant. Try a power law such as $D(x) = \text{const.}\ x^m$, and show that $m = 1$. The tree trunk should look like a slender cone.

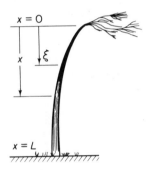

Fig. P7.28 A strong wind blowing on a palm tree.

7.29 In the problem of torsion of a cylindrical bar of mild steel with an elliptic cross section (see Fig. P7.29), it is found that the displacements can be described by the following:

$$u = -\alpha zy, \qquad v = \alpha zx, \qquad w = -\frac{a^2 - b^2}{a^2 + b^2}\alpha xy,$$

where α is the angle of twist in radians per unit length of the bar. Let $a = 2$ cm, $b = 1$ cm. Compute the stresses that act at the point A which is located at the end of the principal minor axis of the elliptic cross section. What is the maximum principal stress at A? What is the maximum shear stress at A? On what planes do the maximum tension and the maximum shear act?

Assume that the material is isotropic and obeys Hooke's law for this problem.

Answer: At A: $\tau_{xz} = -\frac{8}{5}G\alpha$, $\tau_{yz} = 0$, $\sigma_{\max} = -\frac{8}{5}G\alpha$, $\tau_{\max} = -\frac{8}{5}G\alpha$.

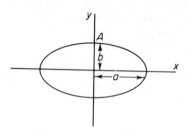

Fig. P7.29   Torsion of an elliptic cylinder.

FURTHER READING

ERINGEN, A. C., *Mechanics of Continnua*, New York: Wiley, (1967), p. 145.

GREEN, A. E., AND J. E. ADKINS, *Large Elastic Deformations*, Oxford: University Press (1960), Chap. 1, esp. pp. 11–35.

LOVE, A. E. H., *A Treatise on the Mathematical Theory of Elasticity* (Cambridge: University Press, 1st ed. (1892), 4th ed. (1927) New York: Dover Publications (1963), Chapter 6, esp. pp. 151–165.

CHAPTER EIGHT

Isotropy

The concept of isotropy is used constantly as a simplifying assumption in continuum mechanics. We shall first define material isotropy and isotropic tensors. The use of isotropic tensors in the constitutive equation of an isotropic elastic material is shown. Then we turn to the determination of isotropic tensors of ranks 2, 3, and 4 and apply them to isotropic materials.

8.1 THE CONCEPT OF MATERIAL ISOTROPY

The class of material whose mechanical property does not depend on the direction is said to be *isotropic*. For example, if we make a tension test on a certain metal, we might find that

(a) The result does not depend on the direction the tension specimen was cut from the ingot.

(b) The lateral contraction is the same in every direction perpendicular to the direction of pulling.

Again, in a torsion experiment, we might find the result to be the same no matter in which orientation the specimen was cut from the ingot. Materials with these properties lead to the concept of isotropy.

To give a precise definition, we make use of the constitutive equation. A material is said to be isotropic if its constitutive equation (the stress-strain-history law) is unaltered under orthogonal transformations of coordinates (Sec. 2.4). For example, if the constitutive equation is $\sigma_{ij} = C_{ijkl}e_{kl}$, we demand that after an orthogonal transformation the law read $\bar{\sigma}_{ij} = C_{ijkl}\bar{e}_{kl}$, where the barred quantities refer to the new coordinates.

Since orthogonal transformations consist of translations, rotations, and reflections of coordinate axes, the definition requires that the mathematical form of the constitutive equation remains unchanged no matter how the axes are translated, rotated, or reflected. In particular, the array of material constants must have the same values in any right-handed or left-handed system of rectangular Cartesian coordinates.

We can think of many materials that are not isotropic—most crystals are not. But if a material were isotropic, its constitutive equation would be greatly simplified. In this chapter we shall show that if a solid obeying the stress-strain relationship

$$(8.1-1) \qquad\qquad \sigma_{ij} = C_{ijkl}e_{kl}$$

is isotropic, then the number of independent elastic constants C_{ijkl} reduces precisely to two. A similar remark applies to the linear viscous fluid.

8.2 ISOTROPIC TENSOR

An isotropic tensor in the Euclidean metric space is defined as a tensor such that its components in any rectangular Cartesian system are unaltered in value by orthogonal transformations of coordinates.

By definition (Sec. 2.4), an orthogonal transformation from x_1, x_2, x_3 to $\bar{x}_1, \bar{x}_2, \bar{x}_3$, is

$$(8.2-1) \qquad\qquad \bar{x}_i = \beta_{ij}x_j + \alpha_i, \qquad (i = 1, 2, 3)$$

where β_{ij} and α_i are constants, under the restriction that

$$(8.2-2) \qquad\qquad \beta_{ik}\beta_{jk} = \delta_{ij}.$$

An orthogonal transformation is said to be *proper* if a right-handed system of coordinate axes is transformed into a right-handed one. For a transformation to be proper, the Jacobian must be positive (see Sec. 2.5). For the orthogonal transformation (8.2-1) the Jacobian is the determinant $|\beta_{ij}|$, which according to Eq. (8.2-2), must have the value ± 1. Hence, for an orthogonal transformation to be proper, we must have

$$(8.2-3) \qquad\qquad |\beta_{ij}| = 1.$$

For example, all rotations of coordinate axes are proper, but a reflection in the x_2x_3-plane

$$(8.2\text{-}4) \quad \begin{cases} \bar{x}_1 = -x_1, \\ \bar{x}_2 = x_2, \\ \bar{x}_3 = x_3, \end{cases} \quad (\beta_{ij}) = \begin{pmatrix} -1 & 0 & 0 \\ 0 & 1 & 0 \\ 0 & 0 & 1 \end{pmatrix}, \quad |\beta_{ij}| = -1$$

is orthogonal, but improper—it turns a right-handed system into a left-handed one.

It is convenient to study isotropic tensors irrespective of stress-strain laws. But there is a connection. We shall prove that if the relation

$$(8.2\text{-}5) \qquad\qquad \sigma_{ij} = C_{ijkl} e_{kl}$$

is isotropic, then C_{ijkl} is an isotropic tensor.

Proof: By the quotient rule (Sec. 2.9), C_{ijkl} is a tensor of rank 4. Hence, C_{ijkl} transforms according to the tensor transformation rule. Now, transforming Eq. (8.2-5) into new coordinates $\bar{x}_i$, we have

$$(8.2\text{-}6) \qquad\qquad \bar{\sigma}_{ij} = \bar{C}_{ijkl} \bar{e}_{kl}.$$

But the definition of material isotropy requires that

$$(8.2\text{-}7) \qquad\qquad \bar{\sigma}_{ij} = C_{ijkl} \bar{e}_{kl}.$$

Hence, by comparing Eq. (8.2-6) and Eq. (8.2-7), we obtain

$$(8.2\text{-}8) \qquad\qquad \bar{C}_{ijkl} = C_{ijkl}.$$

Thus, C_{ijkl} is an isotropic tensor.

Let us now list some isotropic tensors. All scalars are, of course, isotropic. But there is no isotropic tensor of rank 1. For, if the vector A_i were isotropic, then it would have to satisfy the equation

$$(8.2\text{-}9) \qquad\qquad \bar{A}_i = A_i = \beta_{ij} A_j$$

for all possible orthogonal transformations. In particular, for a 180° rotation about the x_1-axis,

$$(8.2\text{-}10) \quad \begin{cases} \bar{x}_1 = x_1, \\ \bar{x}_2 = -x_2, \\ \bar{x}_3 = -x_3, \end{cases} \quad (\beta_{ij}) = \begin{pmatrix} 1 & 0 & 0 \\ 0 & -1 & 0 \\ 0 & 0 & -1 \end{pmatrix}, \quad |\beta_{ij}| = 1,$$

Eq. (8.2-9) becomes

$$A_1 = A_1, \qquad A_2 = -A_2, \qquad A_3 = -A_3.$$

Hence, $A_2 = A_3 = 0$. Similarly, $A_1 = 0$. Thus, the nonexistence of isotropic tensor of rank 1 is proved.

For tensors of rank 2, δ_{ij} is an isotropic tensor, because

$$\begin{aligned}
\bar{\delta}_{ij} &= \beta_{im}\beta_{jn}\delta_{mn} &&\text{(by definition of tensors)}\\
&= \beta_{im}\beta_{jm} &&\text{(since } \delta_{mn} = 0 \text{ if } m \neq n)\\
&= \delta_{ij} &&\text{[by Eq. (8.2-2)].}
\end{aligned}$$

We propose to show that every isotropic tensor of rank 2 may be reduced to the form $p\delta_{ij}$, where p is a scalar.

For proof, we note first that if a tensor B_{ij} is isotropic, it must be diagonal. For, imposing the 180° rotation about the x_1-axis as specified by Eq. (8.2-10), we obtain

$$\bar{B}_{12} = \beta_{1m}\beta_{2n}B_{mn} = -B_{12}.$$

But isotropy requires $\bar{B}_{12} = B_{12}$. Hence, $B_{12} = 0$. Similarly, $B_{ij} = 0$ if $i \neq j$. Hence, B_{ij} is symmetric and diagonal.

Next, let ϵ_{ijk} be the permutation tensor and consider the transformation

$$\bar{x}_j = (\delta_{ij} + d\theta\epsilon_{3ij})x_i,$$

(8.2-11)

$$(\beta_{ij}) = (\delta_{ij} + d\theta\epsilon_{3ij}) = \begin{pmatrix} 1 & d\theta & 0 \\ -d\theta & 1 & 0 \\ 0 & 0 & 1 \end{pmatrix},$$

which represents a rotation about the x_3-axis with an infinitesimal angle of rotation $d\theta$.† The definition of tensors furnishes the relation

(8.2-12)

$$\begin{aligned}
\bar{B}_{ij} &= (\delta_{im} + d\theta\epsilon_{3im})(\delta_{jn} + d\theta\epsilon_{3jn})B_{mn}\\
&= \delta_{im}\delta_{jn}B_{mn} + d\theta(\epsilon_{3im}\delta_{jn}B_{mn} + \epsilon_{3jn}\delta_{im}B_{mn}) + d\theta^2\epsilon_{3im}\epsilon_{3jn}B_{mn}\\
&= B_{ij} + d\theta(\epsilon_{3im}B_{mj} + \epsilon_{3jn}B_{in}) + O(d\theta^2).
\end{aligned}$$

But if B_{ij} is isotropic we must have $\bar{B}_{ij} = B_{ij}$. Hence, for small but arbitrary $d\theta$ we must have

(8.2-13) $\epsilon_{3im}B_{mj} + \epsilon_{3jn}B_{in} = 0.$

---

†See the rotation matrix (2.4-5) and note that when θ is very small, $\cos\theta \doteq 1$, $\sin\theta \doteq \theta$. Identifying the angle θ of Sec. 2.4 with $d\theta$ here furnishes the geometric interpretation of the transformation (8.2-11).

Take $i = 1, j = 1$; then we have

$$\epsilon_{312} B_{21} + \epsilon_{312} B_{12} = B_{21} + B_{12} = 0.$$

But B_{ij} is symmetric as we have shown above. Hence $B_{12} = B_{21}$ and both $= 0$. This agrees with what we learned above, but no new knowledge is gained.
Take $i = 1, j = 2$; then we have

$$\epsilon_{312} B_{22} + \epsilon_{321} B_{11} = B_{22} - B_{11} = 0.$$

Hence, $B_{11} = B_{22}$. It is evident that an entirely similar rotation about the x_1-axis would yield $B_{23} = 0$, $B_{22} = B_{33}$, and a rotation about the x_2-axis would yield $B_{31} = 0$. Hence, the isotropic tensor B_{ij} is reduced to the form $B_{11} \delta_{ij}$. Writing p for B_{11}, we obtain $B_{ij} = p \delta_{ij}$.

Now any rotation from one rectangular Cartesian coordinate system to another can be performed by repeated infinitesimal rotations about coordinate axes. Hence, the conditions examined above are the only conditions imposed by isotropy with respect to proper orthogonal transformations. Hence, $B_{ij} = p \delta_{ij}$ for all proper orthogonal transformations.

For the second-rank isotropic tensor $p \delta_{ij}$, a reflection in the $x_2 x_3$-plane, Eq. (8.2-4), does not change its value. By the argument of arbitrary rotation, we conclude that a reflection in any plane would not affect its value. Hence, the form we just found is isotropic with respect to all orthogonal transformations.   Q.E.D.

Some of the arguments above may be abridged if we note that for an isotropic tensor the coordinate axes may be labeled in an arbitrary order. Thus, a cyclic permutation of the indices 1, 2, 3 cannot affect the values of the components of a tensor which is isotropic with respect to rotation of the coordinate axes. Hence, $B_{12} = 0$ implies $B_{31} = 0$. If the tensor is isotropic also with respect to reflection, then an arbitrary permutation of the indices 1, 2, 3 will not affect the values of the components.

8.3* ISOTROPIC TENSORS OF RANK 3

For tensors of rank 3, we can verify that the permutation tensor ϵ_{ijk} is isotropic with respect to rotation of coordinate axes (proper orthogonal transformations). It is not isotropic with respect to reflection in a coordinate plane, because a reflection such as (8.2-4) turns $\epsilon_{123} = 1$ into $\bar{\epsilon}_{123} = -1$.

We can show that with respect to all rotations of coordinates, the only isotropic tensors of rank 3 are scalar multiples of ϵ_{ijk}. The proof can be constructed similar to that for the second-rank tensor. Let u_{ijk} be an isotropic tensor of rank 3. Let us consider an infinitesimal rotation of an angle $d\theta$ about an arbitrary axis $\boldsymbol{\xi}$ (a vector with components ξ_k) passing through the origin:

(8.3-1) $$\bar{x}_j = (\delta_{ij} + d\theta\, \xi_k\epsilon_{kij})x_i.$$

Then, according to the tensor transformation law,

$$\bar{u}_{ijk} = (\delta_{im} + d\theta\xi_s\epsilon_{sim})(\delta_{jn} + d\theta\xi_s\epsilon_{sjn})(\delta_{kp} + d\theta\xi_s\epsilon_{skp})u_{mnp}$$
$$= u_{ijk} + d\theta\{\xi_s\epsilon_{sim}u_{mjk} + \xi_s\epsilon_{sjn}u_{ink} + \xi_s\epsilon_{skp}u_{ijp}\} + O(d\theta^2).$$

By isotropy, $\bar{u}_{ijk} = u_{ijk}$, hence for small $d\theta$ the quantity in the braces must vanish (we can ignore quantities of the higher order). Thus, for all i, j, k,

(8.3-2) $$\xi_s\epsilon_{sim}u_{mjk} + \xi_s\epsilon_{sjn}u_{ink} + \xi_s\epsilon_{skp}u_{ijp} = 0.$$

Take $i = j = 1$. Then

(8.3-3) $$-\xi_2 u_{31k} + \xi_3 u_{21k} - \xi_2 u_{13k} + \xi_3 u_{12k}$$
$$+ \xi_s\epsilon_{sk1}u_{111} + \xi_s\epsilon_{sk2}u_{112} + \xi_s\epsilon_{sk3}u_{113} = 0.$$

Now put $k = 2$. Then, since ξ_1, ξ_2, ξ_3 are arbitrary, their coefficients must vanish and we obtain

(8.3-4)
$$u_{212} + u_{122} = u_{111},$$
$$u_{312} + u_{132} = 0,$$
$$u_{113} = 0.$$

From the last equation, and by symmetry, $u_{ijk} = 0$ if two of i, j, k are equal and the third unequal. Then, by the first equation of (8.3-4), u_{ijk} is also zero if all of i, j, k are equal. The second shows that

$$u_{ijk} = -u_{jik}.$$

If in Eq. (8.3-3) we put $k = 1$, then every term vanishes, yielding no new information.

Finally, consider the case in which i, j, k are all different in Eq. (8.3-2). We note that u_{mjk} is zero when $m = j$. Then it is seen that Eq. (8.3-2) holds because all the coefficients vanish. It follows that the only isotropic tensors of rank 3 (isotropic with respect to rotations, not reflections) are scalar multiples of ϵ_{ijk}.   Q.E.D.

8.4* ISOTROPIC TENSORS OF RANK 4

Isotropic tensors of rank 4 are of particular interest to the constitutive equations of materials. It is readily seen that since the unit tensor δ_{ij} is isotropic, the tensors

$$(8.4\text{-}1) \qquad \delta_{ij}\delta_{kl}, \qquad \delta_{ik}\delta_{jl} + \delta_{il}\delta_{jk}, \qquad \delta_{ik}\delta_{jl} - \delta_{il}\delta_{jk} = \epsilon_{sij}\epsilon_{skl}$$

are isotropic. We propose to show that if u_{ijkl} is an isotropic tensor of rank 4, then it is of the form

$$(8.4\text{-}2) \qquad \lambda\delta_{ij}\delta_{kl} + \mu(\delta_{ik}\delta_{jl} + \delta_{il}\delta_{jk}) + \nu(\delta_{ik}\delta_{jl} - \delta_{il}\delta_{jk}),$$

where λ, μ, ν are scalars. Furthermore, if u_{ijkl} has the symmetry properties

$$(8.4\text{-}3) \qquad u_{ijkl} = u_{jikl}, \qquad u_{ijkl} = u_{ijlk},$$

then

$$(8.4\text{-}4) \qquad u_{ijkl} = \lambda\delta_{ij}\delta_{kl} + \mu(\delta_{ik}\delta_{jl} + \delta_{il}\delta_{jk}).$$

Proof: We shall establish the results for isotropy with respect to both rotation of coordinate axes and reflections in coordinate planes.

First, we note that the coordinate axes may be labeled in an arbitrary order. Thus, a permutation in the indices 1, 2, 3 cannot affect the values of the components of an isotropic tensor. Hence,

$$(8.4\text{-}5) \qquad \begin{aligned} &u_{1111} = u_{2222} = u_{3333}, \\ &u_{1122} = u_{2233} = u_{3311} = u_{1133} = u_{2211} = u_{3322}, \\ &u_{1212} = u_{2323} = u_{3131} = u_{1313} = u_{2121} = u_{3232}, \\ &u_{1221} = u_{2332} = u_{3113} = u_{2112} = u_{3223} = u_{1331}. \end{aligned}$$

Next, we note that a rotation of $180°$ about the x_1-axis, corresponding to the transformation (8.2-10), changes the sign of any term with an odd number of the index 1. But these terms must not change sign on account of isotropy. Hence, they are zero. For example,

$$(8.4\text{-}6) \qquad u_{1222} = u_{1223} = u_{2212} = 0.$$

By symmetry, this is true for any index i.

These conditions reduce the maximum number of numerically distinct components of the tensor u_{ijkl} to four, namely, $u_{1111}, u_{1122}, u_{1212}, n_{1221}$.

Now let us impose the transformation (8.2-11) corresponding to an infinitesimal rotation about the x_3-axis. The tensor transformation law requires that

$$(8.4\text{-}7)$$
$$\bar{u}_{pqrs} = u_{pqrs} + d\theta\{\epsilon_{3ip}u_{iqrs} + \epsilon_{3iq}u_{pirs} + \epsilon_{3ir}u_{pqis} + \epsilon_{3is}u_{pqri}\} + O(d\theta^2).$$

Since for an isotropic tensor $\bar{u}_{pqrs} = u_{pqrs}$ the terms in the braces must vanish,

(8.4-8) $\epsilon_{3ip}u_{iqrs} + \epsilon_{3iq}u_{pirs} + \epsilon_{3ir}u_{pqis} + \epsilon_{3is}u_{pqri} = 0.$

Since there are only three possible values (1, 2, 3) for each of the four indices *pqrs*, at least two of them must be equal. Hence, we may consider separately the cases where (a) all four are equal, (b) three are equal, (c) two are equal and the other two unequal, and (d) two are equal and the other two equal.

In Case a, take $p = q = r = s = 1$. Then we see that all terms in Eq. (8.4-8) vanish on account of Eq. (8.4-6). Similarly, $p = q = r = s = 2$ or 3 yields no information.

In Case b, take $p = q = r = 1$, $s = 2$. We get

(8.4-9) $-u_{2112} - u_{1212} - u_{1122} + u_{1111} = 0.$

No new information is obtained by setting $p = q = r = 2$, $s = 1$, because this merely amounts to an interchange of indices 1 and 2, which has been considered in Eq. (8.4-5). The case $p = q = r = 3$ is trivial because the ϵ_{3ip} terms vanish.

Cases c and d yield conditions contained in Eq. (8.4-5) and (8.4-6).

Since a rotation from one rectangular coordinate system to another with the same origin can be obtained by repeated infinitesimal rotations about coordinate axes, no additional conditions are imposed on u_{pqrs} by isotropy.

Now let

(8.4-10)
$$u_{1122} = \lambda,$$
$$u_{1212} = \mu + \nu,$$
$$u_{2112} = \mu - \nu.$$

Then Eq. (8.4-9) says

(8.4-11) $u_{1111} = \lambda + 2\mu.$

There appear therefore to be three independent isotropic tensors of order 4, obtainable by taking each of λ, μ, ν in turn equal to 1 and the others to 0.

The tensor obtained by taking $\lambda = 1$, $\mu = \nu = 0$ has components $u_{ijkl} = 1$ if $i = j$, $k = l$, and vanishes in all other cases. Therefore, it is equivalent to

(8.4-12) $u_{ijkl} = \delta_{ij}\delta_{kl}.$

In the tensor obtained by taking $\mu = 1$, $\lambda = \nu = 0$, the component $u_{ijkl} = 1$ if $i = k$, $j = l$, $i \neq j$, and if $i = l$, $j = k$, $i \neq j$; whereas $u_{ijkl} = 2$ if $i = j = k = l$. Other components are zero. This is exactly

(8.4-13) $$u_{ijkl} = \delta_{ik}\delta_{jl} + \delta_{il}\delta_{jk}.$$

The tensor obtained by taking $\lambda = \mu = 0$, $\nu = 1$ has elements $u_{ijkl} = 1$ when $i = k, j = l, i \neq j$; and $u_{ijkl} = -1$ when $i = l, j = k, i \neq j$. All other components are zero. Hence,

(8.4-14) $$u_{ijkl} = \delta_{ik}\delta_{jl} - \delta_{il}\delta_{jk}.$$

The general isotropic tensor of rank 4 is therefore given by (8.4-2). From (8.4-2), Eq. (8.4-4) follows under the symmetry condition (8.4-3).   Q.E.D.

8.5 ISOTROPIC MATERIALS

If an elastic solid is isotropic, the tensor C_{ijkl} in Eq. (8.1-1),

(8.5-1) $$\sigma_{ij} = C_{ijkl}e_{kl},$$

must be isotropic (Sec. 8.2). Furthermore, it has been shown generally that $C_{ijkl} = C_{jikl}$ because the stress tensor is symmetric, and that $C_{ijkl} = C_{ijlk}$ because the strain tensor is symmetric and the sum $C_{ijkl}e_{kl}$ is symmetrizable without loss of generality. Hence, according to Eq. (8.4-4),

(8.5-2) $$C_{ijkl} = \lambda\delta_{ij}\delta_{kl} + \mu(\delta_{ik}\delta_{jl} + \delta_{il}\delta_{jk})$$

and Eq. (8.5-1) becomes

(8.5-3) $$\sigma_{ij} = \lambda e_{kk}\delta_{ij} + 2\mu e_{ij}.$$

This is the most general form of the stress-strain relationship for an isotropic elastic solid for which the stresses are linear functions of the strains. Therefore, an isotropic elastic solid is characterized by two material constants, λ and μ.

Similarly, an isotropic viscous fluid (Sec. 7.3) is governed by the relationship

(8.5-4) $$\sigma_{ij} = -p\delta_{ij} + \lambda V_{kk}\delta_{ij} + 2\mu V_{ij}.$$

8.6 COINCIDENCE OF PRINCIPAL AXES OF STRESS
AND OF STRAIN

An important attribute of isotropy of an elastic body (or a viscous fluid) is that the principal axes of stress and the principal axes of strain (or strain rate) coincide. This follows from the relationship (8.5-3) or (8.5-4), because the direction cosines of the principal axes of stress and strain are, respectively, the solutions of the equations (Sec. 4.4 and 5.7)

(8.6-1) $(\sigma_{ji} - \sigma\delta_{ji})v_j = 0,$ $|\sigma_{ji} - \sigma\delta_{ji}| = 0,$

(8.6-2) $(e_{ji} - e\delta_{ji})v_j = 0,$ $|e_{ji} - e\delta_{ji}| = 0.$

By Eq. (8.5-3), Eq. (8.6-1) becomes

(8.6-3) $(\lambda e_{kk}\delta_{ij} + 2\mu e_{ij} - \sigma\delta_{ji})v_j = 0,$

or

(8.6-4) $2\mu(e_{ji} - \sigma'\delta_{ji})v_j = 0$

if we introduce a new variable σ' defined as follows:

(8.6-5) $$\sigma' = \frac{\sigma - \lambda e_{kk}\delta_{ij}}{2\mu}.$$

But Eq. (8.6-4) is of precisely the same form as Eq. (8.6-2). Thus, although the eigenvalues (principal stresses and strains) are different, the principal directions, given by the solutions v_j, are the same.

There are many other ways of recognizing this fact of coincidence of principal directions of stress and strain. For example, we recognize in the Mohr's circle construction (Sec. 4.8) that the angle between the principal axes and the x-axis does not depend on the location of the center of the circle. The principal angle can be determined if the center is translated to the origin. Such a translation is accomplished by settling $\sigma_{kk} = 0$, $e_{kk} = 0$, under which condition the stress-strain relationship becomes simply

$$\sigma'_{ij} = 2\mu e'_{ij}.$$

The coincidence of principal directions is then evident because 2μ is just a numerical factor.

8.7* OTHER METHODS OF CHARACTERIZING ISOTROPY

There are other ways to characterize isotropy. For example, one may define the property of an elastic body through the strain-energy function, $W(e_{11}e_{12}, \ldots, e_{33})$, which is a function of the strain components and which defines the stress components by the relation

(8.7-1) $$\sigma_{ij} = \frac{\partial W}{\partial e_{ij}}.$$

Then isotropy may be stated as the fact that the strain-energy function depends on the *invariants* of the strain only. For example, using the strain invariants

$$I_1 = e_{ii},$$
$$I_2 = \tfrac{1}{2}e_{ij}e_{ji},$$
$$I_3 = \tfrac{1}{2}e_{ij}e_{jk}e_{ki},$$

we may specify $W(e_{11}, e_{12}, \ldots, e_{33})$ to be a function

(8.7-2) $W(I_1, I_2, I_3)$.

Since the invariants retain their form (and value) under all rotations of coordinates, the same attribute applies to the relationship (8.7-1).

PROBLEMS

8.1 Show that the theorem we proved in Sec. 8.4 may be restated: "The most general isotropic tensor of rank 4 has the form

$$u_{ijkm} = \alpha\delta_{ij}\delta_{km} + \beta\delta_{ik}\delta_{jm} + \gamma\delta_{im}\delta_{jk},$$

where α, β, γ are constants."

8.2 Distinguish the words *homogeneous* and *isotropic*. Consider the atmosphere of the earth.

  (a) If you are concerned with a high-altitude sounding rocket, would you call the atmosphere homogeneous or isotropic?

  (b) If the problem is concerned with the flow around the immediate neighborhood of the rocket which is flying at such a speed that no shock wave is generated, could the air be treated as homogeneous or isotropic?

  (c) If shock waves are generated in part b, what then?

8.3 Do you know of any liquid that is not isotropic?

8.4 Do you know of any solid that is not isotropic?

8.5 If the Apollo astronauts brought back a piece of rock from the moon, what experiments would you devise to determine its mechanical properties, homogeneity, and isotropy? Remember that piece of rock is precious!

8.6 Could the properties of a material change with time and strain-history? Can it change from isotropic to anisotropic and vice versa? Consider a piece of "low carbon" steel as an example. Take into consideration the various possibilities of heat treatment, hardening, tempering, spheroidizing, cold work, etc.

8.7 Can the answers to Prob. 8.6 be divorced from the size of the piece of steel and the purpose of seeking the answer? Give your answers first as a metallurgist creating the steels for an automobile, then as a guidance and control engineer interested in making a high quality gyroscope, and then as a physician who wants to select a material for a heart valve.

8.8 Show that the tensor $\epsilon_{ijk}\delta_{lm}$ is isotropic. Are there other isotropic tensors of rank 5?

8.9 Form some isotropic tensors of rank 6. Generalize to isotropic tensors of even order $2n$.

FURTHER READING

JEFFREYS, H., *Cartesian Tensors*, Cambridge: University Press (1957) Chapter 7.

THOMAS, T. Y., *Concepts from Tensor Analysis and Differential Geometry*, New York: Academic Press (1961) pp. 65–69.

CHAPTER NINE

Mechanical Properties of Fluids and Solids

In Chapter 7 we presented three simple constitutive equations that define three idealized kinds of materials. In Chapter 8 we discussed a further simplification of these constitutive equations if the material is isotropic. In this chapter we shall consider the mechanical properties of real materials in order to see how the idealized constitutive equations fit into the real world. We will see both how restrictive they are and how useful they are. We shall do this by describing a wide variety of materials. In this setting, we shall say a few words about the mechanical properties in relation to the atomic and molecular structure of these materials.

The simplest constitutive equation is that of a nonviscous fluid. There is no nonviscous fluid in the real world, but as is mentioned in Sec. 7.2, air and water may be treated as nonviscous in many problems. The next simple constitutive equation is that of a Newtonian fluid. Air and water are Newtonian fluids, blood and paints are not. Air and water differ in compressibility. Viscosity and compressibility, therefore, are the principal characteristics of fluids, and will be discussed in detail. Our plan is as follows:

In Sec. 9.1 we will consider the relationship between the gaseous and liquid states of a fluid. In Sec. 9.2, the question of whether tensile stress can exist in a liquid is discussed. Section 9.3 treats the viscosity of gases, liquids, and solids. Sections 9.4 and 9.5 treat the compressibility of air and water, respectively. Viscosity of the non-Newtonian type is discussed in Sec. 9.11.

The last half of this chapter deals with solids. The elastic constants of common engineering materials are presented in Sec. 9.6. However, Hooke's law is valid for metals only within a very small range of strain. For larger deformations metals become plastic. Plasticity is discussed in Sec. 9.7. In Sec. 9.8 we raise the question of why the actual strength of engineering metals falls so short of the theoretical strength calculated from the crystal structure. In Sec. 9.9 we discuss the elasticity of softer materials, such as rubber, skin, muscle, and other living tissues. Then we complete the story with a discussion of viscoelasticity in Sec. 9.10, viscoplasticity in Sec. 9.12, and thixotropy in Sec. 9.13.

Thus it is seen that three constitutive equations of Chapter 7 describe only a small part of the real world, but a very important part.

9.1 FLUIDS

Fluids are usually classified as gases or liquids, the chief distinction being density and compressibility. At the "critical point" of a fluid the distinction become confounded and the liquid becomes vapor without a sudden change in properties.

A typical example of the equation of state at a lower temperature and higher pressure is shown for carbon dioxide in Fig. 9.1. The lower curves all pass through a horizontal step—volume change without change of pressure—at a certain value of the pressure. To the left of this step we have the liquid state, wherein it takes a large increase of pressure to produce a small change of volume. To the right is the vapor or gaseous state. A point on this horizontal step (such as AB in Fig. 9.1) actually represents a heterogeneous state consisting of a mixture of liquid and vapor. At 31.05°C, the pause in the CO_2 liquid-vapor isothermal is reduced to zero. At temperatures above this "critical" value the isothermal passes steadily from high to low pressure with no marked division between gas and liquid. At higher temperatures the equation of state becomes better and better approximated by the "perfect gas" law.

The critical points of some substances are given in Table 9.1. The one-atmosphere boiling point is usually about $\frac{2}{3}T_c$ (°K). All the substances shown have roughly the same critical volume, which corresponds to a sphere of radius about 3 Å for each particle.

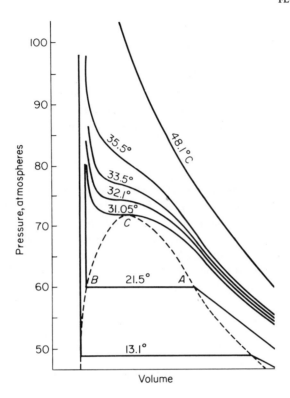

Fig. 9.1 Isothermal curves of volume-pressure relationship for CO_2 near the critical point.

TABLE 9.1

LIQUID-VAPOR-GAS CRITICAL POINTS

Substance	Temperature, T_c (°K)	Pressure, P_c (atm)	Volume, V_c (cm³ mole⁻¹)
Water	647.4	218.3	56
Carbon dioxide	304.2	72.9	94
Ammonia	405.5	111.3	72.5
Oxygen	154.8	50.1	78
Nitrogen	126.2	33.5	90.1
Hydrogen	33.3	12.8	65
Helium 4	5.3	2.26	57.8

Let us say a few words about the properties of fluids from the molecular point of view. Studies of gases led Avogadro to propose the hypothesis that equal volumes of gases contain equal numbers of molecules at the same temperature and pressure. A *mole* of atoms or molecules (a sample whose weight in grams exactly equals the *atomic* or *molecular weight* of the element) contains 6.025×10^{23} particles. This is known as Avogadro's number (N_0). Thus a single oxygen atom weighs $16N_0^{-1}$, i.e., 2.66×10^{-23} g. A sample of an element with density ρ and atomic weight A contains $N_0\rho/A$ atoms/cm³. The volume of 1 mole of gas at the normal temperature and pressure (i.e., 0°C and 760 mm mercury pressure) is 22,400 cm³. Divided into Avogadro's number, this gives *Loschmidt's* number 2.7×10^{19} particles/cm³, which corresponds to an average distance of about 33×10^{-8} cm from one particle to the next. When a vapor is condensed to a liquid or solid, it shrinks to about one-thousandth of its volume at the normal temperature and pressure. (For example, when water vapor is condensed its volume is reduced 1600 to 1.) Thus, the molecular size is of the order of 3×10^{-8} cm or 3 Å.

The kinetic theory interprets the pressure in a gas as the reaction of the gas molecules impinging on a surface. From the consideration of momentum change in molecular impact and rebound, the kinetic theory† derives the equation of state of a perfect gas relating the pressure (p), volume (V), and absolute temperature (T):

(9.1-1) $$pV = RT,$$

where

(9.1-2) $$R = N_0 k.$$

For 1 mole of gas, k and R are *universal* constants, the same for all substances. The constant k = Boltzmann's constant = 1.38×10^{-16} erg deg⁻¹, and R = the gas constant = 8.313×10^7 erg deg⁻¹ = 1.986 cal deg⁻¹ mole⁻¹.

Perhaps we should reflect on how well the idea that pressure is obtained by the bouncing of molecules on a surface fits our definition of stresses as given in Chapter 3: that pressure, an isotropic compressive stress, represents the interaction of molecules on the two sides of any imaginary surface. These two concepts agree perfectly.

Now let us consider the condensed state. The perfect gas law $pV = RT$ obviously fails for highly compressed gases in which most of the space is taken up by the gas molecules themselves. We would expect an equation such as $p(V - \beta) = RT$, where β is proportional to the molecular volume, to describe things better. Van der Waals has proposed the following celebrated equation of state:

(9.1-3) $$\left(p + \frac{\alpha}{V^2}\right)(V - \beta) = RT,$$

---

†See for example, Sir James Jeans, *An Introduction to the Kinetic Theory of Gases*, Cambridge: University Press (1946).

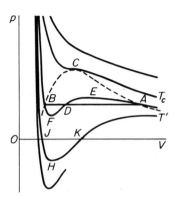

Fig. 9.2   A family of van der Waals isothermals.

in which α/V^2 represents the attractive forces between the gas particles (not accurately except at low gas densities), while β represents the molecular volume. Figure 9.2 shows a family of van der Waals p, V curves. They are like the curves of Fig. 9.1, but the horizontal line AB has become a continuous curve $AEDFB$ with a maximum and a minimum.

It can be argued that the portion of the curve between E and F, where the slope dp/dV is positive, must represent an unstable homogeneous system, because if part of the system expands a little at the expense of the rest, it could continue to expand since its pressure is higher than elsewhere. Simultaneously, the other part would continue to contract because its volume is lower. The segregation ends when the points A and B are reached. The homogeneous system would thus break down into a vapor at A and a liquid at B. The area $AEDA$ and $BFDB$ must be equal, because if this were not so, it would be possible in principle to produce a perpetual motion by taking the system from A to B along the homogeneous path ($AEDFB$) and returning to A along the heterogeneous path (horizontal line AB), or vice versa, obtaining more mechanical work from the system along the one path than is put into it along the other. This would contradict the second law of thermodynamics.

The portions of the curve in AE and FB are stable with respect to infinitesimal disturbances but are unstable against finite disturbances large enough to create *seeds* or *nuclei* of the phase at the other end of the AB line. These portions represent *superheated* liquids or *supersaturated* vapors.

The critical point C is characterized by the coalescence of the minimum and the maximum, thus:

$$(9.1\text{-}4) \qquad \frac{dp}{dV} = \frac{d^2p}{dV^2} = 0$$

on the isothermal at C. Applying these conditions to van der Waals' equation, we obtain, at the critical point,

$$(9.1\text{-}5) \qquad V_c = 3\beta, \qquad p_c = \frac{\alpha}{27\beta^2}, \qquad T_c = \frac{8\alpha}{27R\beta},$$

and hence

$$(9.1\text{-}6) \qquad p_c V_c = \tfrac{3}{8}RT_c.$$

Thus, pV is well below its perfect-gas value at this point. This equation can be used to test the accuracy of van der Waals' equation. From Table 9.1, we find, e.g., for CO_2, the predicted value of $V_c = 3RT_c/8p_c = 130$ cm^3 mole^{-1}, whereas the observed value is 94. This shows the quantitative limitations of van der Waals' equation. Better equations are more complex and can be found in books on thermodynamics.

9.2 TENSILE STRENGTH OF A LIQUID

The pressure in a perfect gas, being directly proportional to the kinetic energy of the molecules, has a lower bound of zero. The pressure in a liquid, however, may be negative (then the liquid is in tension), because the stresses in liquids are derived from short-range cohesive forces between the atoms. In the van der Waals' equation, if the temperature is sufficiently low, such as at T' in Fig. 9.2, the pressure is negative (i.e., the stress is tension) over part of the range, e.g., in JHK. The point of minimum H indicates the *ideal tensile strength* of the liquid.

To estimate this strength, we rewrite van der Waals' equation as

$$(9.2\text{-}1) \qquad p = \frac{RT}{V - \beta} - \frac{\alpha}{V^2}$$

and apply the condition $dp/dV = 0$ at H. This leads to the results that, at H,

$$(9.2\text{-}2) \qquad RTV_H^3 = 2\alpha(V_H - \beta)^2,$$

$$(9.2\text{-}3) \qquad P_H = \alpha\frac{V_H - 2\beta}{V_H^3}.$$

Equation (9.2-3) shows that for p_H to be negative, we must have $V_H < 2\beta$. On the other hand, V_H is greater than β a priori. The roots of (9.2-2) for V_H in the range $\beta \leqslant V_H < 2\beta$, when substituted into (9.2-3), yields the desired answer. Using (9.1-5), we can express (9.2-2), (9.2-3) in the form

$$(9.2\text{-}4) \qquad x^3 - c(x - 1)^2 = 0,$$

$$(9.2\text{-}5) \qquad p_H = 27p_c\frac{x - 2}{x^3},$$

where

$$x = \frac{V_H}{\beta}, \qquad c = \frac{27}{4}\frac{T_c}{T}.$$

For example, according to the data in Table 9.1 we obtain by a simple calculation the limiting tensile strength of water, -1168 atm at $0°C$, -875 atm at $50°C$. This calculation is, of course, very rough. Furthermore, it leaves unanswered the question as to how unstable this ideal state is.

Table 9.2 gives the tensile strength of liquids obtained by various authors using different methods. The tremendous variation in the reported values for water attests to the difficulty in experiments. "Breaks" can be triggered by vapor-nucleating agents, and the liquid can be torn away from the wall of the container while the tension is low. Cavitation such as that due to cosmic rays in Glaser's "bubble chamber"† can cause failure of the liquid.

TABLE 9.2

TENSILE STRENGTH OF LIQUIDS BY VARIOUS METHODS

Liquid and Method	Maximum Negative Pressure (atm)	Ref.	Liquid and Method	Maximum Negative Pressure (atm)	Ref.
Water:			Acetic Acid C-3	288	3
Methods A, B, C.	0.2–150	1	Benzene C-3	150	3
Method C-3.	277	2	Aniline C-3	300	3
Alcohol:			CCl_4 C-3	276	3
Methods A, B, C.	2.4–40	1	Chloroform C-3	317	3
Ether:			Mercury C-3	425	4
Method A.	72	1			
Method B.	2.2	1			

REFERENCES

See *American Institute of Physics Handbook*, edited by D. E. Gray, New York: McGraw-Hill Book Company (1957), pp. 2–170. The specific references listed above are
1. Vincent, R. S.: *Proc. Phys. Soc.* **53** (1941) 141.
2. Briggs, Lyman J.: *J. Appl. Phys.* **21** (1950) 721.
3. Briggs, Lyman J.: *J. Chem. Phys.* **19** (1951) 970.
4. Briggs, Lyman J.: *J. Appl. Phys.* **24** (1953) 488.

METHODS OF MEASURING TENSILE STRENGTH

A. Stress produced by cooling and thus contracting the liquid.
B. Stress produced by expanding the volume of tonometer.
C. Stress produced by centrifugal force.
C-3. Z-shaped capillary tube, open at both ends, rotating in the Z-plane about an axis passing through the center of the Z and perpendicular to the plane. The liquid menisci are located in the bent-back short arms of the Z. The speed of rotation is increased gradually until the liquid in the capillary "breaks."

†D. A. Glaser: *Phys. Rev.*, **87** (1952), 665.

Cavitation can be caused by a variety of reasons; it is an important engineering problem for high-speed vehicles operating in water. For example, flow around a hydrofoil operating at a moderate angle of attack cavitates at a certain critical speed beyond which a vapor bubble grows on the upper side of the hydrofoil, which can cause an increase in drag force and also induce oscillations in the flow. A different kind of cavitation is the formation of a large number of small bubbles. The propellers of seagoing ships require frequent replacement because of surface damage due to cavitation and collapse of such small bubbles. Careful theoretical and experimental examinations of the collapse of these bubbles show that when a bubble collapses, not only is the pressure in the bubble singularly high (because of surface tension, see Sec. 11.3) but also a high speed jet is formed in the bubble due to the convolution of the fluid on the boundary of the bubble. Damage to the propeller surface is presumably due to these jets and pressure centers. Small bubbles of this nature can be formed in the laboratory by focusing ultrasound waves in water. Experiments with ultrasound cavitation in the neighborhood of a metal surface have demonstrated the damaging potential of cavitation. An example due to Dr. A. Ellis is shown in Fig. 9.3, which shows the surface of a stainless steel specimen before and after exposure to cavitation. On the other

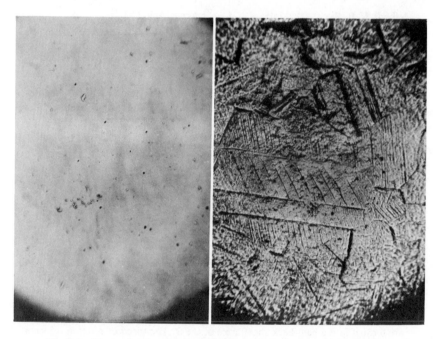

Fig. 9.3 Photomicrograph of a stainless steel specimen before and after exposure to cavitation. × 560. Courtesy A. T. Ellis, *Trans. ASME*, **77**, 7 (Oct. 1955), 1055–1064.

hand, the same principle can be used to advantage industrially for cleansing metal or other surfaces of undesirable materials.

9.3 VISCOSITY

The concept of viscosity in fluid was given by Newton in terms of a shear flow with a uniform velocity gradient as shown in Fig. 9.4. Newton proposed the relationship

$$(9.3\text{-}1) \qquad\qquad \tau = \mu \frac{du}{dy}$$

for the shear stress τ. The coefficient μ is a constant called the *coefficient of viscosity*.

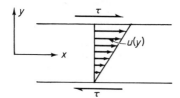

Fig. 9.4 Newtonian concept of viscosity.

Since stress is a force per unit area, it follows that the dimensions of μ are $[ML^{-1}T^{-1}]$. In the centimeter-gram-second system of units, in which the unit of force is the dyne, the unit of μ is called a *poise*, in honor of Poiseuille. In the International units, the unit of viscosity is newton second/meter^2 (ns/m^2). 1 poise is 10 ns/m^2.

The viscosity is small both for air and for water, being 1.8×10^{-4} poise for air and 0.01 poise for water at atmospheric pressure and 20°C. At the same temperature the viscosity of glycerine is about 8.7 poises. In liquids μ diminishes fairly rapidly as the temperature increases. In gases, μ increases as the temperature rises. Tables 9.3 and 9.4 give the values of the viscosity and the kinematic viscosity of air and water. The kinematic viscosity is defined as μ/ρ and is denoted by ν.

An interesting interpretation of the coefficient of viscosity from the kinetic theory of gases was given by Maxwell. Consider a flow with a uniform velocity gradient, and imagine a surface AA as in Fig. 9.5. The shear stress exerted by the gas beneath AA on the gas above is a retarding effect. The shear stress is equal to the time rate of loss of ordered momentum across AA by the random motion of the molecules. A molecule originating at y_1 and moving downward through AA will carry with it a positive momentum

TABLE 9.3

THE VISCOSITY AND KINEMATIC VISCOSITY OF AIR

Temp. (°C)	$10^4\mu$ (g/cm sec)	$10^5\mu$ (lb/ft sec)	v (cm²/sec)	10^3v (ft²/sec)
0	1.709	1.148	0.132	0.142
20	1.808	1.215	0.150	0.161
40	1.904	1.279	0.169	0.181
60	1.997	1.342	0.188	0.202
80	2.088	1.403	0.209	0.225
100	2.175	1.462	0.330	0.248
200	2.582	1.735	0.346	0.372
300	2.946	1.980	0.481	0.518
400	3.277	2.202	0.625	0.673
500	3.583	2.408	0.785	0.845

TABLE 9-4

THE VISCOSITY AND KINEMATIC VISCOSITY OF WATER

Temp. (°C)	100μ (g/cm sec)	$1,000\mu$ (lb/ft sec)	$100v$ (cm²/sec)	10^5v (ft²/sec)
0	1.792	1.204	1.792	1.929
10	1.308	0.879	1.308	1.408
20	1.005	0.675	1.007	1.084
30	0.801	0.538	0.804	0.866
40	0.656	0.441	0.661	0.712
50	0.549	0.369	0.556	0.598
60	0.469	0.315	0.477	0.513
70	0.406	0.273	0.415	0.447
80	0.357	0.240	0.367	0.395
90	0.317	0.213	0.328	0.353
100	0.284	0.191	0.296	0.319

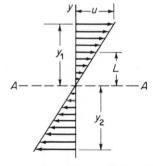

Fig. 9.5   Kinetic interpretation of the coefficient of viscosity of gases.

$m\,(du/dy)\,y_1$ where m is the mass of the molecule, u is the ordered velocity in the x-direction, and du/dy is the vertical velocity gradient, i.e., the rate of shear strain. Similarly, a molecule moving upward through AA and originating at y_2 will carry with it a negative momentum $m\,(du/dy)\,y_2$. Both these excursions represent a loss of ordered momentum from the fluid above AA. The sum of such losses that occur in one second through unit area AA is equal to the shear stress τ.

Let there be N molecules per unit volume. Assume that one-third of the molecules are traveling in each of the three coordinate directions. If the average molecular speed is c, and if one-third of them move perpendicular to AA, $\frac{1}{3}Nc$ molecules will pass through AA each second. Each of these molecules will carry with it a momentum corresponding to the position y at which it originates. Let the average value of the height y be L. Then the shear stress is

$$\tau = \left(\frac{1}{3}Nc\right)m\,\frac{du}{dy}\,L.$$

The product Nm is the density ρ. Therefore,

$$(9.3\text{-}2) \qquad\qquad \tau = \frac{1}{3}\rho cL\,\frac{du}{dy}.$$

A comparison of Eq. (9.3-1) and (9.3-2) shows that

$$(9.3\text{-}3) \qquad\qquad \mu = \tfrac{1}{3}\rho cL.$$

The effective height L is related to the mean free path l (the average distance a molecule travels before colliding with another), and more accurate calculations by Enskog and Chapman show that

$$(9.3\text{-}4) \qquad\qquad \mu = 0.499\,\rho cl.$$

As the density of a gas decreases, the mean free path increases in such a manner that the product ρl almost remains constant. Then μ is proportional to c, which in turn is proportional to the square root of the absolute temperature. Thus the coefficient of viscosity of a gas changes with the temperature but not with the pressure.

For air under standard conditions (sea level, 59°F) the mean free path of the molecules is approximately 8.8×10^{-6} cm.

The argument that leads to the result (9.3-2) can be used on other transport phenomena. When the molecules cross the plane AA, they carry with them not only the momentum of their ordered motion but also their mass and their energy. In a gas with a density gradient the transport of mass corresponds to the phenomenon of diffusion. In a gas with a temperature gradient the transport of energy corresponds to the phenomenon of the conduction of heat. Thus, in the simplest theory the mechanisms of the transport of a component of ordered momentum, of heat energy, and of mass are identical; and as a result it is found that the coefficient of heat conduction k is equal to the product of the viscosity μ and the specific heat at constant volume C_v, while the coefficient of self-diffusion D is equal to the viscosity μ divided by the density ρ. Experiments and more accurate calculations give

$$k = 1.91\mu C_v, \qquad D = 1.2\frac{\mu}{\rho}.$$

Atomic interpretation of viscosity of liquids and solids is different from that of the gases. Solids in the crystal form have *long-range ordered* structures. The atoms are arranged in order by the long-range interaction. On the other hand, gas atoms or molecules interact only when they come "into contact," and the interaction depends on the *short-range* attractive force between two gas atoms or molecules. The liquid state is an interpolation between the gas and the crystal. Generally speaking, other than those properties such as X-ray diffraction, anisotropy, etc., the structure and properties of a liquid, just above the melting point, are fairly similar to those of its crystal. Metals expand only 3 to 5% on melting (bismuth, like ice, contracts), so that the packing of the atoms cannot be too different. It is as if 3 to 5% of the crystal sites became empty and their free volume were broken up and distributed between the particles in such a way as to destroy the long-range order of the structure. A picture of the cause of viscosity for a simple liquid is proposed by Cottrell, as shown in Fig. 9.6.† The elementary movements distort the atomic "cages." The surrounding liquid offers elastic (shear) resistance to such a distortion, but this resistance relaxes as similar movements occur nearby.

On the other hand, the atoms in a crystal are arranged in space lattices. See Fig. 9.7. These lattices may have imperfections, such as dislocations and

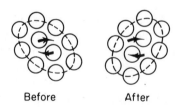

Before   After

Fig. 9.6 Shear due to internal movement in a liquid as suggested by Cottrell as a mechanism for viscosity in a liquid. After A. H. Cottrell, *The Mechanical Properties of Matter* (New York, John Wiley, 1964).

†A. H. Cottrell, *The Mechanical Properties of Matter*, New York: John Wiley & Sons, Inc. (1964), p. 204.

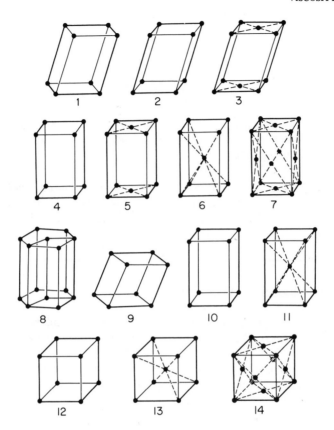

Fig. 9.7 Conventionally chosen unit cells of the fourteen space lattices. 1, Triclinic; 2, monoclinic, simple; 3, monoclinic, base-centered; 4, orthorhombic, simple; 5, orthorhombic, base-centered; 6, orthorhombic, body-centered; 7, orthorhombic, face-centered; 8, hexagonal; 9, rhombohedral; 10, tetragonal, simple; 11, tetragonal, body-centered; 12, cubic, simple; 13, cubic, body-centered; 14, cubic, face-centered. After A. H. Cottrell, *The Mechanical Properties of Matter* (New York, John Wiley, 1964).

vacancies. The movements of these dislocations and vacancies are history-dependent and contribute to viscosity. For example, Fig. 9.8, again taken from Cottrell, pictures the jump of an atom into a vacancy. To make such a jump, an atom must break away from the binding forces of its own site and squeeze between neighbors to enter the vacant site. This represents diffusion on the one hand and viscosity on the other. The much greater viscosity of a crystal as compared with its liquid is due to the fact that the atomic sites in a crystal are defined exactly by the lattices.

Mixtures, colloidal solutions, suspensions, polycrystalline solids, amorphous solids or glass, etc., can have many other relaxation mechanisms which reveal viscosity. In many cases it is not easy to say whether a body

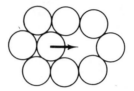

Fig. 9.8 Jump of an atom into a vacancy as a mechanism of viscosity in solid. From A. H. Cottrell, *The Mechanical Properties of Matter* (New York, John Wiley, 1964).

behaves as a fluid or as a solid. Silicone putty can be poured slowly from a cup or bounced quickly like a rubber ball. Conventionally, the distinction between fluids and solids is drawn at a low-stress viscosity of 10^{15} poises. A material with viscosity less than 10^{15} poises is called a fluid, one whose viscosity is greater than 10^{15} poises is called a solid.

9.4 THE COMPRESSIBILITY OF AIR

Air is a nearly perfect gas at normal temperature and pressure (0°C, 760 mm Hg pressure). For a "standard atmosphere" at sea level,

$p = 2116 \, \text{lb/ft}^2 = 1.0332 \, \text{kg/cm}^2 = 1.012 \times 10^6 \, \text{dyne/cm}^2$,
$\rho = 0.002378 \, \text{slug/ft}^3 = 1.225 \, \text{g/liter}$,
$T = 59°F = 519°R = 15.0°C = 288.16°K$,
$a = \text{speed of sound} = 1116 \, \text{ft/sec} = 340.29 \, \text{m/sec}$.

The bulk modulus of air's elasticity, defined as the change of volume per unit volume per unit change of pressure, is

$$(9.4\text{-}1) \qquad K = -\rho \frac{d}{dp}\left(\frac{1}{\rho}\right) = \frac{1}{\rho}\frac{d\rho}{dp} = \frac{1}{\rho a^2},$$

where a is the speed of sound.

It is interesting to note that in the theory of the aerodynamics of the airplane the air pressure at sea level is so high that the air may be treated as an incompressible fluid as long as the velocity of flight is small compared with the velocity of sound. The justification of this obvious disparity with common experience is based on the fact that in such low-subsonic flight the percentage change in the density of air is small because the percentage change of pressure is small. The maximum pressure change associated with motion is of the order of $\rho_0 v^2/2$, where ρ_0 is the density and v is the maximum velocity. Hence the incompressibility approximation will be satisfactory as long as $\rho_0 v^2/2$ is small compared with the undisturbed pressure p_0. Near the ground the atmospheric pressure is about 2100 lb/ft^2, and for a speed of 100 mi/hr $\rho_0 v^2/2$ is about 26 lb/ft^2, or about 1.3% of p_0, which is quite small. Generally, we can say that the condition required is $v \ll \sqrt{p_0/\rho_0}$. Since the velocity of sound is about $1.2\sqrt{p_0/\rho_0}$, the required condition is that v/a should be small. This ratio is called the Mach number and is denoted by M:

$$(9.4\text{-}2) \qquad M = \frac{v}{a} \cdot$$

Hence the condition of negligible compressibility is

$$(9.4\text{-}3) \qquad M \ll 1.$$

9.5 THE COMPRESSIBILITY OF LIQUIDS

Liquids are of course compressible. Table 9.5 shows the volume of pure water as a function of pressure and temperature. Data on many other liquids are known.[†]

Tait (1888)[‡] first discovered a simple equation of state for water, which can be written as

$$(9.5\text{-}1) \qquad V_0^{(P)} = V_0^{(1)} - C \log_{10} \frac{B + P}{B + 1},$$

where V_0 is the specific volume of water (ml/g), $V_0^{(P)}$ is the value of V_0 at pressure P, $V_0^{(1)}$ is that at $P = 1$ (the pressure is measured in bars), and C and B are empirical functions of temperature. According to Y-H. Li,[§] the constants are, in the range of $0 \leqslant t \leqslant 45°C$ and $1 \leqslant P \leqslant 1000$ bars,

$$(9.5\text{-}2) \qquad \begin{aligned} C &= 0.3150 V_0^{(1)}, \\ B &= 2668.0 + 19.867t - 0.311t^2 + 1.778 \times 10^{-3}t^3. \end{aligned}$$

For sea water, the corresponding equations are

$$(9.5\text{-}3) \qquad V^{(P)} = V^{(1)} - (1 - S \times 10^{-2})C \log_{10} \frac{B^* + P}{B^* + 1},$$

$$(9.5\text{-}4) \qquad B^* = 2670.8 + 68.9656S + (19.39 - 0.703178S)t - 0.233t^2$$

[†]See, for example, *The American Institute of Physics Handbook*, New York: McGraw-Hill Book Company (1960). Experimental techniques are thoroughly described by P. W. Bridgman, *The Physics of High Pressure*, London: George Bell & Sons, (1952).

[‡]P. G. Tait, "Report on some of the physical properties of fresh water and sea water," *Report on Scientific Results of Voy. H.M.S., Challenger, Phys. Chem.*, **2** (1888) 1–76.

[§]Yuan-Hui Li, "Equation of state of water and sea water," *J. Geophys. Res.*, **72**, 10 (1967) 2665–2678.

TABLE 9.5

Volume of Pure Air-free H₂O as a Function of Pressure and Temperature

Temperature range—20 to 100°C; pressure range 1–12,000 kg/cm²; specific volume in ml/g

p, kg/cm²	−20°C	−15°C	−10°C	−5°C	0°C	20°C	40°C	60°C	80°C	100°C
1					1.0001	1.0018	1.0079	1.0171	1.0284	1.0435
500					0.9770	0.9819	0.9880	0.9959	1.0063	1.0183
1,000				0.9566	0.9576	0.9632	0.9706	0.9786	0.9883	0.9993
1,500		0.9370	0.9380	0.9394	0.9409	0.9476	0.9550	0.9632	0.9724	0.9826
2,000	0.9203	0.9214	0.9228	0.9246	0.9261	0.9328	0.9408	0.9492	0.9582	0.9679
4,000			0.8771	0.8794	0.8812	0.8888	0.8966	0.9044	0.9126	0.9208
6,000					0.8489	0.8565	0.8645	0.8721	0.8794	0.8871
8,000							0.8396	0.8564	0.8534	0.8604
10,000							0.8186	0.8252	0.8318	0.8385
12,000							0.8006	0.8070	0.8134	0.8199

Based on data of P. W. Bridgman, *J. Chem. Phys*, **3** (1936) 597. See N. F. Dorsey, *Properties of Ordinary Water Substance*, New York; Reinhold Publishing Co. (1948).

in the range $0 \leqslant t \leqslant 20°C$, $1 \leqslant P \leqslant 1000$ bars and $3 \leqslant S \leqslant 4\%$. S is the salinity of the seawater, expressed in percent salt by weight.[†]
The propagation of sound waves is the most familiar revelation of the compressibility of the material. For plane or spherical waves of small amplitude in a liquid for which the density is a unique function of pressure, the phase velocity is given by

$$(9.5\text{-}5) \qquad c = \sqrt{\frac{dp}{d\rho}} = \sqrt{\frac{1}{\rho}\left(-\frac{1}{\rho}\frac{dp}{dV}\right)} = \sqrt{\frac{K}{\rho}},$$

in which p is the pressure, ρ is the density, and $V = 1/\rho$ is the volume per unit mass. $K = -\rho^{-1}(dp/dV)$ is the rate of change of pressure with volume and is known as the bulk modulus of the liquid. Hence, we can compute K if we know c and ρ. Table 9.6 shows the velocity of sound in several liquids.[‡] In the ocean, the velocity of sound varies with the temperature, salinity, and depth (which affects the pressure and density) of the water. Since the ocean

TABLE 9.6

THE VELOCITY OF SOUND c_1 (DILATATIONAL WAVES) IN LIQUIDS

Liquid	t (°C)	c_1 (m/s)	Liquid	t (°C)	c_1 (m/s)
Acetone	25	1170	Mercury	20	1451
Argon	−188	853	Oxygen	−182.9	912
Benzene	25	1295	Seawater (surface	0	1445.5
Carbon tetrachloride	25	930	depth, salinity		
Ether (diethyl)	25	985	35 parts per 1000,		
Ethyl alcohol	20	1177	latitude 30°)*	15	1509.7
Helium	−268.8	179.8	Tin	380	2270
Hydrogen	−256	1187	Water (distilled)§	0	1407
Lead	380	1790			

*In seawater: $c_1 = 1445.5 + 3.92t - 0.024t^2$ m/s, for salinity 3.5% by weight, $t°C$, latitude 30°. The velocity increases by about 18 m/s per km depth and by 13 m/s for a 1% increase in salinity from 3 to 4%. The variation with latitude is zero near the surface, but at 10 km depth the velocity is 0.9 m/s greater at latitude 90° than at the equator.

§The velocity of sound in distilled water is free from dispersion but is anomalous for temperature dependence as shown below:

t (°C)	0	10	20	30	40	50	60	70	80
c_1 (m/s)	1407	1445	1484	1510	1528	1544	1556	1561	1557

†A. Bradshaw and K. E. Schleicher, *Deep Sea Res.* **17** (1970), 691 and F. K. Lepple and F. J. Millero, *Deep Sea Res.* **18** (1971), 1233 presented new data and proposed new equations of state of seawater.

‡Extensive data can be found, for example, in G. W. C. Kaye and T. H. Laby, *Tables of Physical and Chemical Constants*, New York: Longmans, Green & Co., 12th ed. (1960).

water is usually stratified, the medium is not uniform and sound waves do not travel in straight lines. The phenomenon of refraction occurs, which complicates the interpretation of measurements by sonar instruments. The study of such acoustic problems is of great importance to the exploration of ocean engineering.

In most problems of flow, the fluid may be considered incompressible if the velocity of flow is small compared with the speed of sound. This is the case in most problems in hydrodynamics. This is why the reader should familiarize himself with the velocity of sound in the fluid with which he is concerned.

9.6 THE ELASTICITY OF SOLIDS

Hooke's law (7.4-1) contains 21 independent elastic constants for the most general form of anisotropy. A number of anisotropic crystals have had their elastic constants determined.† Most engineering materials are poly-crystalline and can be regarded as isotropic in general applications.

For an *isotropic elastic material* in which there is no change of temperature, Hooke's law may be stated in the form

$$(9.6\text{-}1) \qquad \sigma_{\alpha\alpha} = 3Ke_{\alpha\alpha},$$

$$(9.6\text{-}2) \qquad \sigma'_{ij} = 2Ge'_{ij},$$

where K and G are constants and σ'_{ij} and e'_{ij} are the stress deviation and strain deviation, respectively; i.e.,

$$(9.6\text{-}3) \qquad \sigma'_{ij} = \sigma_{ij} - \tfrac{1}{3}\sigma_{\alpha\alpha}\delta_{ij},$$

$$(9\text{-}6\text{-}4) \qquad e'_{ij} = e_{ij} - \tfrac{1}{3}e_{\alpha\alpha}\delta_{ij}.$$

We have seen before that $\tfrac{1}{3}\sigma_{\alpha\alpha}$ is the mean stress at a point and that, if the strain were infinitesimal, $e_{\alpha\alpha}$ would be the change of volume per unit volume: Both are invariants. Thus, (9.6-1) states that the change in volume of the material is proportional to the mean stress. In the special case of hydrostatic compression,

$$\sigma_{xx} = \sigma_{yy} = \sigma_{zz} = -p, \qquad \sigma_{xy} = \sigma_{yz} = \sigma_{zx} = 0,$$

we have $\sigma_{\alpha\alpha} = -3p$, and Eq. (9.6-1) may be written, in the case of infinitesimal strain, with v and Δv denoting volume and change of volume, respectively,

$$(9\text{-}6\text{-}5) \quad \blacktriangle \qquad \frac{\Delta v}{v} = -\frac{p}{K}.$$

†Data can be found in the *American Institute of Physics Handbook*, New York: McGraw-Hill Book Company (1957), pp. 2-56–2-60.

Thus, the coefficient K is appropriately called the *bulk modulus* of the material.

The strain deviation e'_{ij} describes a deformation without volume change. Equation (9.6-2) states that the stress deviation is simply proportional to the strain deviation. The constant G is called the *modulus of elasticity in shear*, or *shear modulus*, or the *modulus of rigidity*. In the special case in which $e_{xy} \neq 0$ but all other strain components vanish, we have

(9.6-6) ▲ $$\sigma_{xy} = 2Ge_{xy},$$

whereas all other stress components vanish. The coefficient 2 is included because before the tensor concept was introduced, it was customary to define the shear strain as $\gamma_{xy} = 2e_{xy}$.

If we substitute (9.6-3) and (9.6-4) into (9.6-2) and make use of (9.6-1), the result may be written in the form

(9.6-7) ▲ $$\sigma_{ij} = \lambda e_{\alpha\alpha}\delta_{ij} + 2Ge_{ij},$$

or

(9.6-8) ▲ $$e_{ij} = \frac{1+v}{E}\sigma_{ij} - \frac{v}{E}\sigma_{\alpha\alpha}\delta_{ij}.$$

The constants λ and G are called *Lamé's constants* (G. Lamé, 1852). In many books μ is used in place of G. The constant E is called the modulus of elasticity, or Young's modulus (Thomas Young, 1807). The constant v is called Poisson's ratio. The relationships between these constants are

(9.6-9)

$$\lambda = \frac{2Gv}{1-2v} = \frac{G(E-2G)}{3G-E} = K - \frac{2}{3}G = \frac{Ev}{(1+v)(1-2v)}$$

$$= \frac{3Kv}{1+v} = \frac{3K(3K-E)}{9K-E},$$

$$G = \frac{\lambda(1-2v)}{2v} = \frac{3}{2}(K-\lambda) = \frac{E}{2(1+v)} = \frac{3K(1-2v)}{2(1+v)} = \frac{3KE}{9K-E},$$

$$v = \frac{\lambda}{2(\lambda+G)} = \frac{\lambda}{(3K-\lambda)} = \frac{E}{2G}-1 = \frac{3K-2G}{2(3K+G)} = \frac{3K-E}{6K},$$

$$E = \frac{G(3\lambda+2G)}{\lambda+G} = \frac{\lambda(1+v)(1-2v)}{v} = \frac{9K(K-\lambda)}{3K-\lambda}$$

$$= 2G(1+v) = \frac{9KG}{3K+G} = 3K(1-2v),$$

$$K = \lambda + \frac{2}{3}G = \frac{\lambda(1+v)}{3v} = \frac{3G(1+v)}{3(1-2v)} = \frac{GE}{3(3G-E)} = \frac{E}{3(1-2v)}.$$

To these we may add the following combinations that appear frequently.

$$\frac{G}{\lambda + G} = 1 - 2\nu, \qquad \frac{\lambda}{\lambda + 2G} = \frac{\nu}{1 - \nu}.$$

In unabridged notation, Eq. (9.6-8) reads

$$e_{xx} = \frac{1}{E}[\sigma_{xx} - \nu(\sigma_{yy} + \sigma_{zz})], \qquad e_{xy} = \frac{1 + \nu}{E}\sigma_{xy} = \frac{1}{2G}\sigma_{xy},$$

(9.6-10) $$e_{yy} = \frac{1}{E}[\sigma_{yy} - \nu(\sigma_{xx} + \sigma_{zz})], \qquad e_{yz} = \frac{1 + \nu}{E}\sigma_{yz} = \frac{1}{2G}\sigma_{yz},$$

$$e_{zz} = \frac{1}{E}[\sigma_{zz} - \nu(\sigma_{xx} + \sigma_{yy})], \qquad e_{zx} = \frac{1 + \nu}{E}\sigma_{zx} = \frac{1}{2G}\sigma_{zx}.$$

Table 9.7 gives the average values of E, G, and ν at room temperature for several engineering materials which are approximately isotropic.

TABLE 9.7

	E, 10^6 lb/sq in.	G, 10^6 lb/sq in.	ν	Speed of Sound (Dilatational Wave) 10^3 ft/sec
Metals:				
Steels	30	11.5	0.29	16.3
Aluminum alloys	10	2.4	0.31	16.5
Magnesium alloys	6.5	2.4	0.35	16.6
Copper (hot rolled)	15.0	5.6	0.33	—
Plastics:				
Cellulose acetate	0.22			3.6
Vinylchloride acetate	0.46			5.1
Phenolic laminates	1.23		0.25	8.2
Glass	8	3.2	0.25	
Concrete	4		0.2	

In 1829, Poisson advanced arguments, later found untenable, that the value of ν should be $\frac{1}{4}$. The special value of Poisson's ratio $\nu = \frac{1}{4}$ makes

(9.6-11) $$\lambda = G$$

and simplifies the equations of elasticity considerably. Consequently, this assumption is often used, particularly in geophysics, in the study of complicated wave-propagation problems. The special value $\nu = \frac{1}{2}$ implies that

(9.6-12) $$G = \frac{1}{3}E, \qquad \frac{1}{K} = 0,$$

and that, for finite stress,

(9.6-13) $$e_{\alpha\alpha} = 0.$$

Hence, for small deformations, a material with a Poisson's ratio $\frac{1}{2}$ is incompressible. Inasmuch as experience shows that hydrostatic compression induces reduction of volume, $\nu = \frac{1}{2}$ is an upper limit of the Poisson's ratio.

9.7* PLASTICITY OF METALS

If a rod of a ductile metal is pulled in a testing machine at room temperature, the load applied on the test specimen may be plotted against the elongation

$$\epsilon = \frac{l - l_0}{l_0},$$

where l_0 is the original length of the rod and l is the length under load. Numerous experiments show typical load-elongation relationships as indicated in the diagrams of Fig. 9.9. The initial region, appearing as a straight line, is the region in which the law of linear elasticity is expected to hold. Mild steel shows an upper yield point and a flat yield region which is caused

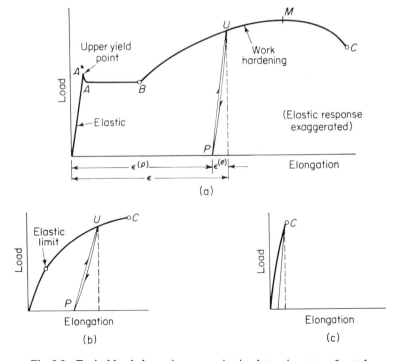

Fig. 9.9 Typical load-elongation curves in simple tension tests of metals.

by many microscopic discontinuous small steps of slip along slip planes of the crystals [Fig. 9.9(a)]. Most other metals do not have such a flat yield region [Fig. 9.9(b)].

Upon unloading at any stage in the deformation, the strain does not retrace the loading curve but is reduced along an elastic unloading line such as the curve *UP* in Fig. 9.9(a) and (b). Reloading retraces the unloading curve with relatively minor deviation and then produces further plastic deformation when approximately the previous maximum stress is exceeded. The test specimen may "neck" at a certain strain, so that its cross-sectional area is reduced in a small region. When *necking* occurs under continued elongation, the load reaches a maximum and then drops down, although the actual average stress in the neck region (load divided by the true area of the neck) continues to increase. The maximum *M* is the ultimate load. Beyond the ultimate load the metal flows. At the point *C* in the curves of Fig. 9.9 the specimen breaks.

Materials like cast iron, titanium carbide, and beryllium, or any rock material which allows minimal plastic deformation before reaching the breaking point, are called brittle materials. The load-strain curve for a brittle material will appear as that in Fig. 9.9(c). The point *C* is the breaking point.

A fact of great importance for geology is that brittle materials such as rocks tend to become ductile when subjected to large hydrostatic pressure (large negative mean stress). This was demonstrated by Theodore von Kármán (1911) in his classical experiments on marbles.

Tests of specimens subjected to simple compression or simple shear lead to load-strain diagrams similar to those of Fig. 9.9.

It is well-known that whereas the elastic moduli of all steels are nearly the same, the yield stress and the ultimate strength vary a great deal depending on the crystal structure (including imperfections, dislocations, vacancies, grain boundaries, twinning, etc.), which can be influenced by small changes in chemical composition, alloying, heat treatment, cold work, etc. In other words, whereas the elastic moduli are "structurally insensitive," the strengths are "structurally sensitive." Table 9.8 shows the elastic and strength constants for some iron and steel alloys.† Note the wide variation in yield stress. For materials without a marked yield point, it is an engineering practice to quote a *yield strength* as the stress at the *proportional limit* which is defined as the point where a tensile strain of 0.2% is reached. Most engineering structures use materials within the proportional limit—hence the strain is truly quite small. For this reason, the linearized theory of elasticity is useful.

Note also how much smaller the ultimate strength is compared with the shear modulus or Young's modulus.

---

†Extensive data can be found in various handbooks: e.g., *The American Institute of Physics Handbook*, New York: McGraw-Hill Book Company (1960).

TABLE 9.8

ELASTIC AND STRENGTH CONSTANTS FOR IRON AND STEEL ALLOYS

%C	Alloy	E Young's mod. (10^9 dyne cm^{-2})	(10^6 lb in.$^{-2}$)	G Shear mod. (10^9 dyne cm^{-2})	(10^9 lb in.$^{-2}$)	ν Poisson's Ratio	σ_{ult} Tensile strength (10^9 dyne cm^{-2})	(10^3 lb in.$^{-2}$)	σ_y Yield strength (10^9 dyne cm^{-2})	(10^3 lb in.$^{-2}$)	Elongation %	τ_{ult} Shear strength (10^9 dyne cm^{-2})	(10^3 lb in.$^{-2}$)
	Cast Iron: 2.50C, 0.79Si, 0.09S, 0.04P	1380	20.0	—	—	—	3.28	47.6	—	—	—	3.07	44.5
	Steel:												
0.03	0.12Mn, 0.005Si, 0.45Cu, 0.07Mo. Hot rolled at 540°C	—	—	—	—	—	3.43	49.7	2.4†	34.8†	35.8	—	—
0.05	0.39Si, 0.25Mn, 0.014P, 0.049S. As rolled	—	—	—	—	—	4.00	58.0	2.79‡	40.5‡	29.5§§	—	—
0.07	18.95Cr, 7.69Ni. Cold rolled ⅜-in. bar	1720	24.9	—	—	—	9.85	142.8	—	—	21††	—	—
0.03	13.47Cr, 0.27V, 0.04P, 0.01S. Hot rolled 3⅜-in. bar	1820	26.4	854	12.4	—	5.68	82.4	—	—	16§	4.72	68.5
0.10	0.5Cr, 0.3Mo, 2.5Ni. Oil quench from 820°C (carburized)	—	—	—	—	—	9.31	135.0	7.84†	113.7†	13‡‡	—	—
0.33	0.78C, 0.24Mo, 0.54Mn, 0.21Si, 0.025P. Wrought, furnace cooled from 1450°F	1970	28.6	827	12.0	0.288	5.28	76.6	2.93	42.5	48	6.02	87.3
0.33	0.78Cr, 0.24Mo, 0.54Mn, 0.21Si, 0.025P. Wrought, oil quenched from 1600°F, tempered at 1100°F	1980	28.7	813	11.8	0.272	8.68	125.9	6.24	90.5	28	7.85	113.9
0.40	1.65Ni, 0.99Cr, 0.51Mn, 0.20Si, 0.028S. Wrought, furnace cooled from 1450°F	1980	28.7	778	11.3	0.299	6.19	89.8	3.02	43.8	40	6.24	90.5
1.27	12.69Mn, 0.12Si. Water quench from 1830°F	—	—	—	—	—	10.2	147.9	5.32	77.2	44	—	—

†At yield point. ‡At 0.2% offset. ††% in 1.5 in. ‡‡% in 3.94 in. §% in 8 in. §§% in 70 mm.
Note: 1 lb/in.^{2} = 6.8947 × 10^4 dyne/cm^2.

9.8* THEORETICAL STRENGTH OF METALS

The theoretical strength of a perfect crystal is of the order of the shear modulus G. This can be demonstrated heuristically as follows. Let a crystal lattice be as shown in Fig. 9.10, with spacing a and b. When a shear stress τ is applied, the lattice distorts. Consider the simultaneous slip of one sheet

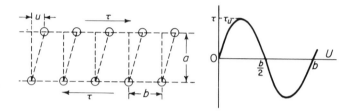

Fig. 9.10 Simultaneous shear of one sheet of atoms in a crystal over another sheet. After A. H. Cottrell, *The Mechanical Properties of Matter* (New York, John Wiley, 1964).

of atoms over the other. The shear stress τ is zero when the atoms stay at the original lattice points; it increases when the displacement u increases; and it will reach a maximum when the shear displacement u is somewhere between 0 and $\tfrac{1}{2}b$, because at $u = \tfrac{1}{2}b$ the arrangement of the atoms will be symmetric and there will be no definite tendency to move either way. If the displacement u exceeds $\tfrac{1}{2}b$, the atoms will be attracted to the nearest lattice position. So the equilibrating shear stress must be negative. At $u = b$ the shear stress must be zero again; and for $u > b$ the pattern must repeat itself. The τ versus u curve must appear roughly as shown in Fig. 9.10 and may be assumed to be of the form

$$(9.8\text{-}1) \qquad \tau = \tau_{\text{ult}} \sin\left(\frac{2\pi u}{b}\right).$$

The initial slope must conform with the shear modulus G. Thus, if e_{xy} is the shear strain which equals $\tfrac{1}{2}(\partial u/\partial y) \doteq \tfrac{1}{2}(u/a)$, we have

$$2G = \left(\frac{d\tau}{de_{xy}}\right)_{u=0} = \left(\frac{d\tau}{du}\right)_{u=0}\left(\frac{du}{de_{xy}}\right) = \frac{2\pi}{b}\tau_{\text{ult}}\cdot 2a;$$

i.e.,

$$(9.8\text{-}2) \qquad \tau_{\text{ult}} = \frac{G}{2\pi}\frac{b}{a}.$$

The quantity τ_{ult} is the largest shear stress that may be sustained; thus it is the ultimate shear strength by definition. It is the *ideal shear strength*. If $a \doteq b$, then $\tau_{\text{ult}} \doteq G/6$.

The values of the ultimate shear strength listed in Table 9.8 are far smaller than the corresponding shear moduli. Thus, the ideal shear strength is not reached. Explanation of this discrepancy leads to the *theory of dislocations*, which is a theory of imperfections in crystals. However, nearly perfect crystals are known in the form of "whisker" crystals, i.e., single crystals in the form of fine fibers. Figure 9.11 shows how a copper whisker attains an unusually high yield stress (about 2% of G), and how it loses its strength as soon as dislocations are made.

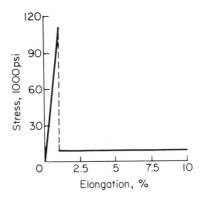

Fig. 9.11   The stress-strain curve of a copper whisker. (From A. H. Cottrell, *ibid*, and after S. Brenner, *J. Appl. Phys.* **24**, (1956) 1484).

The theory of dislocation is one of the most interesting theories in mechanics and in the material sciences. It has received plenty of experimental confirmation and is the principal tool in understanding the structurally sensitive properties of crystalline solids.

9.9* LARGE DEFORMATION. NONLINEAR ELASTICITY

Rubber, the material most qualified to be called "elastic," cannot be described by Hooke's law. Since rubber is capable of large deformation, it is necessary to use the finite-strain analysis, whose details are rather complicated.†

In this section let us describe an example from the living world. Living soft tissues are nonlinearly elastic and somewhat stress-history dependent. For bioengineering it is essential that the stress-strain relationship be known.

Generally speaking, biological solids are multiphase, nonhomogeneous, and anisotropic. A look at any text on anatomy or histology gives the

---

†The reader is referred to the author's book *Foundations of Solid Mechanics*, Chapter 15, for an introduction to the subject.

impression that living tissues are so complex that one may question whether the concept of continuum is applicable. However, the matter really depends on the characteristic scale of length in each problem. If the characteristic length is such that a large number of molecules or cells are enclosed in the smallest dimension concerned, then only the *average* property of the aggregate needs to be considered. For example, if one is concerned with the propagation of long waves in an artery, the artery may be treated as a tube of orthotropic material in the same way that we ignore the individual crystals of metals in structural design and the individual molecules in aerodynamics.

As an example, let us consider a typical connective tissue in an animal: the mesentery of the rabbit. The mesentery is a thin membrane that connects the intestines. It is nearly transparent to the naked eye; it has good, uniform thickness (about 6×10^{-3} cm for the rabbit) and is a favorite of physiologists because its two-dimensional array of small blood vessels is ideal for observation and experimentation. To obtain the gross mechanical property, a strip of uniform width was cut from the mesentery, tied at both ends with fine silk, and tested in simple tension while immersed in a saline solution.†

Figure 9.12 shows a typical load-deflection curve of a specimen when the rate of strain imposed was 0.1 in./min. The ordinate shows the load in grams for a specimen whose cross section in relaxed state is about 10^{-2} cm².

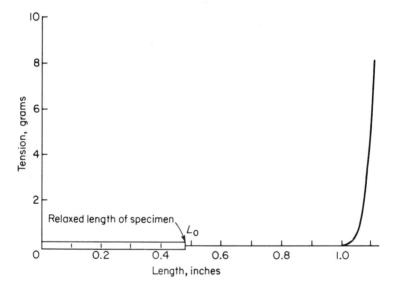

Fig. 9.12   The tension-elongation curve of a mesentery speciman in simple tension.

†See Y. C. Fung, "Elasticity of soft tissues in simple elongation," *American J. of Physiology*, **213**, 6 (1967) 1532–1544.

The abscissa shows the deflection in inches. The completely relaxed length of the specimen was l_0. Note how small the stress was for the first 100% of extension! The load-deflection relationship was definitely nonlinear.

Figure 9.13 shows two typical hysteresis curves of the specimen strained at a rate of ±0.2 and ±0.02 in./min. It is seen that hysteresis exists but is not very large. The effect of strain rate on the hysteresis is seen to be small.

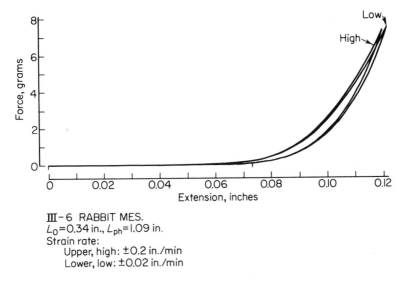

III-6 RABBIT MES.
L_0=0.34 in., L_{ph}=1.09 in.
Strain rate:
    Upper, high: ±0.2 in./min
    Lower, low: ±0.02 in./min

Fig. 9.13   The hysteresis curves for rabbit mesentery in simple elongation.

In reducing the experimental data, the results will be simpler if we use the Lagrangian stress T (obtained by dividing the force by the original cross-sectional area of the specimen in the relaxed state) and the extension ratio λ (the deformed length divided by the relaxed length). The most striking feature of the mechanical behavior of the living mesentery tissue, aside from the extremely small stress in response to a fairly large strain, is revealed when $dT/d\lambda$, the slope of the load-deflection curve, is plotted against the elastic tension T. Figure 9.14 shows such a plot of the slope of the upper curve (loading) in Fig. 9.13. As a first approximation we may fit the experimental curve by a straight line:

$$(9.9\text{-}1) \qquad \frac{dT}{d\lambda} = a(T + \beta), \qquad \text{(for } \lambda < \lambda_y\text{)},$$

where a and β are constants and λ_y indicates the limit of applicability of this equation. The limit λ_y is actually quite low; it is of the order of 2.5 for the mesentery and less than 2 for skin and muscles.

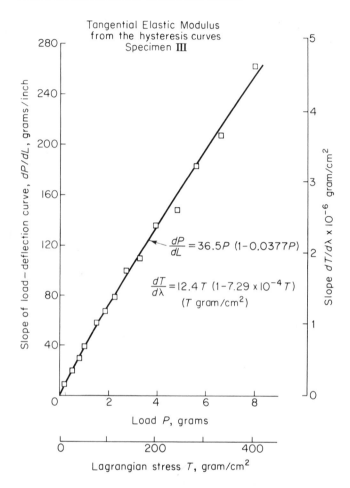

Fig. 9.14 The slope of the load-deflection curve of the rabbit mesentery in simple elongation, plotted against the tension in the specimen.

A simple integration of (9.9-1), together with the condition that the stress must vanish at the natural state so that $T = 0$ when $\lambda = 1$, yields

$$(9.9\text{-}2) \qquad T + \beta = \beta e^{a(\lambda - 1)}, \qquad \text{(for } \lambda = \lambda_y\text{),}$$

which is remarkable indeed. This formula, however, is not convenient for the mesentery for which the constant β is very small and difficult to be ascertained experimentally. For the mesentery, it is more convenient to fix the constant of integration by identifying a point on the stress-strain curve, $T = T^*$ when $\lambda = \lambda^*$, and to calculate a hypothetical constant β which fulfills the condition $T = 0$ when $\lambda = 1$. The result is, for $\lambda < \lambda_y$,

(9.9-3) $T + \beta = (T^* + \beta)e^{a(\lambda - \lambda^*)}, \qquad \beta = \dfrac{T^* e^{-a(\lambda^* - 1)}}{1 - e^{-a(\lambda^* - 1)}}.$

Several other types of soft tissues, such as the skin, the muscles, the ureter, etc., are found to follow similar relationships. Thus, it appears that the exponential type of material is natural in the biological world.

9.10* VISCOELASTICITY

The features of hysteresis, relaxation, and creep are common to many materials. Collectively, they are called the features of viscoelasticity. High polymers show viscoelasticity in a striking way. For example, silicone putty flows when it is left alone but bounces like a rubber ball when loaded quickly. To some extent, metals show these features too. The very fact that all free vibrations in musical instruments ultimately die out is an indication that energy dissipation mechanisms exist. Viscoelasticity of the material is a principal cause of the energy dissipation.

Maxwell (1831–1879) considered viscoelasticity of metals when he designed instruments to measure the viscosity of gases. Lord Kelvin (William Thomson, 1824–1907) studied the damping characteristics of metal wires and rubber rods in torsional oscillations in connection with his effort to formulate the laws of thermodynamics. Zener, Kê, and other solid-state physicists study viscoelasticity to determine the characteristic frequencies associated with various relaxation mechanisms, thus deducing the nature of the crystal grain boundaries, thermal currents, flow of interstitial atoms, vacancies, dislocations, etc. Rocket engineers study viscoelasticity to predict the behavior of solid propellants, including such problems as time-dependent deformation of the propellant grain, stressing under dynamic loading, performance at high temperature, generation of heat in vibrational environment, etc.

Mechanical models are often used to discuss the viscoelastic behavior of materials. In Fig. 9.15 are shown three mechanical models of material behavior, namely, the Maxwell model, the Voigt model, and the "standard linear" model, all of which are composed of combinations of linear springs with spring canstant μ and dashpots with coefficient of viscosity η. A *linear spring* is supposed to produce instantaneously a deformation proportional to

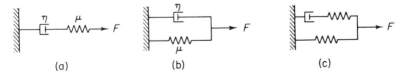

(a)   (b)   (c)

Fig. 9.15 Models of linear viscoelasticity: (a) Maxwell, (b) Voigt, (c) standard linear solid.

the load. A *dashpot* is supposed to produce a velocity proportional to the load at any instant. The relationships between the load F and the deflection u at the point of loading are

(9.10-1)   Maxwell model:

$$\dot{u} = \frac{\dot{F}}{\mu} + \frac{F}{\eta}, \qquad u(0) = \frac{F(0)}{\mu},$$

(9.10-2)   Voigt model:

$$F = \mu u + \eta \dot{u}, \qquad u(0) = 0,$$

(9.10-3)   Standard linear model:

$$F + \tau_\epsilon \dot{F} = E_R(u + \tau_\sigma \dot{u}), \qquad \tau_\epsilon F(0) = E_R \tau_\sigma u(0),$$

where $\tau_\epsilon, \tau_\sigma$ are two constants. When these equations are to be integrated, the initial conditions at $t = 0$ must be prescribed as indicated above.

If we solve Eq. (9.10-1) through (9.10-3) for $u(t)$ when $F(t)$ is a unit-step function $\mathbf{1}(t)$, the result is called the *creep-function*, which represents the elongation produced by a sudden application at $t = 0$ of a constant force of magnitude unity. They are:

(9.10-4)   Maxwell solid:

$$c(t) = \left(\frac{1}{\mu} + \frac{1}{\eta} t \right) \mathbf{1}(t),$$

(9.10-5)   Voigt solid:

$$c(t) = \frac{1}{\mu}(1 - e^{-(\mu/\eta)t}) \mathbf{1}(t),$$

(9.10-6)   Standard linear solid:

$$c(t) = \frac{1}{E_R} \left[1 - \left(1 - \frac{\tau_\epsilon}{\tau_\sigma} \right) e^{-t/\tau_\sigma} \right] \mathbf{1}(t),$$

where the *unit-step function* $\mathbf{1}(t)$ is defined as

(9.10-7) $$\mathbf{1}(t) = \begin{cases} 1 \text{ when } t > 0, \\ \frac{1}{2} \text{ when } t = 0, \\ 0 \text{ when } t < 0. \end{cases}$$

A body which obeys a load-deflection relation like that given by Maxwell's model is said to be a Maxwell solid. Since a dashpot behaves as a piston moving in a viscous fluid, the above-named models are called models of *viscoelasticity*.

Interchanging the roles of F and u, we obtain the *relaxation function* as a response $F(t) = k(t)$ corresponding to an elongation $u(t) = 1(t)$. The relaxation function $k(t)$ is the force that must be applied in order to produce an elongation which changes at $t = 0$ from zero to unity and remains unity thereafter. They are

(9.10-8)   Maxwell solid:

$$k(t) = \mu e^{-(\mu/\eta)t}1(t),$$

(9.10-9)   Voigt solid:

$$k(t) = \eta\delta(t) + \mu 1(t),$$

(9.10-10)   Standard linear solid:

$$k(t) = E_R\left[1 - \left(1 - \frac{\tau_\sigma}{\tau_\epsilon}\right)e^{-t/\tau_\epsilon}\right]1(t).$$

Here we have used the symbol $\delta(t)$ to indicate the *unit-impulse function*, or *Dirac-delta function*, which is defined as a function with a singularity at the origin:

$$\delta(t) = 0 \qquad \text{(for } t < 0, \text{ and } t > 0),$$

(9.10-11)
$$\int_{-\epsilon}^{\epsilon} f(t)\,\delta(t)\,dt = f(0) \qquad (\epsilon > 0),$$

where $f(t)$ is an arbitrary function continuous at $t = 0$. These functions, $c(t)$ and $k(t)$, are illustrated in Fig. 9.16 and 9.17, respectively, for which we add the following comments.

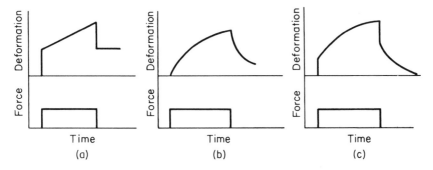

Fig. 9.16 Creep function of (a) Maxwell, (b) Voigt, (c) standard linear solid. A negative phase is superposed at the time of unloading.

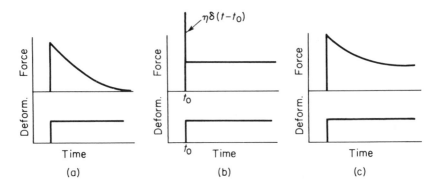

Fig. 9.17 Relaxation functions of (a) Maxwell, (b) Voigt, and (c) standard linear solid.

For the Maxwell solid, a sudden application of a load induces an immediate deflection by the elastic spring, which is followed by "creep" of the dashpot. On the other hand, a sudden deformation produces an immediate reaction by the spring, which is followed by stress relaxation according to an exponential law (9.10-8). The factor η/μ, with dimension of time, may be called a *relaxation time*: It characterizes the rate of decay of the force.

For a Voigt solid, a sudden application of force will produce no immediate deflection because the dashpot, arranged in parallel with the spring, will not move instantaneously. Instead, as shown by Eq. (9.10-5) and Fig. 9.16(b), a deformation will be gradually built up, while the spring takes a greater and greater share of the load. The dashpot displacement relaxes exponentially. Here the ratio η/μ is again a relaxation time: It characterizes the rate of relaxation of the deflection.

For the standard linear solid, a similar interpretation is applicable. The constant τ_ϵ is the time of relaxation of load under the condition of constant deflection [see Eq. (9.10-10)], whereas the constant τ_σ is the time of relaxation of deflection under the condition of constant load [see Eq. (9.10-6)]. As $t \rightarrow \infty$, the dashpot is completely relaxed, and the load-deflection relation becomes that of the springs, as is characterized by the constant E_R in Eq. (9.10-6) and (9.10-10). Therefore, E_R is called the *relaxed elastic modulus*.

Maxwell introduced the model represented by Eq. (9.10-1) with the idea that all fluids are elastic to some extent. Lord Kelvin showed the inadequacy of the Maxwell and Voigt models in accounting for the rate of dissipation of energy in various materials subjected to cyclic loading. Kelvin's model is commonly called the standard linear model because it is the most general relationship to include the load, the deflection, and their first (commonly called "linear") derivatives.

More general models may be built by adding more and more elements to the Kelvin model. Equivalently, we may add more and more exponential terms to the creep function or to the relaxation function.

The most general formulation under the assumption of linearity between cause and effect is due to Boltzmann (1844–1906). In the one-dimensional case, we may consider a simple bar subjected to a force $F(t)$ and elongation $u(t)$. The elongation $u(t)$ is caused by the total history of the loading up to the time t. If the function $F(t)$ is continuous and differentiable, then in a small time interval $d\tau$ at time τ the increment of loading is $(dF/d\tau)\, d\tau$. This increment remains acting on the bar and contributes an element $du(t)$ to the elongation at time t, with a proportionality constant c depending on the time interval $t - \tau$. Hence, we may write

$$(9.10\text{-}12) \qquad du\,(t) = c(t - \tau)\frac{dF(\tau)}{d\tau}\,d\tau.$$

Let the origin of time be taken at the beginning of motion and loading. Then, on summing over the entire history, which is permitted under Boltzmann's hypothesis, we obtain

$$(9.10\text{-}13) \qquad u(t) = \int_0^t c(t - \tau)\frac{dF(\tau)}{d\tau}\,d\tau.$$

A similar argument, with the role of F and u interchanged, gives

$$(9.10\text{-}14) \qquad F(t) = \int_0^t k(t - \tau)\frac{du\,(\tau)}{d\tau}\,d\tau.$$

These laws are linear, since doubling the load doubles the elongation, and vice versa. The functions $c(t - \tau)$ and $k(t - \tau)$ are the *creep* and *relaxation* functions, respectively.

The Maxwell, Voigt, and Kelvin models are special examples of Boltzmann formulation. More generally, we can write the relaxation function in the form

$$(9.10\text{-}15) \qquad k(t) = \sum_{n=0}^{N} \alpha_n e^{-t v_n},$$

which is a generalization of Eq. (9.10-10). If we plot the amplitude α_n associated with each characteristic frequency v_n on a frequency-axis, we obtain a series of lines which resemble an optical spectrum. Hence $\alpha_n(v_n)$ is called a spectrum of the relaxation function. The example shown in Fig. 9.18 is a *discrete* spectrum. A generalization to a continuous spectrum may

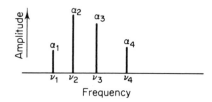

Fig. 9.18   A discrete spectrum of relaxation function.

be desired sometimes. In the case of a living tissue such as mesentery, experimental results on relaxation, creep, and hysteresis cannot be reconciled unless a continuous spectrum is assumed.

9.11* NON-NEWTONIAN FLUIDS

Newton's law of viscosity describes the behavior of water very well, but there are many other fluids that behave differently. Read the advertisements of some paints: "No drip (will not flow on the brush), spreads easily (little resistance to flow), leaves no brush marks (flows to smooth off the surface)." These desirable features for household paints are not Newtonian. Most paints, enamels, and varnishes, are non-Newtonian. Most polymer solutions are non-Newtonian.

Let us illustrate the subject with a fluid that is most important to our lives—blood. The composition of blood is complicated: it contains red blood cells (normally of the order of 45 to 50% by volume), which carry hemoglobin, which in turn is the vehicle for oxygen transport. If a tube of blood is put in a centrifuge and spun for some time, the red cells, being a little heavier, will be packed at the bottom of the tube. Above this lies a thin layer of white cells, whose specific gravity is less than that of the red cells. The upper layer is a clear fluid called plasma. Plasma is about 90% water by weight, 7% plasma protein, 1% inorganic substance, and another 1% of organic substances. There are enough "free" molecules and ions to give a total osmotic pressure of about 8 atmospheres, equivalent to a sodium chloride solution of 0.9% by weight (physiological saline).

The viscosity of the blood depends on the strain rate. Figure 9.19 shows the variation of the coefficient of viscosity with the strain rate as measured with a Couette viscometer by Chien, Usami, Gregersen, et al. The coefficient of viscosity increases as the strain rate decreases below about 100 sec^{-1}. At very low strain rate the blood seems to have a finite "yield" stress; i.e., it seems to be visco-plastic, but the point is yet unsettled.

The world of non-Newtonian fluids is so much larger than that of the Newtonian fluids that the landscape is yet largely unexplored.

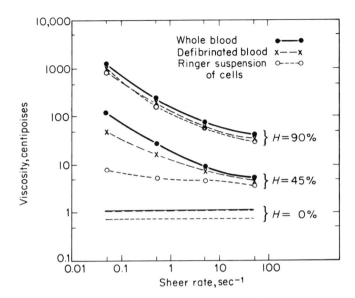

Fig. 9.19 The variation of the coefficient of viscosity with the strain rate in human blood, showing data for whole blood, defibrinated blood, and washed red blood cells resuspended in a Ringer solution, at 45 and 90% hematocrit H, (red cell concentrations by volume). From S. Chien, S. Usami, H. M. Taylor, J. L. Lundberg and M. I. Gregersen, *J. Appl. Physiol.*, **21**, (1966), p. 81, and M. I. Gregersen: "Factors regulating blood viscosity: relation to problems of the microcirculation". *Les Concepts de Claude Bernard sur le Milieu lutérieur* (Paris: Masson, 1967).

9.12* VISCO-PLASTIC MATERIALS

A material obeying Newton's law of viscosity must flow under the slightest shear stress (more precisely, under a nonvanishing stress deviation). Materials such as sour dough, paste, moulding clay, etc., do not follow such a rule. Bingham, who invented the word "rheology" to describe the science of flow (Greek, $\rho\epsilon os$ flow), formulated a law for a class of materials known as visco-plastic, to which sour dough seems to belong. A visco-plastic material is often called a Bingham plastic.

A visco-plastic material can sustain stresses with nonvanishing stress deviator when in a state of rest. Consider first a body subjected to simple shear, i.e., a state in which all components of the tensors of stress and strain rate vanish except $\sigma_{12} = \sigma_{21} = \tau$, and $V_{12} = V_{21} = \dot{e}$. As long as the absolute value of the shearing stress τ is smaller than a certain constant K called the *yield stress*, the material remains rigid, so that $\dot{e} = 0$. As soon as $|\tau|$ exceeds K, however, the material flows, with a strain rate $\dot{e}$ having the same sign as τ

and an absolute value proportional to $|\tau| - K$. Thus,

$$(9.12\text{-}1) \qquad 2\mu\dot{e} = \begin{cases} 0 & \text{if } |\tau| < K, \\ \left(1 - \dfrac{K}{|\tau|}\right)\tau & \text{if } |\tau| > K, \end{cases}$$

where μ is a coefficient of viscosity. The formulation may be written slightly differently with the introduction of a *yield function* F defined as

$$(9.12\text{-}2) \qquad F = 1 - \frac{K}{|\tau|}.$$

Then a visco-plastic material in a state of simple shear is defined by Bingham with the relations

$$(9.12\text{-}3) \qquad 2\mu\dot{e} = \begin{cases} 0 & \text{if } F < 0, \\ F\tau & \text{if } F \geqslant 0. \end{cases}$$

Hohenemser and Prager generalized Bingham's definition to arbitrary states of stress in the following form:

$$(9.12\text{-}4) \qquad 2\mu V_{ij} = \begin{cases} 0 & \text{for } F < 0, \\ F\sigma'_{ij} & \text{for } F \geqslant 0, \end{cases}$$

with

$$(9.12\text{-}5) \qquad F = 1 - \frac{K}{\sqrt{J_2}},$$

where

$\mu =$ coefficient of viscosity,

$V_{ij} =$ strain-rate tensor (see Sec. 6.1),

$\sigma'_{ij} =$ stress-deviator tensor $= \sigma_{ij} - \frac{1}{3}\sigma_{\alpha\alpha}\delta_{ij}$,

$K =$ yield stress,

$J_2 =$ second invariant of the stress deviator

$= \frac{1}{6}[(\sigma_{11} - \sigma_{22})^3 + (\sigma_{22} - \sigma_{33})^2 - (\sigma_{33} - \sigma_{11})^2] + \sigma_{12}^2 + \sigma_{23}^2 + \sigma_{31}^2.$

For simple shear, Eq. (9.12-4) and (9.12-5) reduce to Eq. (9.12-3) and (9.12-2), respectively.

According to Eq. (9.12-4) the rate of deformation tensor in a visco-plastic material is a deviator; i.e., the material is incompressible. When the yield function is negative, the material is rigid. Flow occurs when the yield function has a positive value. The state of stress for which $F = 0$ forms the *yield limit* at which visco-plastic flow sets in or ceases, depending on the sense in which the yield limit is crossed.

Further generalizations of Bingham's equation (9.12-3) are possible. For example, compressibility may be introduced, or other yield criteria may be proposed instead of Eq. (9.12-5). However, it suffices for our purpose here to know that such constitutive laws exist.

A visco-plastic material is a non-Newtonian fluid. Figure 9.20 illustrates the point.

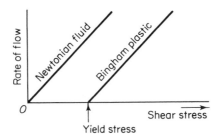

Fig. 9.20 Comparison of the flow rate and stress relationship of a visco-plastic material with that of a Newtonian fluid.

9.13* SOL-GEL TRANSFORMATION.   THIXOTROPY

A colloidal solution may possess rigidity (subjected to shear stress without flow) and be called a *gel*, or it may behave as a fluid without rigidity and be called a *sol*. A gel contains a dispersed component and a dispersion medium, both of which extend continuously throughout the system. The elastic property of the gel may change with its age. The dispersed component of a gel is usually interpreted as forming a three-dimensional network held together by bonds or junction points whose lifetimes are essentially infinite. The junction points may be formed by primary valence bonds, long-range attractive forces, or secondary valence bonds that cause association between segments of polymer chains or formation of submicroscopic crystalline regions. Each of the junctions is a mechanism for relaxation under stress. The statistics of the totality of all these relaxation mechanisms is described by the viscoelasticity of the material.

Certain colloidal systems, such as moderately concentrated polymer solutions, behave as gels but have no yield point and cannot resist shear stress statically. The junction points of the network of the dispersed component are transient. These are viscoelastic materials of the Maxwell type or some of its generalizations.

Gels often can be converted into sols or vice versa by a change of temperature, by agitation, or by chemical action in a process called peptization.

If a reversible gel-sol transformation can be induced isothermally by mechanical vibration, then the material is said to be *thixotropic* according to Freundlich. The gel is transformed into a sol by mechanical agitation, and the sol reverts to a gel when the agitation is discontinued.

Examples of thixotropic substances are paints, printing inks, iron oxide sols, agar, suspensions of kaolin, carbon black, etc. Thixotropic materials pervail in the biological world. The protoplasm in amoeba is perhaps the best known example.

Whether a colloidal system is thixotropic or not may depend on small changes in ionic strength. J. Pryce-Jones has found for mixtures of titanium oxide and various oils that the degree of thixotropy, represented by the variation in viscosity with time after agitation, was much affected by the composition of the oil.† Changes as small as 1 % in the free acid content of the oil can change a free-flowing paint into a highly thixotropic system, the solid content remaining the same.

These examples show the great variety of materials that exist.

PROBLEMS

9.1 Verify that the velocity of sound in the material of construction for an aircraft is a criterion for its safety against such dynamic problems as clear air turbulence, gust encounter, and flutter. For this purpose, let us consider two airplanes identical in geometry and construction but different in material. Simplify the problem to consider only the following four typical parameters: the density of the material σ, the Young's modulus E; the density of air ρ, the velocity of flight of the airplane U. Use dimensional analysis to construct the similarity parameters. Let σ, E, ρ, and U refer to one plane, and σ', E', ρ', and U' refer to the other. Show that for dynamic similarity, we must have

$$\frac{U'}{U} = \sqrt{\frac{E'}{\sigma'}} \bigg/ \sqrt{\frac{E}{\sigma}}.$$

If U represents the limit of the safe flight speed (e.g., the critical flutter speed), then the formula above relates U to the velocity of sound $\sqrt{E/\sigma}$ (speed of longitudinal waves in a rod).

9.2 The velocity of sound in a solid is an important similarity parameter for comparing the rigidity of a flight structure. Suppose you are an airplane designer selecting materials for construction. Using a handbook, list the velocity of sound for the following structural materials: pure aluminum, magnesium, aluminum alloys, magnesium alloys, carbon steels, stainless steels, titanium, titanium carbide, and the rather exotic materials beryllium oxide and pure beryllium. Compare with the plastics lucite and phenolic lami-

†J. Pryce-Jones, *J. Oil and Colour Chem. Assn.*, **295** (1936).

nates and the woods spruce, mahogany, balsa, and bamboo, along the grain. Are you not surprised at the rather small differences between the velocities of sound for many of these materials? What is the best material from this point of view?

9.3 Show that the same conclusion as in Prob. 9.1 would be reached if you consider a suspension bridge which may be induced to vibrate dangerously in wind. [The original Tacoma Narrows Bridge on Puget Sound, Washington, spectacularly failed by flutter on Nov. 7, 1940, four months after the bridge was opened to traffic, in a wind speed of 42 mph. On that morning, the frequency of oscillation of the bridge changed suddenly from 37 to 14 cycles/min, probably because of a failure of a small reinforcing tie rod. The motion grew violently in the torsional mode, and failure occurred half an hour later. If there had not been this aerodynamically induced oscillation (flutter), the bridge should have been able to withstand a steady wind of at least 100 mi/hr.]

9.4 The experimental data on the viscosity of blood measured in a Couette flowmeter (Fig. P3.22, p. 88), as shown in Fig. 9.19, p. 233, can be expressed approximately by Casson's equation

$$\sqrt{\tau} = \sqrt{\tau_y} + b\sqrt{\dot{\gamma}}$$

in which τ is the shear stress, τ_y is a constant that may be identified as a yield stress, and $\dot{\gamma}$ is the shear strain rate (sec^{-1}.) Generalize this result to a constitutive equation for blood that is correct from the point of view of dimensional and tensor analyses.

9.5 Consider Lyman Briggs' experiment on measuring the tensile strength of liquids, described in Ref. 3, p. 205, Table 9.2. See Method C-3 at the bottom of p. 205. If one uses a straight tube which is open at both ends for this experiment, we know that the fluid will fly away and the experiment would not be possible. The bent-back short arms of the Z provide the stability of the fluid. Examine this stability problem and present a theoretical basis for this experiment.

9.6 Assume that no material will expand in volume when it is subjected to a hydrostatic pressure. Show that the maximum value of the Poisson's ratio ν for any isotropic elastic solid obeying Hooke's law is $\frac{1}{2}$.

9.7 Reinforced concrete is concrete poured over steel rods. A vertical, hollow, reinforced concrete column has an internal diameter of 3 ft and thickness of 3 in., with 36 steel rods of 1 sq in. cross-sectional area, uniformly spaced in a circle. The column is subjected to a vertical load, the resultant of which is along the axis of the column. The ratio of Young's modulus of steel to that of concrete is 15. The Poisson's ratio for concrete is 0.4, that for steel is 0.25. Determine the share of the load that is carried by steel at a cross section at some distance from the ends of the column.

9.8 Consider a viscoelastic material characterized as Maxwell's model and described by Eq. (9.10-1), p. 228. Let a sinusoidally varying force $F = a \sin \omega t$ be imposed on the body. What would be the deflection u at steady-state?

Answer:

$$u = \frac{A}{\omega}[\sin{(\omega t - \alpha)} + \sin{\alpha}],$$

where $A = \left[\left(\frac{a\omega}{\mu}\right)^2 + \left(\frac{a}{\eta}\right)^2\right]^{1/2}$, $\tan{\alpha} = \frac{\mu}{\eta\omega}$.

9.9 A liquid flows down a long tube of diameter 1 cm from a reservoir at a rate of 10 cm³/sec. The streamlines are found to be as shown in Fig. P9.9. The principal feature is that the liquid column expands in diameter as it leaves the tube. Can a Newtonian liquid do this? What kind of stress-strain relationship is suggested? [See A. S. Lodge, *Elastic Liquids*, New York: Academic Press (1964), p. 242.]

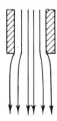

Fig. P9.9   A non-Newtonian fluid coming out of a spout.

9.10 When certain paint was stirred with an electric mixer, it was found that it climbs up the shaft of the mixer. What kind of stress-strain relationship of the paint is revealed by this experiment? (See A. S. Lodge, *ibid*, p. 232.)

9.11 Take a piece of chalk and twist it to failure. Describe the cleavage surface and infer the criterion about the strength of the chalk.

Again, break the chalk by bending and discuss the fracture mechanism.

9.12 Take a piece of nylon thread, pull it to failure, and discuss the failure mechanism of nylon vs. that of the chalk in Prob. 9.11.

9.13 Experiment on a rubber balloon and discuss the phenomena of inflation (pressure-volume relationship) and the final failure.

Take a needle and poke a hole in a rubber balloon before it is inflated. Then stretch the balloon to see if the pin hole has any effect.

Stretch the balloon so that it elongates in one direction (the so-called "uni-axial" tension state) and poke a hole with the needle.

Inflate the balloon with air until it is taut (in a so-called bi-axial tension state) and then poke a hole with the needle. Do you see any difference? Offer an explanation.

FURTHER READING

COTTRELL, A. H., *The Mechanical Properties of Matter*, New York: Wiley (1964).

EIRICH, F., *Rheology*, New York: Academic Press (1956).

FUNG, Y. C., "Biomechanics. Its scope, history, and some problems of continuum in physiology," *Applied Mechanics Reviews*, **21**, 1 (1968), 1–20.

JEANS, SIR JAMES, *An Introduction to the Kinetic Theory of Gases*, Cambridge: University Press (1946).

KINNEY, G. F., *Engineering Properties and Applications of Plastics*, New York: Wiley (1957).

LEVICH, V. G., *Physicochemical Hydrodynamics*, Englewood Cliffs, N.J.: Prentice-Hall (1962).

NADAI, A., *Theory of Flow and Fracture of Solids* (2 vol.), New York: McGraw-Hill (1963).

PAULING, L., *The Chemical Bond. A Brief Introduction to Modern Structural Chemistry*, Ithaca: Cornell University Press (1966).

ZENER, C., *Elasticity and Anelasticity of Metals*, Chicago: University of Chicago Press (1948).

CHAPTER TEN

Derivation
of Field Equations

*In the preceding chapters we have analyzed deformation (strain)
and flow (strain rate) and their relationship with the interaction
(stress) between parts of a material body (continuum). We are
now in a position to use this information to derive differential
equations describing the motion of the continuum under specific
boundary conditions. Our formulation must obey Newton's law
of motion, the principle of conservation of mass, and the laws of
thermodynamics. This chapter is concerned with expressing these
laws in a form suitable for the treatment of a continuum.*

*One may wonder why there is a need for further elaboration
on these well-known laws. The answer may be illustrated in the
following example. If we have a single particle, the principle of
conservation of mass merely states that the mass of the particle is
a constant. However, if we have a large number of particles, such
as the water droplets in a cloud, the situation requires some
thought. For the cloud it is no longer practical to identify the
individual particles. The most convenient way to describe the
cloud is to consider the velocity field, the density distribution,
the temperature distribution, etc. It is the description of the
classical conservation laws in such a circumstance that shall
occupy our attention in this chapter.*

Our approach is based on the fact that these conservation laws must be applicable to the matter enclosed in a volume bounded by an arbitrary closed surface. In such an approach we find that some quantities enter naturally in a surface integral, others in a volume integral. A transformation from a surface integral to a volume integral, and vice versa, is often required. This transformation is embodied in Gauss' theorem, which serves as our mathematical starting point.

10.1 GAUSS' THEOREM

We shall begin with the derivation of Gauss' theorem. Consider a convex region V bounded by a surface S that consists of a finite number of parts whose outer normals form a continuous vector field (e.g., the one shown in Fig. 10.1). Such a region is said to be *regular*. Let a function $A(x_1, x_2, x_3)$ be defined in the volume V and on the surface S. Let A be continuously differentiable in V. Let us consider the volume integral

$$\iiint_V \frac{\partial A}{\partial x_1} \, dx_1 \, dx_2 \, dx_3.$$

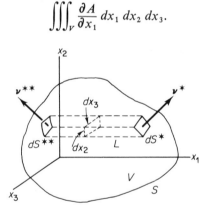

Fig. 10.1 Path of integration illustrating the derivation of Gauss' theorem.

The integrand is the partial derivative of A with respect to x_1. By integrating with respect to x_1 along a line segment L we obtain

(10.1-1) $$\iiint_V \frac{\partial A}{\partial x_1} \, dx_1 \, dx_2 \, dx_3 = \iint_S (A^* - A^{**}) \, dx_2 \, dx_3$$

where A^* and A^{**} are, respectively, the values of A on the surface S at the right and left ends of the line segment L parallel to the x_1-axis. The surface integral on the right-hand side of (10.1-1) may be written more elegantly by the following reasoning. The factors $+dx_2\, dx_3$ and $-dx_2\, dx_3$ are the projections on the $x_2 x_3$-plane of the areas dS^* and dS^{**} at the ends of the line segment L. Let $\mathbf{v} = (v_1, v_2, v_3)$ be the unit vector along the outer normal to the surface S. For the element shown in Fig. 10.1 we see that $v_1^* = \cos(x_1, \mathbf{v}^*)$ is positive, whereas $v_1^{**} = \cos(x_1, \mathbf{v}^{**})$ is negative. It is easy to see that in this case $dx_2\, dx_3 = v_1^*\, dS^*$ at the right end and $-dx_2\, dx_3 = v_1^{**}\, dS^{**}$ at the left end. Therefore, the surface integral in Eq. (10.1-1) can be written as

$$(10.1\text{-}2) \qquad \iint_S (A^*\, dx_2\, dx_3 - A^{**}\, dx_2\, dx_3) = \iint_S (A^* v_1^*\, dS^* + A^{**} v_1^{**}\, dS^{**}).$$

The asterisks may be omitted because they merely indicate the appropriate values of A and v_1 to be taken in a surface integral according to conventional notations. Thus, the right-hand side of Eq. (10.1-1) is reduced to $\int_S A v_1\, dS$. If we write the volume integral on the left-hand side of (10.1-1) as $\int_V (\partial A/\partial x_1)\, dV$, then we have

$$(10.1\text{-}3) \qquad \int_V \frac{\partial A}{\partial x_1}\, dV = \int_S A v_1\, dS,$$

where dV and dS denote the elements of V and S, respectively. A similar argument applies to the volume integral of $\partial A/\partial x_2$ or $\partial A/\partial x_3$. In summary, we obtain Gauss' theorem,

$$(10.1\text{-}4) \quad \blacktriangle \qquad \int_V \frac{\partial A}{\partial x_i}\, dV = \int_S A v_i\, dS, \qquad (i = 1, 2, 3).$$

This formula holds for any convex regular region or for any region that can be decomposed into a finite number of convex regular regions.

Now let us consider a tensor field $A_{jkl\ldots}$. Let the region V with boundary surface S be within the region of definition of $A_{jkl\ldots}$. Let every component of $A_{jkl\ldots}$ be continuously differentiable in V. Then Eq. (10.1-4) is applicable to every component of the tensor, and we obtain the general result

$$(10.1\text{-}5) \qquad \int_V \frac{\partial}{\partial x_i} A_{jkl\ldots}\, dV = \int_S v_i A_{jkl\ldots}\, dS,$$

which is one of the most useful theorems in applied mathematics.

The attribution of a name to this theorem is difficult. In various forms this theorem was given by Lagrange (1762), Gauss (1813), Green (1828), and Ostrogradsky (1831). It is best known in this country as *Green's theorem* or *Gauss' theorem*.

Example 1

Let v_i represent a vector. Then, according to (10.1-5), we have, on identifying $A_i = v_i$,

(10.1-6)
$$\int_V \frac{\partial v_i}{\partial x_i} \, dV = \int_S v_i v_i \, dS.$$

If we write the coordinates x_1, x_2, x_3 as x, y, z; the components v_1, v_2, v_3 as u, v, w; and the direction cosines v_1, v_2, v_3 of the outer normal to the surface S as l, m, n; then

(10.1-7)
$$\iiint_V \left(\frac{\partial u}{\partial x} + \frac{\partial v}{\partial y} + \frac{\partial w}{\partial z} \right) dx \, dy \, dz = \iint_S (lu + mv + nw) \, dS.$$

In another popular notation, we denote the vector by $\mathbf{v}$, the scalar product $v_i v_i$ by $\mathbf{v} \cdot \mathbf{v}$, and define

(10.1-8)
$$\operatorname{div} \mathbf{v} = \frac{\partial u}{\partial x} + \frac{\partial v}{\partial y} + \frac{\partial w}{\partial z}.$$

Then (10.1-7) becomes

(10.1-9)   ▲
$$\int_V \operatorname{div} \mathbf{v} \, dV = \int_S \mathbf{v} \cdot \mathbf{v} \, dS.$$

Equations (10.1-6), (10.1-7) and (10.1-9) are the best known forms of Gauss' theorem.

Example 2

If A is identified with a potential function ϕ, then Eq. (10.1-3) is usually written in the vector form

$$\int_V \operatorname{grad} \phi \, dV = \int_S \mathbf{v} \phi \, dS.$$

Example 3

Let e_{ijk} be the permutation tensor. Then

$$\int e_{ijk} u_{k,j} \, dV = e_{ijk} \int u_{k,j} \, dV = e_{ijk} \int u_k v_j \, dS = \int e_{ijk} u_k v_j \, dS;$$

i.e.

$$\int \operatorname{curl} \mathbf{u} \, dV = \int \mathbf{v} \times \mathbf{u} \, dS.$$

10.2 MATERIAL DESCRIPTION OF THE MOTION
OF A CONTINUUM

Let a fixed frame of reference $O\text{-}x_1x_2x_3$ be chosen. Let the location of a material particle be $x_1 = a_1$, $x_2 = a_2$, $x_3 = a_3$ when time $t = t_0$. We shall use (a_1, a_2, a_3) as the label for that particle. As time goes on the particle moves: its location has the history

$$x_1 = x_1(a_1, a_2, a_3, t), \qquad x_2 = x_2(a_1, a_2, a_3, t), \qquad x_3 = x_3(a_1, a_2, a_3, t)$$

referred to the same coordinate system or, in short,

$$(10.2\text{-}1) \qquad x_i = x_i(a_1, a_2, a_3, t), \qquad (i = 1, 2, 3).$$

If such an equation is known for every particle in the body, then we know the history of motion of the entire body. Mathematically, Eq. (10.2-1) defines the *transformation*, or *mapping*, of a domain $D(a_1, a_2, a_3)$ into a domain $D'(x_1, x_2, x_3)$, with t as a parameter. An example is shown in Fig. 10.2. If the

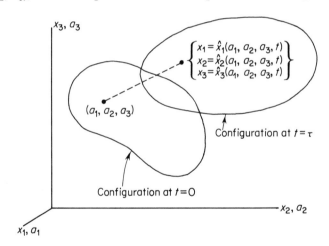

Fig. 10.2 Labeling of particles.

mapping is continuous and one-to-one; i.e., for every point (a_1, a_2, a_3) there is one and only one point (x_1, x_2, x_3) and vice versa, and neighboring points in $D(a_1, a_2, a_3)$ are mapped into neighboring points in $D'(x_1, x_2, x_3)$, then the functions $x_i(a_1, a_2, a_3, t)$ must be single-valued, continuous, and continuously differentiable, and the Jacobian must not vanish in the domain D.

The mapping (10.2-1) is said to be a *material description* of the motion of the body. In a material description, the velocity and acceleration of the particle (a_1, a_2, a_3) are, respectively,

$$(10.2\text{-}2) \qquad v_i(a_1, a_2, a_3, t) = \frac{\partial x_i}{\partial t}\bigg|_{(a_1, a_2, a_3)},$$

$$(10.2\text{-}3) \qquad \alpha_i(a_1, a_2, a_3, t) = \frac{\partial v_i}{\partial t}\bigg|_{(a_1, a_2, a_3)} = \frac{\partial^2 x_i}{\partial t^2}\bigg|_{(a_1, a_2, a_3)}.$$

Conservation of mass may be expressed as follows. Let $\rho(\mathbf{x})$ be the density of the material at location $\mathbf{x}$, where the symbol $\mathbf{x}$ stands for (x_1, x_2, x_3). Let $\rho_0(\mathbf{a})$ be the density at the piont (a_1, a_2, a_3) when $t = 0$. Then the mass of the material enclosed in a volume V is $\int_D \rho_0(\mathbf{a})\, da_1\, da_2\, da_3$ at $t = 0$, and $\int_{D'} \rho(\mathbf{x})\, dx_1\, dx_2\, dx_3$ at time t. Thus, conservation of mass is expressed by the formula

$$(10.2\text{-}4) \qquad \int_{D'} \rho(\mathbf{x})\, dx_1\, dx_2\, dx_3 = \int_D \rho_0(\mathbf{a})\, da_1\, da_2\, da_3,$$

where the integrals extend over the same particles. But

$$(10.2\text{-}5) \qquad \int_{D'} \rho(\mathbf{x})\, dx_1\, dx_2\, dx_3 = \int_D \rho(\mathbf{x})\left|\frac{\partial x_i}{\partial a_j}\right| da_1\, da_2\, da_3,$$

where $|\partial x_i/\partial a_j|$ is the Jacobian of the transformation, i.e., the determinant of the matrix $(\partial x_i/\partial a_j)$:

$$(10.2\text{-}6) \qquad \left|\frac{\partial x_i}{\partial a_j}\right| = \begin{vmatrix} \partial x_1/\partial a_1 & \partial x_1/\partial a_2 & \partial x_1/\partial a_3 \\ \partial x_2/\partial a_1 & \partial x_2/\partial a_2 & \partial x_2/\partial a_3 \\ \partial x_3/\partial a_1 & \partial x_3/\partial a_2 & \partial x_3/\partial a_3 \end{vmatrix}.$$

Identifying the right-hand sides of (10.2-4) and (10.2-5) and realizing that the result must hold for any arbitrary domain D, we see that the integrands must be equal:

$$(10.2\text{-}7) \qquad \rho_0(\mathbf{a}) = \rho(\mathbf{x})\left|\frac{\partial x_i}{\partial a_j}\right|.$$

Similarly,

$$(10.2\text{-}8) \qquad \rho(\mathbf{x}) = \rho_0(\mathbf{a})\left|\frac{\partial a_i}{\partial x_j}\right|.$$

These equations relate the density in different configurations of the body to the transformation that leads from one configuration to another.

Thus, the material description of a continuum follows the method used in particle mechanics.

10.3 SPATIAL DESCRIPTION OF THE MOTION
OF A CONTINUUM

In the material description every particle is identified by its coordinates at a given instant of time t_0. This is not always convenient. When we describe the flow of water in a river, we do not desire to identify the location from which every particle of water comes. Instead, we are generally interested in the instantaneous velocity field and its evolution with time. This leads to the *spatial description* traditionally used in hydrodynamics. The location (x_1, x_2, x_3) and the time t are taken as independent variables. It is natural for hydrodynamics because measurements are more easily made and directly interpreted in terms of what happens at a certain place, rather than following the particles.

In a spatial description, the instantaneous motion of the continuum is described by the velocity vector field $v_i(x_1, x_2, x_3, t)$ which, of course, is the velocity of a particle instantaneously located as (x_1, x_2, x_3) at time t. We shall show that the instantaneous acceleration of the particle is given by the formula

$$(10.3\text{-}1) \quad \blacktriangle \qquad \dot{v}_i(\mathbf{x}, t) = \frac{\partial v_i}{\partial t}(\mathbf{x}, t) + v_j \frac{\partial v_i}{\partial x_j}(\mathbf{x}, t),$$

where $\mathbf{x}$ again stands for the variables x_1, x_2, x_3, and every quantity in this formula is evaluated at $(\mathbf{x}, t)$. The proof follows from the fact that a particle located at (x_1, x_2, x_3) at time t is moved to a point with coordinates $x_i + v_i\, dt$ at the time $t + dt$ and that, according to Taylor's theorem, and by omitting the higher order infinitesimal terms as $dt \rightarrow 0$,

$$\dot{v}_i(\mathbf{x}, t)\, dt = v_i(x_j + v_j\, dt, t + dt) - v_i(x_j, t)$$

$$= v_i + \frac{\partial v_i(\mathbf{x}, t)}{\partial t}\, dt + \frac{\partial v_i(\mathbf{x}, t)}{\partial x_j} v_j\, dt - v_i,$$

which reduces to (10.3-1). The first term in (10.3-1) may be interpreted as arising from the time dependence of the velocity field; the second term as the contribution of the motion of the particle in the nonhomogeneous velocity field. Accordingly, these terms are called the *local* and the *convective* parts of the acceleration, respectively.

The reasoning that led to (10.3-1) is applicable to any function $F(x_1, x_2, x_3, t)$ that is attributable to the moving particles, such as the temperature. A convenient terminology is the *material derivative*, which is denoted by a dot or the symbol D/Dt. Thus, the material derivative of F is

(10.3-2) ▲ $\qquad \dot{F} = \dfrac{DF}{Dt} \equiv \left(\dfrac{\partial F}{\partial t}\right)_{\mathbf{x}=\text{const.}} + v_1 \dfrac{\partial F}{\partial x_1} + v_2 \dfrac{\partial F}{\partial x_2} + v_3 \dfrac{\partial F}{\partial x_3}.$

On the other hand, if $F(x_1, x_2, x_3, t)$ is transformed into $F(a_1, a_2, a_3, t)$ through the transformation (10.2-1), then $F(a_1, a_2, a_3, t)$ is indeed the value of F attached to the particle (a_1, a_2, a_3). Hence, the material derivative F indeed means the rate of change of the property F of the particle (a_1, a_2, a_3). Formally,

(10.3-3) $\qquad\qquad\qquad \dot{F} = \dfrac{\partial F(a_1, a_2, a_3, t)}{\partial t}\bigg|_{\mathbf{a}}.$

On regarding $F(x_1, x_2, x_3, t)$ as an implicit function of a_1, a_2, a_3, t, we have

(10.3-4) $\quad \dot{F} = \dfrac{\partial F}{\partial t}\bigg|_{\mathbf{x}} + \dfrac{\partial F}{\partial x_1}\bigg|_{t} \dfrac{\partial x_1}{\partial t}\bigg|_{\mathbf{a}} + \dfrac{\partial F}{\partial x_2}\bigg|_{t} \dfrac{\partial x_2}{\partial t}\bigg|_{\mathbf{a}} + \dfrac{\partial F}{\partial x_3}\bigg|_{t} \dfrac{\partial x_3}{\partial t}\bigg|_{\mathbf{a}},$

which reduces to Eq. (10.3-2) by virtue of Eq. (10.2-2).

10.4 THE MATERIAL DERIVATIVE OF A VOLUME INTEGRAL

Let $I(t)$ be a volume integral of a continuously differentiable function $A(\mathbf{x}, t)$ defined over a spatial domain $V(x_1, x_2, x_3, t)$ occupied by a given set of material particles:

(10.4-1) $\qquad\qquad I(t) = \iiint_V A(\mathbf{x}, t)\, dx_1\, dx_2\, dx_3.$

Here again we write $\mathbf{x}$ for x_1, x_2, x_3. The function $I(t)$ is a function of the time t because both the integrand $A(\mathbf{x}, t)$ and the domain $V(\mathbf{x}, t)$ depend on the parameter t. As t varies, $I(t)$ varies also, and we ask: What is the rate of change of $I(t)$ with respect to t? This rate, denoted by DI/Dt and called the *material derivative* of I, is defined for a given set of material particles.

The phrase "for a given set of particles," is of primary importance. The question is how fast the material body itself sees the value of I is changing. To evaluate this rate, note that the boundary S of the body at the instant t will have moved at time $t + dt$ to a neighboring surface S', which bounds the domain V' (Fig. 10.3). The material derivative of I is defined as

(10.4-2) $\quad \dfrac{DI}{Dt} = \lim\limits_{dt \to 0} \dfrac{1}{dt}\left[\int_{V'} A(\mathbf{x}, t + dt)\, dV - \int_V A(\mathbf{x}, t)\, dV \right].$

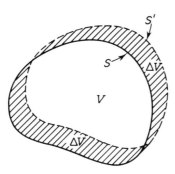

Fig. 10.3 Continuous change of the boundary of a region.

Attention is drawn to the difference in domains V' and V. Let ΔV be the domain $V' - V$. We note that ΔV is swept out by the motion of the surface S in the small time interval dt. Since $V' = V + \Delta V$, we can write (10.4-2) as

(10.4-3)

$$\frac{DI}{Dt} = \lim_{dt \to 0} \frac{1}{dt} \left[\int_V A(\mathbf{x}, t + dt) \, dV + \int_{\Delta V} A(\mathbf{x}, t + dt) \, dV - \int_V A(\mathbf{x}, t) \, dV \right]$$

$$= \lim_{dt \to 0} \left\{ \frac{1}{dt} \int_V [A(\mathbf{x}, t + dt) - A(\mathbf{x}, t)] \, dV + \frac{1}{dt} \int_{\Delta V} A(\mathbf{x}, t + dt) \, dV \right\}.$$

For a continuously differentiable function $A(\mathbf{x}, t)$, the first term on the right-hand side contributes the value $\int_V \frac{\partial A}{\partial t} \, dV$ to DI/Dt. The last term in (10.4-3) may be evaluated by noting that for an infinitesimal dt the integrand may be taken as $A(\mathbf{x}, t)$ on the boundary surface S [because of the assumed continuity of $A(\mathbf{x}, t)$] and that the integral is equal to the sum of $A(\mathbf{x}, t)$ multiplied by the volume swept by the particles situated on the boundary S in the time interval dt. If v_i is the unit vector along the outer normal of S, then, since the displacement of a particle on the boundary is $v_i \, dt$, the volume swept by particles occupying an element of area dS on the boundary S is $dV = v_i v_i \, dS \cdot dt$. On ignoring infinitesimal quantities of the second or higher order, we see that the contribution of this element to DI/Dt is $A v_i v_i \, dS$. The total contribution is obtained by an integration over S. Therefore,

(10.4-4) ▲ $$\frac{D}{Dt} \int_V A \, dV = \int_V \frac{\partial A}{\partial t} \, dV + \int_S A v_j v_j \, dS.$$

Transforming the last integral by Gauss' theorem and using Eq. (10.3-2) we have

$$(10.4\text{-}5) \quad \blacktriangle \quad \frac{D}{Dt} \int_V A \, dV = \int_V \frac{\partial A}{\partial t} \, dV + \int_V \frac{\partial}{\partial x_j} (Av_j) \, dV$$

$$= \int_V \left(\frac{\partial A}{\partial t} + v_j \frac{\partial A}{\partial x_j} + A \frac{\partial v_j}{\partial x_j} \right) dV$$

$$= \int_V \left(\frac{DA}{Dt} + A \frac{\partial v_j}{\partial x_j} \right) dV.$$

This important formula will be used repeatedly in the sections which follow. It should be noted that according to (10.4-5) the *operation of forming the material derivative and that of spatial integration is noncommutative in general.*

10.5 THE EQUATION OF CONTINUITY

The law of conservation of mass was discussed in Sec. 10.2. With the results of Sec. 10.4, we can now give some alternative forms.

The mass contained in a domain V at a time t is

$$(10.5\text{-}1) \qquad\qquad m = \int_V \rho \, dV,$$

where $\rho = \rho(\mathbf{x}, t)$ is the density of the continuum at the location $\mathbf{x}$ at time t. Conservation of mass requires that $Dm/Dt = 0$. The derivative Dm/Dt is given by Eq. (10.4-4) or (10.4-5), if A is replaced by ρ. Since the result must hold for an arbitrary domain V, the integrand must vanish. Hence, we obtain the following alternative forms of the law of conservation of mass.

$$(10.5\text{-}2) \quad \blacktriangle \qquad\qquad \int_V \frac{\partial \rho}{\partial t} \, dV + \int_S \rho v_j v_j \, dS = 0.$$

$$(10.5\text{-}3) \quad \blacktriangle \qquad\qquad \frac{\partial \rho}{\partial t} + \frac{\partial \rho v_j}{\partial x_j} = 0.$$

$$(10.5\text{-}4) \quad \blacktriangle \qquad\qquad \frac{D\rho}{Dt} + \rho \frac{\partial v_j}{\partial x_j} = 0.$$

These are called the *equations of continuity.* The integral form (10.5-2) is useful when the differentiability of ρv_j cannot be assumed.

In problems of statics, these equations are identically satisfied. Then the conservation of mass must be expressed by Eq. (10.2-7) or Eq. (10.2-8).

10.6 THE EQUATIONS OF MOTION

Newton's laws of motion state that in an inertial frame of reference the material rate of change of the linear momentum of a body is equal to the resultant of applied forces.

At an instant of time t, the linear momentum of all the particles contained in a domain V is

(10.6-1)
$$\mathcal{P}_i = \int_V \rho v_i \, dV.$$

If the body is subjected to surface tractions $\overset{v}{T}_i$ and body force per unit volume X_i, the resultant force is

(10.6-2)
$$\mathcal{F}_i = \int_S \overset{v}{T}_i \, dS + \int_V X_i \, dV.$$

According to the Cauchy's formula, Eq. (3.4-2), the surface traction may be expressed in terms of the stress field σ_{ij}, so that $\overset{v}{T}_i = \sigma_{ji} v_j$, where v_j is the unit vector along the outer normal to the boundary surface S of the domain V. On substituting $\sigma_{ij} v_j$ for $\overset{v}{T}_i$ into (10.6-2) and transforming the surface integral into a volume integral by Gauss' theorem, we have

(10.6-3)
$$\mathcal{F}_i = \int_V \left(\frac{\partial \sigma_{ij}}{\partial x_j} + X_i \right) dV.$$

Newton's law states that

(10.6-4)
$$\frac{D}{Dt} \mathcal{P}_i = \mathcal{F}_i.$$

Hence, according to Eq. (10.4-5), with A identified with ρv_i, we have

(10.6-5)
$$\int_V \left[\frac{\partial \rho v_i}{\partial t} + \frac{\partial}{\partial x_j} (\rho v_i v_j) \right] dV = \int_V \left(\frac{\partial \sigma_{ij}}{\partial x_j} + X_i \right) dV.$$

Since this equation must hold for an arbitrary domain V, the integrands on the two sides must be equal. Thus,

(10.6-6)
$$\frac{\partial \rho v_i}{\partial t} + \frac{\partial}{\partial x_j} (\rho v_i v_j) = \frac{\partial \sigma_{ij}}{\partial x_j} + X_i.$$

The left-hand side of (10.6-6) is equal to

$$v_i \left(\frac{\partial \rho}{\partial t} + \frac{\partial \rho v_j}{\partial x_j} \right) + \rho \left(\frac{\partial v_i}{\partial t} + v_j \frac{\partial v_i}{\partial x_j} \right).$$

The quantity in the first parentheses vanishes according to the equation of continuity (10.5-3), while that in the second is the acceleration Dv_i/Dt. Hence, we obtain the celebrated *Eulerian equation of motion* of a continuum.

$$(10.6\text{-}7) \quad \blacktriangle \qquad\qquad \rho \frac{Dv_i}{Dt} = \frac{\partial \sigma_{ij}}{\partial x_j} + X_i.$$

The equation of equilibrium discussed in Sec. 3.5 is a special case which can be obtained by setting all velocity components v_i equal to zero.

10.7* MOMENT OF MOMENTUM

An application of the law of balance of angular momentum to the particular case of *static equilibrium* leads to the conclusion that stress tensors are symmetric tensors (see Sec. 3.5.) We shall now show that no addditional restriction to the motion of a continuum is introduced in dynamics by the angular momentum postulate, which states that the material rate of change of the moment of momentum with respect to an origin is equal to the resultant moment of all the applied forces about the same origin.

At an instant of time t, a body occupying a regular region V of space with boundary S has the moment of momentum [Eq. (3.3-2)]

$$(10.7\text{-}1) \qquad\qquad \mathcal{H}_i = \int_V e_{ijk} x_j \rho v_k \, dV$$

with respect to the origin of coordinates. If the body is subjected to surface traction $\overset{v}{T}_i$ and body force per unit volume X_i, the resultant moment about the origin is

$$(10.7\text{-}2) \qquad\qquad \mathcal{L}_i = \int_V e_{ijk} x_j X_k \, dV + \int_S e_{ijk} x_j \overset{v}{T}_k \, dS.$$

Introducing Cauchy's formula, $\overset{v}{T}_k = \sigma_{lk} v_l$, into the last integral and transforming the result into a volume integral by Gauss' theorem, we obtain

$$(10.7\text{-}3) \qquad\qquad \mathcal{L}_i = \int_V e_{ijk} x_j X_k \, dV + \int_V (e_{ijk} x_j \sigma_{lk})_{,l} \, dV.$$

Euler's law states that, for any region V,

$$(10.7\text{-}4) \qquad\qquad \frac{D}{Dt} \mathcal{H}_i = \mathcal{L}_i.$$

Evaluating the material derivative of $\mathcal{H}_i$ according to (10.4-5) and using (10.7-3), we obtain

$$(10.7\text{-}5) \qquad e_{ijk} x_j \frac{\partial}{\partial t} (\rho v_k) + \frac{\partial}{\partial x_l} (e_{ijk} x_j \rho v_k v_l) = e_{ijk} x_j X_k + e_{ijk} (x_j \sigma_{lk})_{,l}.$$

The second term in (10.7-5) can be written as

$$e_{ijk}\rho v_j v_k + e_{ijk} x_j \frac{\partial}{\partial x_l}(\rho v_k v_l) = 0 + e_{ijk} x_j \frac{\partial}{\partial x_l}(\rho v_k v_l)$$

because e_{ijk} is antisymmetric and $v_j v_k$ is symmetric with respect to j, k. The last term in (10.7-5) can be written as $e_{ijk}\sigma_{jk} + e_{ijk}x_j\sigma_{lk,l}$. Hence, Eq. (10.7-5) becomes

$$(10.7\text{-}6) \qquad e_{ijk} x_j \left[\frac{\partial}{\partial t}(\rho v_k) + \frac{\partial}{\partial x_l}(\rho v_k v_l) - X_k - \sigma_{lk,l} \right] - e_{ijk}\sigma_{jk} = 0.$$

The sum in the square brackets vanishes by the equation of motion (10.6-6). Hence, Eq. (10.7-6) is reduced to

$$(10.7\text{-}7) \qquad\qquad\qquad e_{ijk}\sigma_{jk} = 0;$$

i.e., $\sigma_{jk} = \sigma_{kj}$. Thus, if the stress tensor is symmetric, the law of balance of moment of momentum is identically satisfied.

10.8* THE BALANCE OF ENERGY

The motion of a continuum must be governed further by the law of conservation of energy. If mechanical energy alone is of interest in a problem, then the energy equation is merely the first integral of the equation of motion. If thermal process is significant, then the equation of energy becomes an independent equation to be satisfied.

The law of conservation of energy is the first law of thermodynamics. Its expression for a continuum can be derived as soon as all forms of energy and work are listed. Let us consider a continuum for which there are three forms of energy: the kinetic energy K, the gravitational energy G, and the internal energy E:

$$(10.8\text{-}1) \qquad\qquad \text{Energy} = K + G + E.$$

The *kinetic energy* contained in a regular domain V at a time t is

$$(10.8\text{-}2) \qquad\qquad K = \int \tfrac{1}{2}\rho v_i v_i\, dV,$$

where v_i are the components of the velocity vector of a particle occupying an element of volume dV and ρ is the density of the material. The *gravitational* energy depends on the distribution of mass and may be written as

$$(10.8\text{-}3) \qquad\qquad G = \int \rho\phi(x)\, dV,$$

where ϕ is the graviatational potential per unit mass. In the important special case of a uniform gravitational field, we have

$$(10.8\text{-}4) \qquad G = \int \rho g z \, dV,$$

where g is the gravitational acceleration and z is a distance measured from a certain plane in a direction opposite to the gravitational field. The *internal energy* is written in the form

$$(10.8\text{-}5) \qquad E = \int \rho \varepsilon \, dV,$$

where ε is the *internal energy per unit mass*. The first law of thermodynamics states that the energy of a system can be changed by absorption of heat Q and by work done on the system, W:

$$(10.8\text{-}6) \qquad \Delta \text{ energy} = Q + W.$$

Expressing this in terms of rates, we have

$$(10.8\text{-}7) \qquad \frac{D}{Dt}(K + G + E) = \dot{Q} + \dot{W},$$

where $\dot{Q}$ and $\dot{W}$ are the rate of change of Q and W per unit time.

Now, the heat input into the body must be imparted through the boundary. To describe the heat flow, a *heat flux* vector $\mathbf{h}$ (with components h_1, h_2, h_3) is defined as follows. Let dS be a surface element in the body, with unit outer normal ν_i. Then the rate at which heat is transmitted across the surface dS in the direction of ν_i is assumed to be representable as $h_i \nu_i \, dS$. If the medium is moving, we insist that the surface element dS be composed of the same particles. The rate of heat input is, therefore,

$$(10.8\text{-}8) \qquad \dot{Q} = -\int_S h_i \nu_i \, dS = -\int_V \frac{\partial h_i}{\partial x_i} \, dV.$$

The rate at which work is done on the body by the body force per unit volume F_i in V and surface traction $\overset{\nu}{T}_i$ in S is the *power*

$$(10.8\text{-}9) \qquad \dot{W} = \int F_i v_i \, dV + \int \overset{\nu}{T}_i v_i \, dS$$

$$= \int F_i v_i \, dV + \int \sigma_{ij} v_j v_i \, dS$$

$$= \int F_i v_i \, dV + \int (\sigma_{ij} v_i)_{,j} \, dV.$$

Since in Eq. (10.8-7) the gravitational energy is included in the term G, the power W must be evaluated with the gravitational force excluded from the body force F_i. Substituting Eq. (10.8-2), (10.8-3), (10.8-5), (10.8-8), and (10.8-9) into the first law of thermodynamics (10.8-7), using formula (10.4-5) to compute the material derivatives, we obtain the following result after some calculation:

(10.8-10)
$$\frac{1}{2}\rho\frac{Dv^2}{Dt} + \frac{v^2}{2}\frac{D\rho}{Dt} + \frac{v^2}{2}\rho\,\text{div}\,\mathbf{v} + \rho\frac{D\mathcal{E}}{Dt} + \mathcal{E}\frac{D\rho}{Dt}$$

$$+ \mathcal{E}\rho\,\text{div}\,\mathbf{v} + \rho\frac{D\phi}{Dt} + \phi\frac{D\rho}{Dt} + \phi\rho\,\text{div}\,\mathbf{v}$$

$$= -\frac{\partial h_i}{\partial x_i} + F_i v_i + \sigma_{ij,j}v_i + \sigma_{ij}v_{i,j}.$$

This equation can be simplified greatly if we make use of the equations of continuity and motion:

(10.8-11)
$$\frac{D\rho}{Dt} + \rho\,\text{div}\,\mathbf{v} = 0, \qquad \rho\frac{Dv_i}{Dt} = X_i + \sigma_{ij,j}.$$

Here X_i is the total body force per unit mass. The difference between X_i and F_i is the gravitational force and is, by definition,

(10.8-12)
$$X_i - F_i = -\rho\frac{\partial\phi}{\partial x_i}$$

Since

$$\frac{D\phi}{Dt} = \frac{\partial\phi}{\partial t} + v_i\frac{\partial\phi}{\partial x_i}$$

and $\partial\phi/\partial t = 0$ for a gravitational field that is independent of time, we have, for such a field and with Eq. (10.8-11) and (10.8-12),

(10.8-13)
$$\frac{1}{2}\rho\frac{Dv^2}{Dt} + \rho\frac{D\mathcal{E}}{Dt} = -\frac{\partial h_i}{\partial x_i} + \rho v_i\frac{Dv_i}{Dt} + \sigma_{ij}v_{i,j}.$$

But

(10.8-14)
$$\rho v_i\frac{Dv_i}{Dt} = \frac{1}{2}\rho\frac{Dv^2}{Dt},$$

and

(10.8-15)
$$\sigma_{ij}v_{i,j} = \sigma_{ij}[\tfrac{1}{2}(v_{i,j} + v_{j,i}) + \tfrac{1}{2}(v_{i,j} - v_{j,i})] = \sigma_{ij}V_{ij} + 0,$$

where

(10.8-16)
$$V_{ij} = \tfrac{1}{2}(v_{i,j} + v_{j,i})$$

is the *strain-rate tensor*. The last term in (10.8-15) vanishes because it is the contraction of the product of a symmetric tensor σ_{ij} with an antisymmetric one. Hence, Eq. (10.8-13) can be simplified, and we obtain the final form of the energy equation

$$(10.8\text{-}17) \qquad \rho \frac{D\varepsilon}{Dt} = -\frac{\partial h_i}{\partial x_i} + \sigma_{ij}V_{ij}.$$

Specialization

(A) If all the nonmechanical transfer of energy consists of heat conduction, which obeys Fourier's law,

$$(10.8\text{-}18) \qquad h_i = -J\lambda \frac{\partial T}{\partial x_i},$$

where J is the mechanical equivalent of heat, λ is the conductivity, and T is the absolute temperature, then the energy equation becomes

$$(10.8\text{-}19) \qquad \rho \frac{D\varepsilon}{Dt} = J \frac{\partial}{\partial x_i}\left(\lambda \frac{\partial T}{\partial x_i}\right) + \sigma_{ij}V_{ij}.$$

(B) The usual equation of heat conduction in a continuum at rest is obtained by deleting the terms involving ϕ, v_i and V_{ij}, and setting

$$(10.8\text{-}20) \qquad \varepsilon = JcT,$$

where c is the specific heat for the vanishing rate of deformation. Then Eq. (10.8-19) becomes

$$(10.8\text{-}21) \qquad \rho c \frac{\partial T}{\partial t} = \frac{\partial}{\partial x_i}\left(\lambda \frac{\partial T}{\partial x_i}\right).$$

10.9* THE EQUATIONS OF MOTION AND CONTINUITY IN POLAR COORDINATES

In Sec. 3.7 and 5.8 we considered the stress and strain components in polar coordinates. The corresponding equations of motion and continuity can be derived in the same manner: by the method of general tensor analysis, by transformation from the Cartesian coordinates, or by direct ad hoc derivation from first principles. Illustrations of the last two approaches follow.

The basic equations for transformation between Cartesian coordinates x, y, z to polar coordinates r, θ, z are given in Sec. 5.8. If we substitute Eq. (3.7-5) into the equation of equilibrium,

$$(10.9\text{-}1) \qquad \frac{\partial \sigma_{ij}}{\partial x_j} = 0;$$

i.e.,

$$\frac{\partial \sigma_{xx}}{\partial x} + \frac{\partial \sigma_{xy}}{\partial y} + \frac{\partial \sigma_{xz}}{\partial z} = 0,$$

etc., and use Eq. (5.8-3) to transform the derivatives, we obtain

$$(10.9\text{-}2) \quad \left(\frac{\partial \sigma_{rr}}{\partial r} + \frac{1}{r} \frac{\partial \sigma_{r\theta}}{\partial \theta} + \frac{\sigma_{rr} - \sigma_{\theta\theta}}{r} + \frac{\partial \sigma_{rz}}{\partial z} \right) \cos \theta$$

$$- \left(\frac{1}{r} \frac{\partial \sigma_{\theta\theta}}{\partial \theta} + \frac{\partial \sigma_{r\theta}}{\partial r} + 2\frac{\sigma_{r\theta}}{r} + \frac{\partial \sigma_{\theta z}}{\partial z} \right) \sin \theta = 0.$$

Since this equation must hold for all values of θ, we must have, at $\theta = 0$ and at $\theta = \pi/2$, respectively,

$$(10.9\text{-}3)$$

$$\frac{\partial \sigma_{rr}}{\partial r} + \frac{1}{r} \frac{\partial \sigma_{r\theta}}{\partial \theta} + \frac{\sigma_{rr} - \sigma_{\theta\theta}}{r} + \frac{\partial \sigma_{rz}}{\partial z} = 0,$$

$$\frac{1}{r} \frac{\partial \sigma_{\theta\theta}}{\partial \theta} + \frac{\partial \sigma_{r\theta}}{\partial r} + \frac{2\sigma_{r\theta}}{r} + \frac{\partial \sigma_{\theta z}}{\partial z} = 0.$$

But the choice of the x-direction is arbitrary, so that Eq. (10.9-3) must be valid for all values of θ. Similarly, from Eq. (10.9-1) with $i = 3$ we obtain the third equation of equilibrium,

$$(10.9\text{-}4) \quad \frac{\partial \sigma_{zz}}{\partial z} + \frac{1}{r} \frac{\partial \sigma_{z\theta}}{\partial \theta} + \frac{\partial \sigma_{zr}}{\partial r} + \frac{\sigma_{rz}}{r} = 0.$$

If the continuum is subjected to an acceleration and a body force, then the equation of motion (10.6-7) is

$$(10\text{-}9\text{-}5) \quad \frac{\partial \sigma_{ij}}{\partial x_j} + X_i = \rho \frac{Dv_i}{Dt} = \rho a_i.$$

The body force per unit volume may be resolved into components F_r, F_θ, F_z along the r-, θ-, and z-directions. The acceleration $Dv_i/Dt = a_i$ must be considered carefully. The component of acceleration in the x-direction in the rectangular coordinates is

$$(10.9\text{-}6) \quad a_x = \frac{\partial v_x}{\partial t} + v_x \frac{\partial v_x}{\partial x} + v_y \frac{\partial v_x}{\partial y} + v_z \frac{\partial v_x}{\partial z}.$$

The components of acceleration a_x, a_y, a_z and of velocity v_x, v_y, v_z are related to the components a_r, a_θ, a_z and v_r, v_θ, v_z in polar coordinates by the same Eq. (5.8-4) which relate the displacements, provided that u be replaced by a or v, respectively. Hence, by substitution of Eq. (5.8-3) and (5.8-4) into (10.9-6), we obtain

(10.9-7)

$$a_x = \frac{\partial}{\partial t}(v_r \cos\theta - v_\theta \sin\theta)$$

$$+ (v_r \cos\theta + v_\theta \sin\theta)\left(\cos\theta \frac{\partial}{\partial r} - \frac{\sin\theta}{r}\frac{\partial}{\partial\theta}\right)(v_r \cos\theta - v_\theta \sin\theta)$$

$$+ (v_r \sin\theta + v_\theta \cos\theta)\left(\sin\theta \frac{\partial}{\partial r} + \frac{\cos\theta}{r}\frac{\partial}{\partial\theta}\right)(v_r \cos\theta - v_\theta \sin\theta)$$

$$+ v_z \frac{\partial}{\partial z}(v_r \cos\theta - v_\theta \sin\theta)$$

$$= \cos\theta\left(\frac{\partial v_r}{\partial t} + v_r\frac{\partial v_r}{\partial r} + \frac{v_\theta}{r}\frac{\partial v_r}{\partial\theta} - \frac{v_\theta^2}{r} + v_z\frac{\partial v_r}{\partial z}\right)$$

$$- \sin\theta\left(\frac{\partial v_\theta}{\partial t} + v_r\frac{\partial v_\theta}{\partial r} + \frac{v_\theta}{r}\frac{\partial v_\theta}{\partial\theta} + \frac{v_r v_\theta}{r} + v_z\frac{\partial v_\theta}{\partial z}\right).$$

Comparing (10.9-7) with the equation

(10.9-8) $$a_x = a_r \cos\theta - a_\theta \sin\theta,$$

we obtain the components of acceleration

(10.9-9)
$$a_r = \frac{\partial v_r}{\partial t} + v_r\frac{\partial v_r}{\partial r} + \frac{v_\theta}{r}\frac{\partial v_r}{\partial\theta} - \frac{v_\theta^2}{r} + v_z\frac{\partial v_r}{\partial z},$$

$$a_\theta = \frac{\partial v_\theta}{\partial t} + v_r\frac{\partial v_\theta}{\partial r} + \frac{v_\theta}{r}\frac{\partial v_\theta}{\partial\theta} + \frac{v_r v_\theta}{r} + v_z\frac{\partial v_\theta}{\partial z}.$$

Similarly,

(10.9-10) $$a_z = \frac{\partial v_z}{\partial t} + v_r\frac{\partial v_z}{\partial r} + \frac{v_\theta}{r}\frac{\partial v_z}{\partial\theta} + v_z\frac{\partial v_z}{\partial z}.$$

The full equations of motion are

(10.9-11)
$$\rho a_r = \frac{\partial\sigma_{rr}}{\partial r} + \frac{1}{r}\frac{\partial\sigma_{r\theta}}{\partial\theta} + \frac{\sigma_{rr} - \sigma_{\theta\theta}}{r} + \frac{\partial\sigma_{rz}}{\partial z} + F_r,$$

$$\rho a_\theta = \frac{1}{r}\frac{\partial\sigma_{\theta\theta}}{\partial\theta} + \frac{\partial\sigma_{r\theta}}{\partial r} + \frac{2\sigma_{r\theta}}{r} + \frac{\partial\sigma_{r\theta}}{\partial z} + F_\theta,$$

$$\rho a_z = \frac{\partial\sigma_{zz}}{\partial z} + \frac{1}{r}\frac{\partial\sigma_{z\theta}}{\partial\theta} + \frac{\partial\sigma_{zr}}{\partial r} + \frac{\sigma_{rz}}{r} + F_z.$$

These derivations are again straightforward but not very instructive from the physical point of view. A second derivation based on an examination of the balance of forces acting on an element may supply further insight into these equations. Figure 10.4 shows the free-body diagram for an isolated element with the stress pattern indicated. The equation of motion indicates that the acceleration in the radial direction is equal to the sum of all the forces

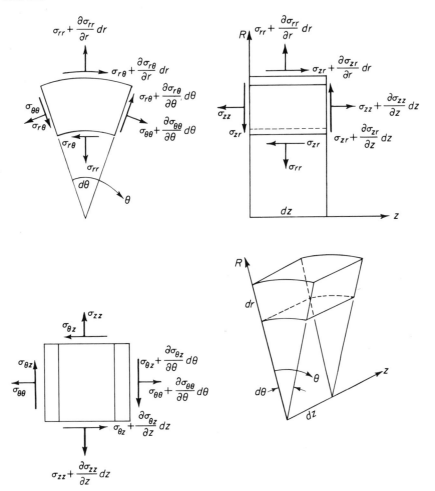

Fig. 10.4 Stress field in cylindrical polar coordinates.

acting in the radial direction. Thus,

$$
(10.9\text{-}12) \quad \rho a_r \, dr \, dz \left[\frac{r \, d\theta + (r + dr) \, d\theta}{2} \right] = F_r \, dr \, dz \left[\frac{r \, d\theta + (r + dr) \, d\theta}{2} \right]
$$

$$
+ \left(\sigma_{rr} + \frac{\partial \sigma_{rr}}{\partial r} \, dr \right)(r + dr) \, d\theta \, dz - \sigma_{rr} r \, d\theta \, dz
$$

$$
- \sigma_{\theta\theta} \, dr \, dz \sin \frac{d\theta}{2} - \left(\sigma_{\theta\theta} + \frac{\partial \sigma_{\theta\theta}}{\partial \theta} \, d\theta \right) dr \, dz \sin \frac{d\theta}{2}
$$

$$
+ \left(\sigma_{r\theta} + \frac{\partial \sigma_{r\theta}}{\partial \theta} \, d\theta \right) dr \, dz - \sigma_{r\theta} \, dr \, dz
$$

$$
+ \left(\sigma_{rz} + \frac{\partial \sigma_{rz}}{\partial z} \, dz - \sigma_{rz} \right) \left[\frac{r \, d\theta + (r + dr) \, d\theta}{2} \right] dr.
$$

Expanding, dropping higher order infinitesimal quantities, and dividing through by r, we obtain the first equation of Eq. (10.9-11). The other equations can be obtained in a similar manner. We note that in the equation for radial equilibrium the term $-\sigma_{\theta\theta}/r$ is a radial pressure in the nature of hoop stress; the term σ_{rr}/r is the contribution due to the larger area of the outer surface at $r + dr$ than that at radius r. The term σ_{rz}/r in the equation for axial equilibrium is due to the same reason. The term $2\sigma_{r\theta}/r$ in the tangential equation has two origins: One is due to the same reason as above, that the outer surface is larger; the other arises from the fact that the radial surfaces at θ and $\theta + d\theta$ are not parallel but make an angle $d\theta$.

A similar graphical interpretation can be made of the individual terms in the expressions for acceleration. The term $-v_\theta^2/r$ in a_r is of the nature of centripetal acceleration. The term $v_\theta v_r/r$ in a_θ arises from the rotation of the radial velocity vector v_r, thus contributing a tangential component of acceleration.

A similar treatment can be used to transform the equation of continuity (10.5-3) into polar coordinates. But here it is perhaps most instructive to study the balance of mass flow in an element, as shown in Fig. 10.5. With the area through which the mass flow takes place accounted for properly, we obtain

$$(10.9\text{-}13) \qquad \frac{1}{r}\frac{\partial}{\partial r}(\rho r v_r) + \frac{1}{r}\frac{\partial \rho v_\theta}{\partial \theta} + \frac{\partial \rho v_z}{\partial z} + \frac{\partial \rho}{\partial t} = 0.$$

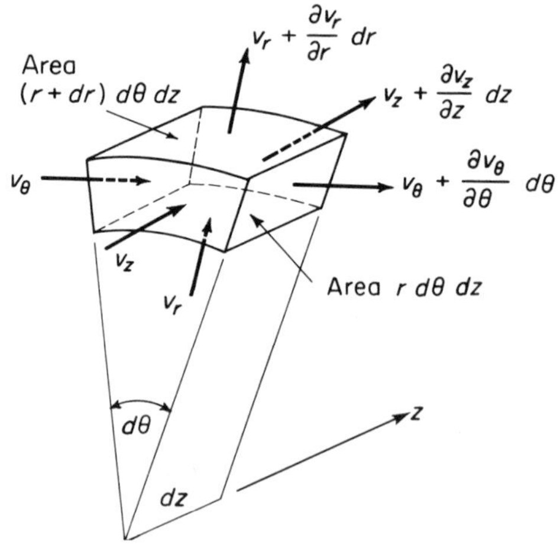

Fig. 10.5 Conservation of mass in cylindrical polar coordinates.

PROBLEMS

10.1 State the definitions of (a) a line integral, (b) a surface integral, (c) a volume integral.

10.2 State the mathematical conditions under which Eq. (10.1-4), (10.1-5), (10.4-4), and (10.4-5) are valid.

10.3 Evaluate the line integral

$$\oint_c y^2 \, dx + x^2 \, dy,$$

where C is a triangle with vertices $(1,0)$, $(1, 1)$, $(0, 0)$. See Fig. P10.3. *Answer:* $\frac{1}{3}$.

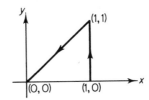

Fig. P.10.3 Path of integration.

10.4 Evaluate $\oint_c (x^2 - y^2) \, ds$, where C is the circle $x^2 + y^2 = 4$.

10.5 Derive Green's theorem: Let D be a domain of the xy-plane and let c be a piece-wise smooth simple closed curve in D whose interior is also in D. Let $P(x, y)$ and $Q(x, y)$ be functions defined and continuous and having continuous first partial derivatives in D. Then

$$\oint_c P \, dx + Q \, dy = \iint_R \left(\frac{\partial Q}{\partial x} - \frac{\partial P}{\partial y} \right) dx \, dy,$$

where R is the closed region bounded by c.

10.6 Interpret Green's theorem vectorially to derive the following theorems:

(a) $\oint_c u_T \, ds = \iint_R \text{curl}_z \, \mathbf{u} \, dx \, dy,$

(b) $\oint_c v_n \, ds = \iint_R \text{div} \, \mathbf{v} \, dx \, dy,$

where $\mathbf{u}$, $\mathbf{v}$ are vector fields, u_T is the tangential component of $\mathbf{u}$ (tangent to the curve c), ds is the arc length, v_n is the normal component of $\mathbf{v}$ on c. Equation (a) is a special case of Stokes' theorem. Equation (b) is the two-dimensional form of Gauss' theorem.

10.7 A rubber spherical balloon is quickly blown up in an angry sea by a ditched pilot. Let a particle on the balloon be located at

$$x = x(t), \qquad y = y(t), \qquad z = z(t).$$

Let the surface of the balloon be described by the equation

$$F(t) = (x - \lambda)^2 + (y - \mu)^2 + (z - \nu)^2 - a^2 = 0,$$

where $\lambda(t)$, $\mu(t)$, $\nu(t)$, which define the center of the sphere, and $a(t)$, the radius, are functions of time (see Fig. P10.7).

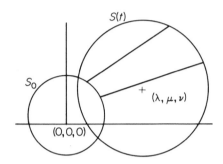

Fig. P10.7   Expanding balloon.

Show that $DF/Dt = 0$.

Derive the boundary conditions for the air and water moving about the balloon.

Solution: The equation $F(t) = 0$ representing the surface of the balloon is true at all times. Therefore its derivative with respect to t must vanish. Since x, y, z are coordinates of the particles, and $F(t)$ is associated with the balloon at all times, the time derivative is the material derivative, i.e., DF/Dt, which is zero.

Conversely, from the equation $DF/Dt = 0$, we conclude that $F = $ const. for a given set of particles. In particular, if the set of particles is defined by the equation $F = 0$, they remain the same set. If $F = 0$ defines the balloon at $t = 0$, it defines the balloon at any t.

The equation becomes more significant if we consider the fluid (air and water) around the balloon. Fluid particles once in contact with the balloon remain in contact with it (the so-called "no-slip" condition of a viscous fluid in contact with a solid body). Hence the boundary conditions of the flow field are $F = 0$ and $DF/Dt = 0$.

10.8 The surface of a flag fluttering in the wind is described by the equation

$$F(x, y, z, t) = 0.$$

Write down analytically the constraints imposed by the flag on the airstream. In other words, given the shape of the boundary surface $F = 0$, derive the boundary condition for the flow. For this problem, consider the air as a nonviscous fluid.

What difference would it make if the air is taken to be a viscous fluid?

Solution: As in Prob. 10.7, the boundary condition of the airstream on the flag surface $F = 0$ is

$$\frac{\partial F}{\partial t} + u_x \frac{\partial F}{\partial x} + u_y \frac{\partial F}{\partial y} + u_z \frac{\partial F}{\partial z} = 0 \tag{1}$$

where $\mathbf{u}$ (u_x, u_y, u_z) is the velocity vector. For the surface $F(x, y, z, t) = 0$, the vector $\mathbf{n}$ with the components

$$\frac{\partial F}{\partial x}, \frac{\partial F}{\partial y}, \frac{\partial F}{\partial z}$$

is normal to the surface. Hence Eq. (1) may be written

$$\frac{\partial F}{\partial t} + \mathbf{u \cdot n} = 0. \tag{2}$$

It means that the normal velocity must be equal to $-\partial F/\partial t$ on the flag surface.

For a viscous fluid, the *no-slip* condition requires, in addition, $F = 0$. See discussion in Sec. 11.2, p. 268.

10.9 Two components of the velocity field of a fluid are known in the region $-2 \leq x, y, z \leq 2$:

$$u = (1 - y^2)(a + bx + cx^2), \qquad w = 0.$$

The fluid is incompressible. What is the velocity component v in the direction of the y-axis?

10.10 Let the temperature field of the fluid described in Prob. 10.9 be

$$T = T_0 e^{-kt} \sin \alpha x \cos \beta y.$$

Find the material rate of change of the temperature of a particle located at the origin $x = y = z = 0$. Find the same for a particle at $x = y = z = 1$.

10.11 For an isotropic Newtonian viscous fluid, derive an equation of motion expressed in terms of the velocity components.

CHAPTER ELEVEN

Field Equations and Boundary Conditions in Fluids

We have acquired enough basic equations to deal with a broad range of problems. Most objects on a scale that we can see are continua. Their motion follows the laws of conservation of mass, momentum, and energy. With the proper constitutive equations and boundary conditions, we can describe many physical problems mathematically. In this chapter we illustrate the formulation of some problems on the flow of fluids.

11.1 THE NAVIER-STOKES EQUATIONS

Let us derive the basic equations governing the flow of a Newtonian viscous fluid. Let x_1, x_2, x_3 or x, y, z be rectangular Cartesian coordinates. Let the velocity components along the x-, y-, z-axis directions be denoted by v_1, v_2, v_3 or u, v, w, respectively. Let p denote pressure; σ_{ij}, or σ_{xx}, σ_{xy}, etc., be the stress components; and μ be the coefficient of viscosity. Then the stress-strain-rate relationship is given by Eq. (7.3-3):

(11.1-1)

$$\sigma_{ij} = -p\delta_{ij} + \lambda V_{kk}\delta_{ij} + 2\mu V_{ij} = -p\delta_{ij} + \lambda \frac{\partial v_k}{\partial x_k}\delta_{ij} + \mu\left(\frac{\partial v_i}{\partial x_j} + \frac{\partial v_j}{\partial x_i}\right);$$

i.e.,

(11.1-1a)

$$\sigma_{xx} = -p + 2\mu\frac{\partial u}{\partial x} + \lambda\left(\frac{\partial u}{\partial x} + \frac{\partial v}{\partial y} + \frac{\partial w}{\partial z}\right),$$

$$\sigma_{yy} = -p + 2\mu\frac{\partial v}{\partial y} + \lambda\left(\frac{\partial u}{\partial x} + \frac{\partial v}{\partial y} + \frac{\partial w}{\partial z}\right),$$

$$\sigma_{zz} = -p + 2\mu\frac{\partial w}{\partial z} + \lambda\left(\frac{\partial u}{\partial x} + \frac{\partial v}{\partial y} + \frac{\partial w}{\partial z}\right),$$

$$\sigma_{xy} = \mu\left(\frac{\partial u}{\partial y} + \frac{\partial v}{\partial x}\right), \qquad \sigma_{yz} = \mu\left(\frac{\partial v}{\partial z} + \frac{\partial w}{\partial y}\right), \qquad \sigma_{zx} = \mu\left(\frac{\partial w}{\partial x} + \frac{\partial u}{\partial z}\right).$$

Substituting these into the equation of motion (10.6-7), we obtain the Navier-Stokes equations:

(11.1-2) $$\rho\frac{Dv_i}{Dt} = \rho X_i - \frac{\partial p}{\partial x_i} + \frac{\partial}{\partial x_i}\left(\lambda\frac{\partial v_k}{\partial x_k}\right) + \frac{\partial}{\partial x_k}\left(\mu\frac{\partial v_k}{\partial x_i}\right) + \frac{\partial}{\partial x_k}\left(\mu\frac{\partial v_i}{\partial x_k}\right),$$

where X_i stands for the body force per unit mass.

The velocity components must satisfy the equation of continuity (10.5-3) derived from the conservation of mass:

(11.1-3) $$\frac{\partial\rho}{\partial t} + \frac{\partial(\rho v_k)}{\partial x_k} = 0.$$

These equations are to be supplemented by the equations of thermal state, balance of energy, and heat flow.

If the fluid is *incompressible*, then

(11.1-4) $$\rho = \text{const.},$$

and all thermodynamic considerations need not be introduced explicitly. Limiting ourselves to an incompressible homogeneous fluid, we see that the equation of continuity becomes

(11.1-5) $$\frac{\partial v_k}{\partial x_k} = 0, \quad \text{or} \quad \frac{\partial u}{\partial x} + \frac{\partial v}{\partial y} + \frac{\partial w}{\partial z} = 0,$$

and the Navier-Stokes equation is simplified to

(11.1-6)
$$\rho \frac{Dv_i}{Dt} = \rho X_i - \frac{\partial p}{\partial x_i} + \mu \frac{\partial^2 v_i}{\partial x_k \partial x_k}.$$

Written out *in extenso*, these are

(11.1-7)
$$\frac{Du}{Dt} = X - \frac{1}{\rho} \frac{\partial p}{\partial x} + \nu \nabla^2 u,$$
$$\frac{Dv}{Dt} = Y - \frac{1}{\rho} \frac{\partial p}{\partial y} + \nu \nabla^2 v,$$
$$\frac{Dw}{Dt} = Z - \frac{1}{\rho} \frac{\partial p}{\partial z} + \nu \nabla^2 w,$$

where $\nu = \mu/\rho$ is the *kinematic viscosity* and ∇^2 is the *Laplacian operator*

(11.1-8)
$$\nabla^2 = \frac{\partial^2}{\partial x^2} + \frac{\partial^2}{\partial y^2} + \frac{\partial^2}{\partial z^2}.$$

Equations (11.1-5) and (11.1-7) comprise four equations for the four variables u, v, w, and p occurring in an incompressible viscous flow.

The solution of Navier-Stokes equation is the central problem in fluid mechanics. This equation embraces a tremendous range of physical phenomena and has many applications to science and engineering. The equation is nonlinear and is, in general, very difficult to solve.

To complete the formulation of a problem we must specify the boundary conditions. In Sec. 11.2 we consider the no-slip condition on a solid-fluid interface. In Sec. 11.3 the condition at a "free," or a fluid-fluid, interface is considered, where surface tension plays an important role. Then a dimensional analysis is presented to illustrate the significance of the Reynolds number. We shall then consider the laminar flow in a channel or a tube as an example of a simplified solution when the nonlinear terms can be ignored. As a warning that such a solution may not be valid in all cases, we use the classical experiments of Reynolds in Sec. 11.5.

In some instances the viscosity of a fluid may be ignored completely, and we deal with the idealized world of "perfect fluids." In association with this idealization the boundary conditions must be changed: The order of the differential equation would be too low to permit the satisfaction of all the boundary conditions of a viscous fluid. We relinquish the no-slip condition at the solid-fluid interface and ignore any shear gradient requirement at a free surface. As a consequence, sometimes the resulting simpler mathematical problems lead to difficulties in physical interpretations.

11.2 BOUNDARY CONDITIONS AT A
SOLID-FLUID INTERFACE

One of the boundary conditions that must be satisfied at a solid-fluid interface is that the fluid must not penetrate the solid if it is impermeable to the fluid. Most containers of fluids are of this nature. Mathematically, this requires that the relative velocity component of the fluid *normal* to the solid surface must vanish.

The specification of the tangential component of velocity of the fluid relative to the solid requires much greater care. It is customary to assume that the *no-slip condition* prevails at an interface between a viscous fluid and a solid boundary. In other words, on the solid-fluid boundary the velocities of the fluid and the solid are exactly equal. This conviction was realized only after a long historical development.

If the solid boundary is stationary, the no-slip condition requires that the velocity changes continuously from zero at the surface to the free-stream value some distance away. This boundary condition is in drastic contrast to that which is required of a nonviscous fluid, for which we can only specify that no fluid shall penetrate the solid surface; but the fluid must be permitted to slide over the solid so that their tangential velocities can be different. This is a penalty for the idealization of complete absence of viscosity. Figure 11.1 illustrates the difference. On the left, the flow of a nonviscous fluid over a stationary solid object is shown: At the interface the fluid slips over the solid with a large tangential velocity. On the right, it is shown that for a viscous fluid the velocity must vanish on the interface.

Since the no-slip condition must be imposed for all real fluid no matter how small the viscosity, the illustration shown on the right-hand side of Fig. 11.1 must prevail for all real fluids.

For the airfoil shown in Fig. 11.1, it is known from wind-tunnel measurements that the flow field is well represented by the left-hand figure; i.e.,

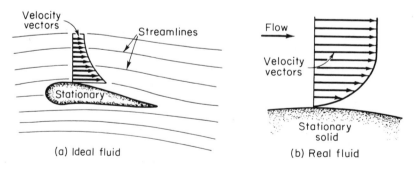

(a) Ideal fluid   (b) Real fluid

Fig. 11.1 The difference in boundary conditions for flows of ideal and real fluids over a solid body.

except for the immediate neighborhood of the solid boundary, the flow can be obtained as though air had no viscosity. Yet we know that air has viscosity, even though very small. Therefore, the no-slip condition must prevail. How can we resolve this conflict?

The answer to this question and the resolution of this conflict is a triumph of modern fluid mechanics. The modern view is that the illustration shown in Fig. 11.1(b) is an indication of what happens in the immediate neighborhood of a solid boundary. We should consider Fig. 11.1(b) as an enlargement of what happens in a very small region of a flow next to an interface. This region is the *boundary layer*. Beyond the boundary layer the flow is practically nonviscous. The dramatic importance of this boundary layer will be seen at the sharp trailing edge of the airfoil. It dictates the condition that the flow must leave the sharp trailing edge smoothly, with no discontinuity in the velocity field. If we insist on idealized nonviscous flow, the tangential velocity could differ on the top and bottom sides of the trailing edge. In the theory of nonviscous fluids such a discontinuity can be eliminated either by permitting the flow to round the sharp corner with an infinite velocity or by introducing an exact amount of circulation so that the trailing edge becomes a stagnation point. The latter condition was proposed by the German mathematician Kutta (1902) and the Russian mathematician Joukowski (1907) and is known as *Kutta-Joukowski hypothesis*, which is the basis for our modern theory of flight. Thus we see that the fluid viscosity, no matter how small, has a profound influence on flow.

But how can we believe the no-slip condition? On what basis is this condition established? The molecular theory of gases does not provide a firm answer. From the same molecular hypotheses which led him to the equations of motion of a viscous fluid, Navier deduced (1823) that there is slipping at a solid boundary, and this slipping is resisted by a force proportional to the relative velocity. This is equivalent to the boundary condition $\beta u = \mu \, \partial u/\partial n$ for flow in one direction along a solid wall, where u is the velocity, $\partial u/\partial n$ is the derivative along the normal away from the wall, β is a constant, and μ is the coefficient of viscosity. The ratio μ/β is a length. This length is zero if there is no slip. Maxwell (1879) calculated that the length μ/β is a moderate multiple of the mean free path L of the gas molecule—probably about $2L$ if there are no inequalities of temperature. Thus, at atmospheric pressure the slip would be negligible. For rarefied gases, however, it would be considerable. (In the atmosphere at an altitude of 50 miles the mean free path is about 1 inch.) This later deduction is in agreement with experimental evidence. Recent molecular theories of rarefied gas flow substantiate this conclusion.

Experiments on liquids and gases at atmospheric pressure (not rarefied), however, support the no-slip condition conclusively. Daniel Bernoulli (1738), on discovering significant discrepancies between the results he had calculated for a perfect fluid and those he had measured with a real one, attributed the difference to the boundary conditions. Coulomb (1800) found that the resistance of an oscillating metallic disk in water was scarcely altered when the disk was smeared with grease or when the surface was covered with powdered sandstone, so the nature of the surface had little influence on the resistance. Poiseuille, in his memoir on blood flow (1841), observed directly with a microscope that the flow at the wall of a tube was stagnant. Hagen, who obtained the laws of flow in capillary tubes experimentally a short time before Poiseuille (1839), stated that the velocity at the wall is zero. After Stokes and others carried out calculations of flow in tubes (1845) based on the no-slip condition, it was found that the results not only agreed with Poiseuille's experiments on the flow of water, but also—in disagreement with earlier experiments carried out by Poiseuille himself—with the flow of mercury in glass tubes (which are not wetted by the mercury). Other experimenters, such as Whetham (1890), Couette (1890), etc., came to the same conclusion. Fage and Townsend (1932) used an ultramicroscope to examine the flow of water containing small particles and confirmed the no-slip condition. In addition, there is agreement between theory and experiment on Stokes's and Oseen's theories of motion at small Reynolds numbers, on Taylor's calculations and observations on the stability of flow between rotating cylinders, etc. All these experiences, taken together, support the conclusion that for a real fluid the slip, if it takes place on a solid boundary, is too small to be observed or to make any sensible difference in the results of theoretical deductions.

11.3 SURFACE TENSION AND THE BOUNDARY
CONDITIONS AT A FREE SURFACE

If an interface between two flows can be identified and is stable, then the normal component of velocity must be the same on the two sides of the interface. This normal-velocity condition is the same as that required at a solid boundary discussed in the previous section. However, the stress boundary condition (Sec. 3.8) has a new element introduced by surface tension.

A certain amount of energy is required to create an interface between two fluids, in particular, between a gas and a liquid. The energy per unit area, of the dimentions [force·length/length²] or [force/length], is called the *surface tension* and is denoted by γ. In a static condition, surface tension can be identified as a tension of intensity γ per unit length across any imaginary line in the interface and is tangent to the interface.

A soap bubble is shown in Fig. 11.2. To create the bubble one must blow and create an internal pressure p, which is balanced by the tension in the soap

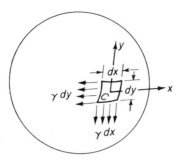

Fig. 11.2 A soap bubble.

film. Let C be a small closed rectangular curve of sides dx and dy drawn on the surface of the bubble (Fig. 11.2). The tensions acting on the sides are shown in the figure. To compute the pressure required to balance the tensions, let us consider two cross-sectional views: one in the xz-plane (z being normal to the soap film), another in the yz-plane. The former is shown in Fig. 11.3 where the tensile forces $\gamma\, dy$ are active at each end. Since these forces are tangent to the surface, they have a resultant $\gamma\, dy\, d\theta$ normal to the surface.

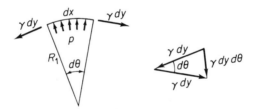

Fig. 11.3 Equilibrium of membrane forces acting on an element of the soap bubble.

But $d\theta = dx/R_1$, where R_1 is the radius of curvature for the soap film. Hence the normal force is $\gamma\, dx\, dy/R_1$. Similarly, the tensions active on the other sides of the rectangle contribute a resultant $\gamma\, dx\, dy/R_2$. Since the soap film has two air-liquid interfaces (inside and outside), the total resultant force due to surface tension acting on the curve C is normal to the soap film and is equal to $2\gamma\, dx\, dy/R_1 + 2\gamma\, dx\, dy/R_2$. This force is balanced by the pressure multiplied by the area $dx\, dy$. On equating these forces, we obtain for the soap film the celebrated equation named after Laplace (1805), although it was actually obtained a year earlier by Thomas Young (1804):

$$(11.3\text{-}1) \qquad\qquad 2\gamma\left(\frac{1}{R_1} + \frac{1}{R_2}\right) = p.$$

If the soap bubble is spherical, then $R_1 = R_2$. If the bubble is not spherical, we note that the sum

(11.3-2) $$\frac{1}{R_1} + \frac{1}{R_2} = \text{mean curvature}$$

is invariant with respect to the rotation of coordinates on any surface. Hence the directions chosen for the x- and y-axes are immaterial.

As a particular case, let us consider soap films formed by boundary curves under zero pressure difference. Then the surface is the so-called *minimal surface*, governed by the equation

(11.3-3) $$\frac{1}{R_1} + \frac{1}{R_2} = 0.$$

Equation (11.3-1) indicates that the pressure required to balance the surface tension becomes very large if the radii R_1, R_2 become very small. For a constant γ, if $R_1, R_2 \to 0$ the pressure p would tend toward infinity. A collapsing vapor bubble in water thus induces a nucleus of high pressure that can cause damage if there is a metal surface nearby, which will become pitted and corroded. Propellers on ships are known to suffer severely from such cavitation damages. (See Fig. 9.3 and the discussion on p. 206.)

Surface tension is very important in such chemical engineering problems as foaming, in such mechanical engineering problems as fracture of metals and rocks, and in such physiological problems as the stability and atelectasis of the pulmonary alveoli. Sometimes the structure of the interface may be quite complex, so that its property cannot be uniquely characterized simply by the energy per unit area named above. For example, the alveolar surface in our lungs is moist and the surface tension is modulated by the presence of "surfactants," lipids such as lecithin. The arrangement of these polar molecules on the interface depends on the concentration of these molecules, the rate at which the surface is strained, and the history of strain, so that the surface tension-area relationship has a huge hysteresis loop when the surface is subjected to a periodic strain. Figure 11.4 shows the experimental results obtained by J. A. Clements by means of a surface balance of the Wilhelmy type. It shows the surface tension-area relationship between air and pure water, blood plasma, 1 % Tween 20 detergent, and a saline extract of a normal lung. The loops of water and detergent are exaggerated schematically to show the cylic nature of the strain history.

When there is an interface, there is a question of permeability of the fluid moving through it. In the case of a solid boundary, the permeability may change the boundary condition with respect to the normal component of velocity. In the case of a free boundary between two fluids in relative motion, the interface condition can be rather subtle and a variety of possible situations

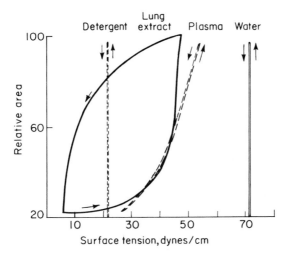

Fig. 11.4 The variation of surface tension with strain for several fluids. From J. A. Clements "Surface Phenomena in Relation to Pulmonary Function". *The Physiologist*, **5**. 1 (1962). 11–28.

may arise. Thus, a certain amount of mass transfer, laminar or turbulent mixing, etc., may occur. If the free surface is covered with a thin layer of another material (such as a layer of oil on a lake or a flag between two airstreams), it may be necessary also to account for interfacial elasticity and viscosity.

11.4* DYNAMIC SIMILARITY AND
REYNOLDS NUMBER

Let us put the Navier-Stokes equation in dimensionless form. For simplicity, we shall consider a homogeneous incompressible fluid. Choose a characteristic velocity V and a characteristic length L. For example, if we investigate the flow of air around an airplane wing, we may take V to be the airplane speed and L to be the wing chord length. If we investigate the flow in a tube, V may be taken as the mean flow speed and L, the tube diameter. For a falling sphere, we may take the speed of falling to be V, the diameter of the sphere to be L, and so on. Having chosen these characteristic quantities, we introduce the dimensionless variables:

$$
x' = \frac{x}{L}, \qquad y' = \frac{y}{L}, \qquad z' = \frac{z}{L}, \qquad u' = \frac{u}{V},
$$

(11.4-1)

$$
v' = \frac{v}{V}, \qquad w' = \frac{w}{V}, \qquad p' = \frac{p}{\rho V^2}, \qquad t' = \frac{Vt}{L},
$$

and the parameter

$$(11.4\text{-}2) \qquad \text{Reynolds number} = R_N = \frac{VL\rho}{\mu} = \frac{VL}{\nu}.$$

Equation (11.1-7) for an incompressible fluid can then be put into the form

$$(11.4\text{-}3) \quad \frac{\partial u'}{\partial t'} + u'\frac{\partial u'}{\partial x'} + v'\frac{\partial u'}{\partial y'} + w'\frac{\partial u'}{\partial z'} = -\frac{\partial p'}{\partial x'} + \frac{1}{R_N}\left(\frac{\partial^2 u'}{\partial x'^2} + \frac{\partial^2 u'}{\partial y'^2} + \frac{\partial^2 u'}{\partial z'^2}\right)$$

and two additional equations obtainable from Eq. (11.4-3) by changing u' into v', v' into w', w' into u' and x' into y', y' into z', z' into x'. The equation of continuity (11.1-5) can also be put in dimensionless form:

$$(11.4\text{-}4) \qquad \frac{\partial u'}{\partial x'} + \frac{\partial v'}{\partial y'} + \frac{\partial w'}{\partial z'} = 0.$$

Since Eq. (11.4-3) and (11.4-4) constitute the complete set of field equations for an incompressible fluid, it is clear that only one physical parameter, the Reynolds number R_N, enters into the field equations of the flow.

Consider two geometrically similar bodies immersed in a moving fluid. The bodies are similar (same shape but different size); the boundary conditions are identical (no-slip). The two flows will be identical (in the dimensionless variables) if the Reynolds numbers for the two bodies are the same, because two geometrically similar bodies having the same Reynolds number will be governed by identical differential equations and boundary conditions (in dimensionless form). Therefore, *flows about geometrically similar bodies at the same Reynolds numbers are completely similar in the sense that the functions* $u'(x', y', z', t)$, $v'(x', y', z', t)$, $w'(x', y', z', t)$, $p'(x', y', z', t)$ *are the same for the various flows.* Thus the Reynolds number is said to govern the *dynamic similarity.* For this statement to be useful and true we must notice that the similarity of the bodies must involve not only the shaps but also the surface roughness and that the flows must also be similar as regards turbulence.

The wide range of Reynolds numbers that occurs in practical porblems is illustrated in the following examples.

PROBLEMS

11.1 Smokestacks are known to sway in the wind if they are not rigid enough. The wind force depends on the Reynolds number of the flow. Let the wind speed be 30 mi/hr, the smokestack diameter be 20 ft. Compute the Reynolds number of the flow.

Answer: 5.46×10^6.

The coefficient of viscosity of air at 20°C is $\mu = 1.808 \times 10^{-4}$ poise (g/cm sec), the kinematic viscosity ν is 0.150 Stokes (cm²/sec).

11.2 Compute the Reynolds number for a submarine periscope of diameter 16 in. at 15 knots.

Answer: 2.4×10^6.

For water at 10°C, $\mu = 1.308 \times 10^{-2}$ g/cm sec, $\nu = 1.308 \times 10^{-2}$ cm²/sec. 1 knot = 1 nautical mile per hour, or 6080.20 ft/hr.

11.3 Suppose that in a cloud chamber experiment designed to determine the charge of an electron (Robert Millikan's experiment) the water droplet diameter is 5 micra (i.e., 5×10^{-4} cm), which moves in air at 0°C at a speed of 2 mm/sec. What is the Reynolds number?

Answer: 7.6×10^{-4}.

For air at 0°C, $\nu = 0.132$ cm²/sec.

11.4 For blood plasma to flow in a capillary blood vessel of diameter 10 micra (i.e., 10^{-3} cm), at a speed of 2 mm/sec, what is the Reynolds number?

Answer: 1.4×10^{-2}.

For blood plasma at body temperature, μ is about 1.4 centipoise (1.4 $\times 10^{-2}$ g/cm sec).

11.5 Compute the Reynolds number for a large airplane wing with a chord length of 10 ft flying at 600 mi/hr at an altitude of 7500 ft, (0°C).

Answer: 6.2×10^7.

Let us now consider the physical interpretation of the Reynolds number. First, it will be shown that the Reynolds number expresses the ratio of inertial force to the shear stress. In a flow the inertial force due to convective acceleration arises from terms such as ρu^2, whereas the shear stress arises from terms such as $\mu \, \partial u / \partial y$. The orders of magnitude of these terms are, respectively,

$$\text{inertial force:} \quad \rho V^2,$$

$$\text{shear stress:} \quad \frac{\mu V}{L}.$$

The ratio is

(11.4-5) $$\frac{\text{inertial force}}{\text{shear stress}} = \frac{\rho V^2}{\mu V / L} = \frac{\rho V L}{\mu} = \text{Reynolds number.}$$

A large Reynolds number signals a preponderant inertial effect. A small Reynolds number signals a predominant shear effect.

11.5* LAMINAR FLOW IN A HORIZONTAL
CHANNEL OR TUBE

Navier-Stokes equations are not easy to solve. If, however, one can find a special problem in which the nonlinear terms disappear, then the solution can be obtained by standard methods. A particularly simple problem of this nature is the steady flow of an incompressible fluid in a horizontal channel of width $2h$ between two parallel planes as shown in Fig. 11.5.

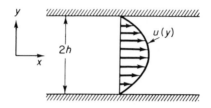

Fig. 11.5 Laminar flow in a parallel channel.

We search for a flow

$$(11.5\text{-}1) \qquad u = u(y), \qquad v = 0, \qquad w = 0$$

that satisfies the Navier-Stokes equations, the equation of continuity, and the no-slip conditions on the boundaries $y = \pm h$:

$$(11.5\text{-}2) \qquad u(h) = 0, \qquad u(-h) = 0.$$

Obviously (11.5-1) satisfies the equation of continuity (11.1-3) exactly, whereas Eq. (11.1-7) becomes

$$(11.5\text{-}3) \qquad 0 = -\frac{\partial p}{\partial x} + \mu \frac{d^2 u}{dy^2},$$

$$(11.5\text{-}4) \qquad 0 = \frac{\partial p}{\partial y},$$

$$(11.5\text{-}5) \qquad 0 = \frac{\partial p}{\partial z}.$$

Equations (11.5-4) and (11.5-5) show that p is a function of x only. If we differentiate Eq. (11.5-3) with respect to x and use Eq. (11.5-1), we obtain $\partial^2 p/\partial x^2 = 0$. Hence $\partial p/\partial x$ must be a constant, say $-\alpha$. Equation (11.5-3) then becomes

$$(11.5\text{-}6) \qquad \frac{d^2 u}{dy^2} = -\frac{\alpha}{\mu},$$

which has a solution

$$(11.5\text{-}7) \qquad u = A + By - \frac{\alpha}{\mu}\frac{y^2}{2}.$$

The two constants A and B can be determined by the boundary conditions (11.5-2) to yield the final solution

$$(11.5-8) \qquad u = \frac{\alpha}{2\mu}(h^2 - y^2).$$

Thus, the velocity profile is a parabola.

A corresponding problem is the flow through a horizontal circular cylindrical tube of radius a (see Fig. 11.6). We search for a solution

$$u = u(y, z), \qquad v = 0, \qquad w = 0.$$

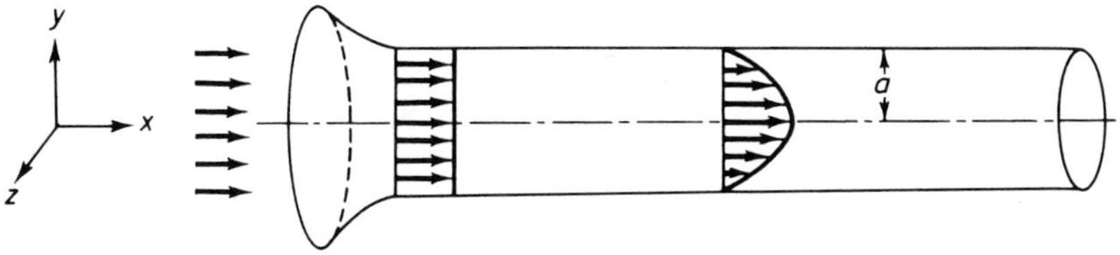

Fig. 11.6 Laminar flow in a circular cylindrical tube.

In analogy with Eq. (11.5-6), the Navier-Stokes equation becomes

$$(11.5-9) \qquad \frac{\partial^2 u}{\partial y^2} + \frac{\partial^2 u}{\partial z^2} = -\frac{\alpha}{\mu}.$$

It is convenient to transform from Cartesian coordinates x, y, z to the cylindrical polar coordinates x, r, θ, with $r^2 = y^2 + z^2$ (see Sec. 5.8). Then Eq. (11.5-9) becomes

$$(11.5-10) \qquad \frac{\partial^2 u}{\partial y^2} + \frac{\partial^2 u}{\partial z^2} = \frac{1}{r}\frac{\partial}{\partial r}\left(r\frac{\partial u}{\partial r}\right) + \frac{1}{r^2}\frac{\partial^2 u}{\partial \theta^2} = -\frac{\alpha}{\mu}.$$

Let us assume that the flow is symmetric so that u is a function of r only; then $\partial^2 u/\partial\theta^2 = 0$, and the equation

$$(11.5-11) \qquad \frac{1}{r}\frac{d}{dr}\left(r\frac{du}{dr}\right) = -\frac{\alpha}{\mu}$$

can be integrated immediately to yield

$$(11.5-12) \qquad u = -\frac{\alpha}{\mu}\frac{r^2}{4} + A\log r + B.$$

The constants A and B are determined by the conditions of no-slip at $r = a$ and symmetry on the center line, $r = 0$:

$$(11.5-13) \qquad u = 0 \quad \text{at} \quad r = a.$$

$$(11.5-14) \qquad \frac{du}{dr} = 0 \quad \text{at} \quad r = 0.$$

The final solution is

(11.5-15)
$$u = \frac{\alpha}{4\mu}(a^2 - r^2).$$

This is the famous parabolic velocity profile of the Hagen-Poiseuille flow; the theoretical solution was worked out by Stokes.

The classical solution of Hagen-Poiseuille flow has been subjected to innumerable experimental observations. It is not valid near the entrance to a tube. It is satisfactory at a sufficiently large distance from the entrance but is again invalid if the tube is too large or if the velocity is too high. The difficulty at the entry region is due to the transitional nature of the flow in that region, so that our assumption $v = 0$, $w = 0$ is not valid. The difficulty with too large a Reynolds number, however, is of a different kind. The flow becomes turbulent!

Osborne Reynolds demonstrated the transition to turbulent flow in a classical experiment in which he examined an outlet through a small tube from a large water tank. At the end of the tube there was a stopcock used to vary the speed of water through the tube. The junction of the tube with the tank was nicely rounded, and a filament of colored fluid was introduced at the mouth. When the speed of water was slow, the filament remained distinct through the entire length of the tube. When the speed was increased, the filament broke up at a given point and diffused throughout the cross section (see Fig. 11.7). Reynolds identified the governing parameter $u_m\, d/v$—the

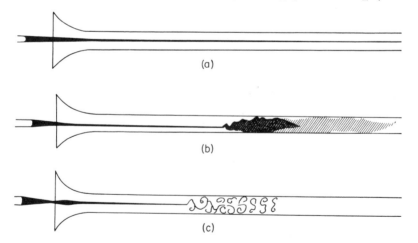

Fig. 11.7   Reynolds' turbulence experiment: (a) laminar flow; (b) and (c), transition from laminar to turbulent flow. After Osborne Reynolds, "An experimental investigation of the circumstances which determine whether the motion of water shall be direct or sinuous, and of the law of resistance in parallel channels," *Phil. Trans., Roy. Soc.*, **174**, (1883), 935–982.

Reynolds number—where u_m is the mean velocity, d is the diameter, and v is the kinematic viscosity. The point at which the color diffuses throughout the tube is the transition point from laminar to turbulent flow in the tube. Reynolds found that transition occurred at Reynolds numbers between 2000 and 13,000, depending on the smoothness of the entry conditions. When extreme care is taken, the transition can be delayed to Reynolds numbers as high as 40,000. On the other hand, a value of 2000 appears to be about the lowest value obtainable on a rough entrance.

Turbulence is one of the most important and most difficult problems in fluid mechanics. Not only is it technically important because turbulence affects the skin friction, the resistance to flow, the heat generation and transfer, diffusion, etc., but also because it is widespread. One might say that the normal mode of fluid flow is turbulent. The water in the ocean, the air above the earth, and the state of motion in the sun are turbulent. The theory of turbulence will greet you when you study fluid mechanics in greater depth.

PROBLEM 11.6 From the basic solution (11.5-15), show that the rate of mass flow through the tube is

$$(11.5\text{-}16) \qquad\qquad Q = \frac{\pi a^4 \rho}{8\mu}\, \alpha,$$

that the mean velocity is

$$(11.5\text{-}17) \qquad\qquad u_m = \frac{a^2}{8\mu}\, \alpha,$$

and that the skin friction coefficient is

$$(11.5\text{-}18) \qquad c_f = \frac{\text{shear stress}}{\text{mean dynamic pressure}} = \frac{-\mu\,(\partial u/\partial r)_{r=a}}{\tfrac{1}{2}\rho u_m^2} = \frac{16}{R_N},$$

where $R_N = 2au_m/v$.

11.6* BOUNDARY LAYER

If we let $R_N \to \infty$ in the dimensionless Navier-Stokes equation (11.4-3) for a homogeneous incompressible fluid, namely,

$$(11.6\text{-}1) \qquad \frac{Du_i'}{Dt'} = -\frac{\partial p'}{\partial x_i'} + \frac{1}{R_N}\nabla^2 u_i', \qquad (i = 1, 2, 3),$$

the last term would drop out unless the second derivatives become very large. In a general flow field in which the velocity and its derivatives are finite, the effect of viscosity would disappear when the Reynolds number tends toward infinity. Near a solid wall, however, a rapid transition takes place for the velocity to vary from that of the free stream to that of the solid because of the no-slip condition. If this transition layer is very thin, the last term cannot be dropped even though the Reynolds number is very large.

We shall define the boundary layer as *the region of a fluid in which the effect of viscosity is felt even though the Reynolds number is very large.* In the boundary layer the flow is such that the shear-stress term—the last term in Eq. (11.6-1)—is of the same order of magnitude as the convective force term. Based on the observation that in a high-speed flow the boundary layer is very thin, Prandtl (1904) simplified the Navier-Stokes equation into a much more tractable boundary-layer equation.

To see the nature of the boundary-layer equation, let us consider a two-dimensional flow over a fixed flat plate. See Fig. 11.8. We take the x'-axis in the direction of flow along the surface and the y'-axis normal to it. The velocity component w along the z'-axis is assumed to vanish. Then Eq. (11.6-1) becomes

$$(11.6\text{-}2) \qquad \frac{\partial u'}{\partial t'} + u' \frac{\partial u'}{\partial x'} + v' \frac{\partial u'}{\partial y'} = -\frac{\partial p'}{\partial x'} + \frac{1}{R_N}\left(\frac{\partial^2 u'}{\partial x'^2} + \frac{\partial^2 u'}{\partial y'^2}\right),$$

$$(11.6\text{-}3) \qquad \frac{\partial v'}{\partial t'} + u' \frac{\partial v'}{\partial x'} + v' \frac{\partial v'}{\partial y'} = -\frac{\partial p'}{\partial y'} + \frac{1}{R_N}\left(\frac{\partial^2 v'}{\partial x'^2} + \frac{\partial^2 v'}{\partial y'^2}\right).$$

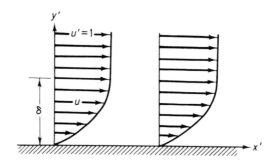

Fig. 11.8 A boundary layer of flow.

If we take the free-stream velocity as the characteristic velocity, then the dimensionless velocity u' is equal to 1 in the free stream (outside the boundary layer). The velocity u' varies from 0 on the solid surface $y' = 0$ to $u' = 1$ at $y' = \delta$ where δ denotes the boundary-layer thickness (which is dimensionless and is numerically small). We can now estimate the order of magnitude of the terms occurring in Eq. (11.6-2) as follows. We write $u' = O(1)$ to mean that u' is *at most* of the order unity. We notice that $O(1) + O(1) = O(1)$, $O(1) \cdot O(1) = O(1)$, $O(1) + O(\delta) = O(1)$, $O(1) \cdot O(\delta) = O(\delta)$. Then, since the variation of u' with respect to t' and x' is finite, we have

$$u' = O(1), \qquad \frac{\partial u'}{\partial x'} = O(1),$$

(11.6-4)

$$\frac{\partial^2 u'}{\partial x'^2} = O(1). \qquad \frac{\partial u'}{\partial t'} = O(1).$$

By the equation of continuity (11.4-4), we have

(11.6-5)
$$\frac{\partial u'}{\partial x'} = -\frac{\partial v'}{\partial y'} = O(1).$$

Hence,

(11.6-6)
$$v' = \int_0^\delta \frac{\partial v'}{\partial y'} dy' \sim \int_0^\delta O(1)\, dy = O(\delta).$$

Thus, the vertical velocity is at most of the order of δ, which is numerically small:

(11.6-7)
$$\delta \ll 1.$$

Since $v' = O(\delta)$ while $\partial v'/\partial y' = O(1)$ according to Eq. (11.6-5), we see that a differentiation with respect to y' in the boundary layer increases the order of magnitude by $1/\delta$. Then

$$\frac{\partial u'}{\partial y'} = O\left(\frac{1}{\delta}\right), \qquad \frac{\partial^2 u'}{\partial y'^2} = O\left(\frac{1}{\delta^2}\right),$$

(11.6-8)

$$\frac{\partial v'}{\partial x'} = O(\delta), \qquad \frac{\partial^2 v'}{\partial x'^2} = O(\delta),$$

$$\frac{\partial v'}{\partial t'} = O(\delta), \qquad \frac{\partial^2 v'}{\partial y'^2} = O\left(\frac{1}{\delta}\right).$$

Now, by definition, the shear stress term is of the same order of magnitude as the inertial force term in the boundary layer. But the terms on the left-hand side of (11.6-2) are all $O(1)$; hence those on the right-hand side must be also $O(1)$; in particular,

$$O(1) = \frac{\partial p'}{\partial x'},$$

(11.6-9)

$$O(1) = \frac{1}{R_N}\left(\frac{\partial^2 u'}{\partial x'^2} + \frac{\partial^2 u'}{\partial y'^2}\right) = \frac{1}{R_N}\left[O(1) + O\left(\frac{1}{\delta^2}\right)\right].$$

Since the first term in the bracket is much smaller than the second term, we have

$$O(1) = \frac{1}{R_N} O\left(\frac{1}{\delta^2}\right).$$

Hence,

(11.6-10)
$$R_N = O\left(\frac{1}{\delta^2}\right).$$

Thus, we obtain an estimate of the boundary-layer thickness,

(11.6-11)
$$\delta = O\left(\frac{1}{\sqrt{R_N}}\right).$$

Substituting Eq. (11.6-4), (11.6-8), (11.6-10) into Eq. (11.6-3), we see that all terms involving v' are $O(\delta)$; hence the remaining term $\partial p'/\partial y'$ must also be $O(\delta)$. Thus,

(11.6-12)
$$\frac{\partial p'}{\partial y'} = O(\delta) \sim 0.$$

In other words, *the pressure is approximately constant through the boundary layer.* By retaining only terms of order 1, the Navier-Stokes equations are reduced to

(11.6-13)
$$\frac{\partial u'}{\partial t'} + u'\frac{\partial u'}{\partial x'} + v'\frac{\partial u'}{\partial y'} = -\frac{\partial p'}{\partial x'} + \frac{1}{R_N}\frac{\partial^2 u'}{\partial y'^2},$$

and Eq. (11.6-12). Equation (11.6-13) is Prandtl's boundary-layer equation; it is subjected to the boundary conditions

(11.6-14)
$$u' = v' = 0 \qquad \text{for } y' = 0,$$
$$u' = 1 \qquad \text{for } y' = \delta.$$

PROBLEM 11.7 Estimate the boundary-layer thickness of air flowing over a plate 10 ft long at 100 ft/sec.

Answer: $\delta = O(0.05 \text{ in.})$.

11.7* LAMINAR BOUNDARY LAYER OVER A
FLAT PLATE

To apply Prandtl's boundary-layer theory, let us consider an incompressible fluid flowing over a flat plate as in Fig. 11.9, in which the vertical scale is magnified to make the picture clearer. The velocity outside the boundary layer is assumed constant, $\bar{u}$. We shall seek a steady-state solution so that $\partial u/\partial t = 0$. An additional assumption will be made, to be justified *a posteriori*, that the pressure gradient $\partial p/\partial x$ is negligible compared to the other terms in the boundary-layer equation. Then Eq. (11.6-13) becomes

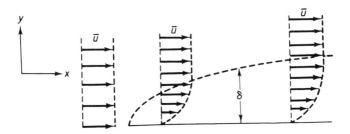

Fig. 11.9 Laminar boundary layer over a flat plate, showing the growth in thickness.

$$(11.7\text{-}1) \qquad u\frac{\partial u}{\partial x} + v\frac{\partial u}{\partial y} = v\frac{\partial^2 u}{\partial y^2}.$$

Here we return to the physical quantities and drop the primes. The equation of continuity is

$$(11.7\text{-}2) \qquad \frac{\partial u}{\partial x} + \frac{\partial v}{\partial y} = 0.$$

Equation (11.7-2) is satisfied identically if u, v are derived from a stream function $\psi(x, y)$:

$$(11.7\text{-}3) \qquad u = -\frac{\partial \psi}{\partial y}, \qquad v = \frac{\partial \psi}{\partial x}.$$

Then Eq. (11.7-1) becomes

$$(11.7\text{-}4) \qquad \frac{\partial \psi}{\partial x}\frac{\partial^2 \psi}{\partial y^2} - \frac{\partial \psi}{\partial y}\frac{\partial^2 \psi}{\partial x\,\partial y} = v\frac{\partial^3 \psi}{\partial y^3}.$$

The boundary conditions are (a) no-slip on the plate and (b) continuity at the free stream outside the boundary layer; i.e.,

$$(11.7\text{-}5) \qquad u = v = 0 \quad \text{or} \quad \frac{\partial \psi}{\partial x} = \frac{\partial \psi}{\partial y} = 0 \qquad \text{for } y = 0,$$

$$(11.7\text{-}6) \qquad u = \bar{u} \quad \text{or} \quad -\frac{\partial \psi}{\partial y} = \bar{u} \qquad \text{for } y = \delta.$$

Following Blasius,[†] we seek a "similarity" solution. Consider the transformation

$$(11.7\text{-}7) \qquad \bar{x} = \alpha x, \qquad \bar{y} = \beta y, \qquad \bar{\psi} = \gamma\psi,$$

[†] H. Blasius, "Grenzschichten in Flüssigkeiten mit kleiner Reibung," *Zeitschrift f. Math. u. Phys.*, **56** (1908), 1.

in which α, β, and γ are constants. A substitution of (11.7-7) into (11.7-4) shows that the equation for the function $\bar{\psi}(\bar{x}, \bar{y})$ has the same form as (11.7-4) if we choose $\gamma = \alpha/\beta$. A similar substitution into (11.7-6) shows that $-\partial\bar{\psi}/\partial\bar{y} = \bar{u}$ if we choose $\gamma = \beta$. Hence, $\beta = \alpha/\beta$ or $\beta = \sqrt{\alpha}$. With this choice, we have

(11.7-8)
$$\frac{\bar{y}}{\sqrt{\bar{x}}} = \frac{y}{\sqrt{x}}, \qquad \frac{\bar{\psi}}{\sqrt{\bar{x}}} = \frac{\psi}{\sqrt{x}}.$$

These relations suggest that there are solutions of the form

(11.7-9) $\qquad \psi = -f(\xi)\sqrt{\nu\bar{u}x}, \qquad \xi = \sqrt{\dfrac{\bar{u}}{\nu}}\,\dfrac{y}{\sqrt{x}}.$

Substitution of (11.7-9) into (11.7-4) yields the ordinary differential equation

(11.7-10) $\qquad\qquad 2f''' + ff'' = 0,$

were the primes indicate differentiation with respect to ξ. This equation has been solved numerically to a high degree of accuracy under the boundary conditions

(11.7-11) $\qquad f(0) = 0, \qquad f'(0) = 0, \qquad f'(\infty) = 1,$

which express the conditions that $u = 0$, $v = 0$ at the plate, and $u \to \bar{u}$, the free-stream velocity, outside the boundary layer. From Eq. (11.7-9) it is seen that for fixed x/L, $\xi \to \infty$ means that y/L is large compared with the boundary-layer thickness $\sqrt{\nu/L\bar{u}}$, or δ. The velocity distribution, yielded by the solution of Eq. (11.7-10) and (11.7-11), agrees closely with experimental evidence,[†] as seen in Fig. 11.10, except very near the leading edge of the plate where the boundary-layer approximation breaks down, and far downstream where the flow becomes turbulent.

The flow corresponding to the solution given above is a laminar flow. At a sufficient distance downstream from the leading edge the flow becomes turbulent and the Blasius solution fails. The transition occurs when a Reynolds number based on the boundary layer thickness,

$$R = \frac{\bar{u}\delta}{\nu},$$

reaches a critical value. Generally the value of the critical transitional

---

[†]J. Nikuradse, *Laminare Reibungsschichten an der längsangeströmten platte.* Monograph, Zentrale f. Wiss. Berichtswesen, Berlin, 1942. See H. Schlichting: *Boundary Layer Theory*, translated by J. Kestin, New York: McGraw-Hill Book Company (1960), p. 124.

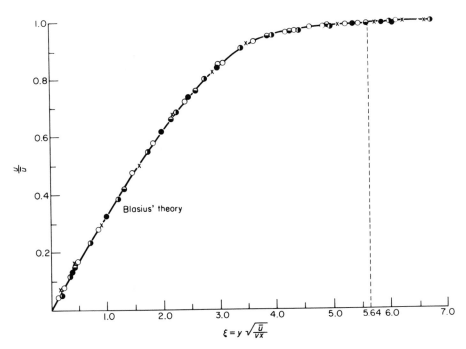

Fig. 11.10 Blasius' solution of velocity distribution in a laminar boundary layer on a flat plate at zero incidence, and comparison with Nikuradse's measurements.

Reynolds number is of the order 3000, but the exact value depends on the surface roughness, curvature, Mach number, etc.

There is a tremendous difference between a laminar boundary layer and a turbulent one with respect to heat transfer, skin friction, heat generation, etc. In our space age, the question of laminar-turbulent transition is of supreme importance with respect to reentry vehicles. As a satellite reenters the atmosphere, the heat generated by skin friction in the boundary layer is tremendous—but a turbulent boundary layer generates much more than a laminar one. For most reentry vehicles, survival is possible if the boundary layer over the nose cone is laminar: But if the flow were turbulent, the nose cone would be burned out.

11.8* NONVISCOUS FLUID

A great simplification is obtained if the coefficient of viscosity vanishes exactly. Then the stress tensor is isotropic,

$$(11.8\text{-}1) \qquad\qquad \sigma_{ij} = -p\delta_{ij},$$

and the equation of motion can be simplified into

$$(11.8\text{-}2) \qquad \rho \frac{Dv_i}{Dt} = \rho X_i - \frac{\partial p}{\partial x_i}.$$

Here ρ is the density of the fluid; p is the pressure; v_1, v_2, v_3 are the velocity components, and X_1, X_2, X_3 are the body force per unit mass.

If, in addition, the fluid is homogeneous and incompressible, then its density is a constant, and the equation of continuity (10.4-4) is reduced to the form

$$(11.8\text{-}3) \qquad \frac{\partial u}{\partial x} + \frac{\partial v}{\partial y} + \frac{\partial w}{\partial z} = 0 \quad \text{or} \quad \frac{\partial u_i}{\partial x_i} = 0.$$

A vector field satisfying Eq. (11.8-3) is said to be *solenoidal*. According to the general theory of potential, a solenoidal field can be derived from another vector field. This can be illustrated in the simple case of a *two-dimensional flow field* for which $w = 0$ and u, v are independent of z, and for which the equation of continuity is

$$(11.8\text{-}4) \qquad \frac{\partial u}{\partial x} + \frac{\partial v}{\partial y} = 0.$$

Then it is obvious that if we take an arbitrary function $\psi(x, y)$ and derive u, v according to the rules

$$(11.8\text{-}5) \qquad u = \frac{\partial \psi}{\partial y}, \qquad v = -\frac{\partial \psi}{\partial x},$$

the equation (11.8-4) will be satisfied identically. Such a function ψ is called a *stream function*.

Substituting (11.8-5) into the equation of motion (11.8-2), we obtain the governing equations (for the two-dimensional flow)

$$(11.8\text{-}6) \qquad \begin{aligned} \frac{\partial^2 \psi}{\partial t \, \partial y} + \frac{\partial \psi}{\partial y} \frac{\partial^2 \psi}{\partial x \, \partial y} - \frac{\partial \psi}{\partial x} \frac{\partial^2 \psi}{\partial y^2} &= X - \frac{1}{\rho} \frac{\partial p}{\partial x}, \\ -\frac{\partial^2 \psi}{\partial t \, \partial x} - \frac{\partial \psi}{\partial y} \frac{\partial^2 \psi}{\partial x^2} + \frac{\partial \psi}{\partial x} \frac{\partial^2 \psi}{\partial x \, \partial y} &= Y - \frac{1}{\rho} \frac{\partial p}{\partial y}. \end{aligned}$$

If the body force is zero, an elimination of p yields

$$(11.8\text{-}7) \qquad \frac{\partial}{\partial t} \nabla^2 \psi + \psi_y \nabla^2 \psi_x - \psi_x \nabla^2 \psi_y = 0,$$

in which

$$\nabla^2 = \frac{\partial^2}{\partial x^2} + \frac{\partial^2}{\partial y^2},$$

and subscripts indicate partial differentiation.

PROBLEM 11.8 Show that for a two-dimensional flow of an incompressible viscous fluid the governing equation for the stream function defined by (11.8-5) is

(11.8-8) $\dfrac{\partial}{\partial t}\nabla^2\psi + \psi_y\nabla^2\psi_x - \psi_x\nabla^2\psi_y = \nu\nabla^2\nabla^2\psi + \dfrac{\partial X}{\partial y} - \dfrac{\partial Y}{\partial x}.$

11.9* VORTICITY AND CIRCULATION

The concept of circulation and vorticity is of great importance in fluid mechanics. The *circulation* $I(\mathcal{C})$ in any closed circuit $\mathcal{C}$ is defined by the line integral

(11.9-1) $I(\mathcal{C}) = \displaystyle\int_{\mathcal{C}} \mathbf{v} \cdot \mathbf{dl} = \int_{\mathcal{C}} v_i\,dx_i,$

where $\mathcal{C}$ is any closed curve in the fluid, and the integrand is the scalar product of the velocity vector $\mathbf{v}$ and the vector $\mathbf{dl}$, which is tangent to the curve $\mathcal{C}$ and of length dl (Fig. 11.11). Clearly, the circulation is a function of both the velocity field and the chosen curve $\mathcal{C}$.

By means of Stokes's theorem, if $\mathcal{C}$ encloses a simply connected region, the line integral can be transformed into a surface integral

(11.9-2) $I(\mathcal{C}) = \displaystyle\int_{S} (\nabla \times \mathbf{v})_n\,dS = \int_{S} (\text{curl }\mathbf{v})_i v_i\,dS,$

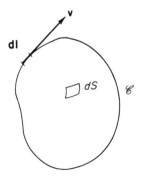

Fig. 11.11   Circulation. Notations.

where S is any surface in the fluid bounded by the curve $\mathcal{C}$, v_i is the unit normal to the surface and curl $\mathbf{v} = e_{ijk}v_{j,k}$. The curl $\mathbf{v}$ is called the *vorticity* of the velocity field.

The law of change of circulation with time, when the circuit $\mathcal{C}$ is a *fluid line*, i.e., a curve $\mathcal{C}$ formed by the same set of fluid particles as time changes, is given by the *theorem of Lord Kelvin:* If the fluid is nonviscous and the body force is conservative, then

(11.9-3)
$$\frac{DI}{Dt} = -\int_{\mathcal{C}} \frac{dp}{\rho}.$$

If, in addition to the conditions above, the density ρ is a unique function of the pressure, then the fluid is called *barotropic*, and the last integral vanishes because the integral would be single-valued and $\mathcal{C}$ is a closed curve. We then have the *Helmholtz theorem that*

(11.9-4)
$$\frac{DI}{Dt} = 0.$$

To prove the theorems above, we note that since $\mathcal{C}$ is a fluid line composed always of the same particles, the order of differentiation and integration may be interchanged in the following:

(11.9-5) $$\frac{D}{Dt} \int_{\mathcal{C}} v_i \, dx_i = \int_{\mathcal{C}} \frac{D}{Dt}(v_i \, dx_i) = \int_{\mathcal{C}} \left(\frac{Dv_i}{Dt} dx_i + v_i \frac{D \, dx_i}{Dt}\right).$$

But $D \, dx_i/Dt$ is the rate at which dx_i is increasing as a consequence of the motion of the fluid; hence it is equal to the difference of the velocities parallel to x_i at the ends of the element, i.e., dv_i. Substituting Dv_i/Dt from the equation of motion (11.8-2) and replacing $D \, dx_i/Dt$ by dv_i, we obtain

(11.9-6)

$$\frac{DI}{Dt} = \int_{\mathcal{C}} \left[\left(-\frac{1}{\rho}\frac{\partial p}{\partial x_i} + X_i\right) dx_i + v_i \, dv_i\right] = -\int_{\mathcal{C}} \frac{dp}{\rho} + \int_{\mathcal{C}} X_i \, dx_i + \int_{\mathcal{C}} dv^2.$$

Of the terms on the right-hand side, the last vanishes because v^2 is single-valued in the flow field; the second vanishes if the body force X_i is conservative. Hence Kelvin's theorem is proved. Helmholtz's theorem follows immediately as a special case because the integral on the right-hand side vanishes if the fluid is barotropic.

In the clear cut conclusion of Helmholtz theorem lies its importance. For if we limit our attention to a barotropic fluid, then we have $I = $ constant. Hence if the circulation vanishes at one instant of time, it must vanish for all times. If this is so for any arbitrary fluid lines in a field, then according to Eq. (11.9-2) the vorticity vanishes in the whole field. This leads to a great simplification which will be discussed in Sec. 11.10, namely, the irrotational flow. To appreciate the importance of this simplification, one need to remark only that the vast majority of the classical literature on fluid mechanics deals with irrotational flows.

Note that the circulation around a fluid line *does not* have to remain constant if the density ρ depends on other variables in addition to pressure. Into this category fall most geophysical problems in which the temperature enters as a parameter affecting both ρ and p. Also, in stratified flows ρ is a function of location, not necessarily as a function of p alone.

The significance of the word *fluid line* in the theorems of Kelvin and Helmholtz may be seen by considering the problem of a thin airfoil moving in the air. The conditions of the Helmholtz theorem are satisfied. Hence, the circulation I about any fluid line never changes with time. Since the motion of the fluid is caused by the motion of the airfoil and since at the beginning the fluid is at rest and $I = 0$, it follows that I vanishes at all times. Note, however, that the volume occupied by the airfoil is exclusive of the fluid. A fluid line $\mathcal{C}$ enclosing the boundary of the airfoil becomes elongated when the airfoil moves forward as shown in Fig. 11.12. According to the Helmholtz theorem, the circulation about $\mathcal{C}$ is zero, so that the total vorticity inside $\mathcal{C}$ vanishes, but one cannot conclude that the vorticity actually vanishes everywhere inside $\mathcal{C}$. In the region occupied by the airfoil, *and* in the wake behind the airfoil, vorticity does exist. However, the Helmholtz theorem applies to the region outside the airfoil and its wake, and the vanishing of circulation about every possible fluid line clearly shows that the flow is irrotational outside the airfoil *and* its wake.

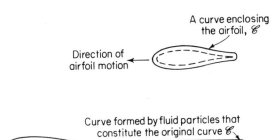

Fig. 11.12 Fluid line enclosing an airfoil and its wake.

11.10* IRROTATIONAL FLOW

A flow is said to be *irrotational* if the vorticity vanishes everywhere, i.e., if

$$(11.10\text{-}1) \qquad \nabla \times \mathbf{v} = \operatorname{curl} \mathbf{v} = 0,$$

or

$$e_{ijk}v_{j,k} = 0.$$

For a two-dimensional irrotational flow, we must have

(11.10-2)
$$\frac{\partial u}{\partial y} - \frac{\partial v}{\partial x} = 0.$$

If the fluid is incompressible and a stream function defined by (11.8-5) is introduced, then a substitution of (11.8-5) into (11.10-2) yields the equation

(11.10-3)
$$\frac{\partial^2 \psi}{\partial x^2} + \frac{\partial^2 \psi}{\partial y^2} = 0.$$

This is the famous *Laplace equation*, whose solution is the concern of many books on applied mathematics.

We can show that an irrotational flow of an incompressible fluid is governed by a Laplace equation even in the three-dimensional case, because by the definition of irrotationality the following three equations must hold:

(11.10-4) $\dfrac{\partial u}{\partial y} - \dfrac{\partial v}{\partial x} = 0,$ $\dfrac{\partial v}{\partial z} - \dfrac{\partial w}{\partial y} = 0,$ $\dfrac{\partial w}{\partial x} - \dfrac{\partial u}{\partial z} = 0.$

These equations can be satisfied identically if the velocities u, v, w are derived from a *potential function* $\Phi(x, y, z)$ according to the rule

(11.10-5) $u = \dfrac{\partial \Phi}{\partial x},$ $v = \dfrac{\partial \Phi}{\partial y},$ $w = \dfrac{\partial \Phi}{\partial z}.$

If, in addition, the fluid is *incompressible*, then a substitution of (11.10-5) into (11.1-5) yields the *Laplace equation*

(11.10-6)
$$\frac{\partial^2 \Phi}{\partial x^2} + \frac{\partial^2 \Phi}{\partial y^2} + \frac{\partial^2 \Phi}{\partial z^2} = 0.$$

Since Φ is a potential function, this equation is also called a *potential equation*.

The incompressible potential flow is governed by the Laplace equation. If a solution can be found that satisfies all the boundary conditions, then the Eulerian equation of motion yields the pressure gradient and the problem is solved. The nonlinear convective acceleration, which causes the central difficulty of fluid mechanics, does not hinder the solution of potential flows of an incompressible fluid. This is why the potential theory is so simple and so important.

To realize the usefulness of the potential theory, we must quote the

Helmholtz theorem (see Sec. 11.9). If the motion of any portion of a fluid mass is irrotational at any one instant of time, it will continue to be irrotational at all times, provided that the body forces are conservative and that the fluid is *barotropic* (i.e., its density is a function of pressure alone). These conditions are met in most problems. If a solid body is immersed in a fluid and suddenly set in motion, the motion generated in a nonviscous fluid would be irrotational.† Hence, a whole class of technologically important problems are irrotational.

11.11* COMPRESSIBLE NONVISCOUS FLUIDS

If the fluid is compressible, the equation of continuity (10.5-3) is

$$\frac{\partial \rho}{\partial t} + \frac{\partial \rho v_j}{\partial x_j} = 0. \tag{11.11-1}$$

If the fluid is nonviscous, the Eulerian equation of motion is

$$\frac{\partial v_i}{\partial t} + v_j \frac{\partial v_i}{\partial x_j} = -\frac{1}{\rho} \frac{\partial p}{\partial x_i} + X_i. \tag{11.11-2}$$

The density is uniquely related to the pressure only if the temperature T is explicitly accounted for. Thus, if the temperature is known to be constant (isothermal), we have for an ideal gas

$$\frac{p}{\rho} = \text{const.,} \qquad T = \text{const.,} \tag{11.11-3}$$

whereas if the flow is isentropic (adiabatic and reversible), we have

$$\frac{p}{\rho^\gamma} = \text{const.,} \qquad \frac{T}{\rho^{\gamma-1}} = \text{const.,} \tag{11.11-4}$$

where γ is the ratio of specific heats at constant pressure C_p and constant volume C_v; i.e., $\gamma = C_p/C_v$. Either of these two cases is *barotropic*.

In other cases, it is necessary to introduce the temperature explicitly as a variable. Then we must introduce also the equation of state relating p, ρ, and T and the *caloric* equation of state relating C_p, C_v, and T.

Let us consider, as an example, the propagation of small disturbances in a barotropic fluid in the absence of body force. Let us write

$$c^2 = \frac{dp}{d\rho}. \tag{11.11-5}$$

---

†See H. Lamb, *Hydrodynamics*, New York: Dover Publications, 6th ed. (1945), pp. 10, 11.

The velocity of flow will be assumed to be so small that the second-order terms may be neglected in comparison with the first-order term. Correspondingly, the disturbances in the density ρ and the pressure p and the derivatives of ρ and p are also first-order infinitesimal quantities. Then, on neglecting the body force X_i and all small quantities of the second or higher order, Eq. (11.11-1) and (11.11-2) are linearized to

$$(11.11\text{-}6) \qquad \frac{\partial \rho}{\partial t} + \rho \frac{\partial v_j}{\partial x_j} = 0,$$

$$(11.11\text{-}7) \qquad \frac{\partial v_i}{\partial t} = -\frac{1}{\rho} \frac{\partial p}{\partial x_i} = -\frac{1}{\rho} \frac{dp}{d\rho} \frac{\partial \rho}{\partial x_i} = -\frac{c^2}{\rho} \frac{\partial \rho}{\partial x_i}.$$

Differentiating Eq. (11.11-6) with respect to t and Eq. (11.11-7) with respect to x_i, again neglecting the second-order terms and eliminating the sum $\rho \, \partial^2 v_j / \partial t \, \partial x_j$, we obtain

$$(11.11\text{-}8) \qquad \frac{1}{c^2} \frac{\partial^2 \rho}{\partial t^2} = \frac{\partial^2 \rho}{\partial x_i \, \partial x_i};$$

i.e.,

$$\frac{1}{c^2} \frac{\partial^2 \rho}{\partial t^2} = \frac{\partial^2 \rho}{\partial x^2} + \frac{\partial^2 \rho}{\partial y^2} + \frac{\partial^2 \rho}{\partial z^2}.$$

This is the *wave equation for the propagation of small disturbances.* It is the basic equation of acoustics.

By the same linearization procedure, and because the change of pressure is proportional to the change in density, $dp = c^2 \, d\rho$, we see that the same wave equation governs the pressure disturbance:

$$(11.11\text{-}9) \qquad \frac{1}{c^2} \frac{\partial^2 p}{\partial t^2} = \frac{\partial^2 p}{\partial x_k \, \partial x_k}.$$

Further, from Eq. (11.11-7) and (11.11-8) or (11.11-9), we deduce that

$$(11.11\text{-}10) \qquad \frac{1}{c^2} \frac{\partial^2 v_i}{\partial t^2} = \frac{\partial^2 v_i}{\partial x_k \, \partial x_k}.$$

Hence, in the linearized theory, ρ, p, v_1, v_2, v_3 are governed by the same wave equation.

Let us apply these equations to the problem of a source of disturbance (sound) located at the origin and radiating symmetrically in all directions. We may visualize a spherical siren. Because of the radial symmetry, we have

$$(11.11\text{-}11) \qquad \frac{\partial^2}{\partial x^2} + \frac{\partial^2}{\partial y^2} + \frac{\partial^2}{\partial z^2} = \frac{\partial^2}{\partial r^2} + \frac{2}{r} \frac{\partial}{\partial r}.$$

Hence, Eq. (11.11-8) becomes

$$(11.11\text{-}12) \qquad \frac{1}{c^2}\frac{\partial^2 \rho}{\partial t^2} = \frac{\partial^2 \rho}{\partial r^2} + \frac{2}{r}\frac{\partial \rho}{\partial r}.$$

It can be verified by direct substitution that a general solution of this equation is the sum of two arbitrary functions f and g:

$$(11.11\text{-}13) \qquad \rho = \rho_0 + \frac{1}{r} f(r - ct) + \frac{1}{r} g(r + ct),$$

Here ρ_0 is a constant (undisturbed density of the field.) The f term represents a wave radiating out from the origin, and the g term, one converging toward the origin. Perhaps the clearest way to see this is to consider a special case in which the function $f(r - ct)$ is a step function: $f(r - ct) = \epsilon\mathbf{1}(r - ct)$, where ϵ is small and $\mathbf{1}(r - ct)$ is the unit-step function, which is zero when $r - ct < 0$ and is 1 when $r - ct > 0$. The disturbance is, therefore, a small jump across a line of discontinuity described by the equation $r - ct = 0$. At time $t = 0$, the disturbance is located at the origin. At time t, the line of discontinuity is moved to $r = ct$. Thus, c is the speed of propagation of the disturbance. The general case follows by the principle of superposition. In acoustics, c is called the *velocity of sound.*

The velocity of sound $c = (dp/d\rho)^{1/2}$ depends on the relationship between pressure and density. If we are concerned with an ideal gas and the condition is isentropic, we have, from Eq. (11.11-4),

$$(11.11\text{-}14) \qquad c = \sqrt{\frac{\gamma p}{\rho}}.$$

In the history of mechanics there was a long story about the propagation of sound in the air. The first theoretical investigation of the velocity of sound was made by Newton (1642–1727), who assumed (11.11-3) and obtained $c = \sqrt{p/\rho}$ in a publication in 1687. It was found that the sound speed calculated from Newton's formula falls short of the experimental value of sound speed by a factor of approximately one-sixth. This discrepancy was not explained until Laplace (1749–1827) pointed out that the rate of compression and expansion in a sound wave is so fast that there is no time for any appreciable heat interchange by conduction; thus the process must be considered as adiabatic. This argument becomes plausible if we think of the step wave discussed in the preceding paragraph. For a step wave the sudden changes in ρ and p that take place as the wave front sweeps by must be accomplished at the wave front in an infinitesimal region of space and time. Heat transfer in such a small time interval is negligible. Hence, the gas flows isentropically across the discontinuity. As a general sound wave is a superposition of such step waves, the entire flow is isentropic. Therefore,

Eq. (11.11-4) applies and (11.11-14) results. Experiments have verified that Laplace was right.

Generally, then, the wave equations (11.11-8), et seq. are associated with isentropic flows. To apply these equations, conditions that guarantee isentropy, such as the absence of strong shock waves and the smallness of thermal diffusivity, must be observed.

11.12* SUBSONIC AND SUPERSONIC FLOW

The basic wave equation (11.11-8) is referred to a frame of reference which is at rest relative to the fluid at infinity. It imposes no restriction on where and how the disturbances are generated. The sources of disturbances may be moving or changing with time; the same equation holds. The nature of the sources would appear only in the boundary conditions and initial conditions.

A flying aircraft is a source of disturbance in still air. The disturbances come to us as sound waves governed by the wave equation. As we all know, the nature of disturbances changes drastically as the aircraft's flight speed changes from subsonic to supersonic. In the latter case we hear the sonic boom.

It is convenient to study the nature of flow about an aircraft in a wind tunnel. We shall therefore write down the wave-propagation equation for disturbances in the air flowing in a tunnel as they appear to us standing on the ground.

Consider a body of fluid coming from, say, the left, with a uniform velocity U at infinity. If the disturbances are indicated by a prime, we assume the velocity components to be

$$(11.12\text{-}1) \qquad u = U + u', \qquad v = v', \qquad w = w', \qquad U = \text{const.}$$

and the pressure and density

$$(11.12\text{-}2) \qquad p = p_0 + p', \qquad \rho = \rho_0 + \rho'.$$

The whole investigation would be simplified if we could assume that the disturbances are infinitesimal quantities of the first order; i.e.,

$$(11.12\text{-}3) \qquad u', v', w' \ll U, \qquad p' \ll p_0, \qquad \rho' \ll \rho_0.$$

Under these assumptions the basic equations (11.11-1) and (11.11-4) may be linearized as before. In fact, repeating the steps as in Sec. 11.11 with our new assumptions, we obtain the equation of continuity

$$\frac{\partial \rho'}{\partial t} + \rho_0\left(\frac{\partial U}{\partial x} + \frac{\partial u'}{\partial x} + \frac{\partial v'}{\partial y} + \frac{\partial w'}{\partial z}\right) + (U + u')\frac{\partial \rho'}{\partial x} + v'\frac{\partial \rho'}{\partial y} + w'\frac{\partial \rho'}{\partial z} = 0,$$

which is linearized into

(11.12-4) $$\frac{\partial \rho'}{\partial t} + \rho_0 \left(\frac{\partial u'}{\partial x} + \frac{\partial v'}{\partial y} + \frac{\partial w'}{\partial z}\right) + U\frac{\partial \rho'}{\partial x} = 0.$$

Similarly, the equations of motion are linearized into

(11.12-5)
$$\frac{\partial u'}{\partial t} + U\frac{\partial u'}{\partial x} = -\frac{1}{\rho_0}\frac{\partial p'}{\partial x} = -\frac{c^2}{\rho_0}\frac{\partial \rho'}{\partial x},$$

$$\frac{\partial v'}{\partial t} + U\frac{\partial v'}{\partial x} = -\frac{c^2}{\rho_0}\frac{\partial \rho'}{\partial y},$$

$$\frac{\partial w'}{\partial t} + U\frac{\partial w'}{\partial x} = -\frac{c^2}{\rho_0}\frac{\partial \rho'}{\partial z}.$$

Differentiating the three equations (11.12-5) with respect to x, y, z, respectively, adding, and again neglecting the second-order terms, we obtain

$$\frac{\partial}{\partial t}\left(\frac{\partial u'}{\partial x} + \frac{\partial v'}{\partial y} + \frac{\partial w'}{\partial z}\right) + U\frac{\partial}{\partial x}\left(\frac{\partial u'}{\partial x} + \frac{\partial v'}{\partial y} + \frac{\partial w'}{\partial z}\right)$$

$$= -\frac{c^2}{\rho_0}\left(\frac{\partial^2 \rho'}{\partial y^2} + \frac{\partial^2 \rho'}{\partial y^2} + \frac{\partial^2 \rho'}{\partial z^2}\right).$$

Hence, on eliminating $\partial u'/\partial x + \partial v'/\partial y + \partial w'/\partial z$ with (11.12-4), we have

(11.12-6) $$\frac{\partial^2 \rho'}{\partial t^2} + 2U\frac{\partial^2 \rho'}{\partial x\,\partial t} + U^2\frac{\partial^2 \rho'}{\partial x^2} = c^2\left(\frac{\partial^2 \rho'}{\partial x^2} + \frac{\partial^2 \rho'}{\partial y^2} + \frac{\partial^2 \rho'}{\partial z^2}\right).$$

This is the basic equation for compressible flow in aerodynamics. The pressure p', velocity components v_i', and the velocity potential Φ, $v_i' = \Phi_{,i}$ are governed by the same equation.

Let us examine the basic equation above in some simpler cases. Consider a steady flow around a model at rest. Then all derivatives with respect to time t vanish, and the velocity potential Φ is governed by the equation

(11.12-7) $$\frac{U^2}{c^2}\frac{\partial^2 \Phi}{\partial x^2} = \frac{\partial^2 \Phi}{\partial x^2} + \frac{\partial^2 \Phi}{\partial y^2} + \frac{\partial^2 \Phi}{\partial z^2}.$$

This equation now depends on only one dimensionless parameter, U/c, which is called the *Mach number* and is denoted by

(11.12-8) $$M = \frac{U}{c}.$$

The nature of the solution to Eq. (11.12-7) changes tremendously depending on whether M is greater or less than 1. We call a flow *subsonic* if $M < 1$ or *supersonic* if $M > 1$. We write, for a subsonic flow,

$$(11.12\text{-}9) \qquad (1 - M^2)\frac{\partial^2\Phi}{\partial x^2} + \frac{\partial^2\Phi}{\partial y^2} + \frac{\partial^2\Phi}{\partial z^2} = 0 \qquad (M < 1),$$

whereas, for a supersonic flow,

$$(11.12\text{-}10) \qquad (M^2 - 1)\frac{\partial^2\Phi}{\partial x^2} - \frac{\partial^2\Phi}{\partial y^2} - \frac{\partial^2\Phi}{\partial z^2} = 0 \qquad (M > 1).$$

Equation (11.12-9) is a partial differential equation of the *elliptic type*. Equation (11.12-10) is one of the *hyperbolic type*. Let us consider an example showing the difference between these equations.

Let a very thin plate with a small sinusoidal wavy profile be placed in a steady flow, with the mean chord of the plate parallel to the velocity U at infinity. See Figs. 11.13 and 11.14. The waves of the plate shall be described by the equation

$$(11.12\text{-}11) \qquad\qquad z = a \sin\frac{\pi x}{L}.$$

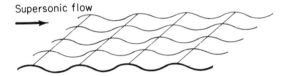

Fig. 11.13   A wavy plate in a steady supersonic flow.

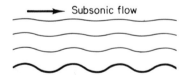

Fig. 11.14   A wavy plate in a steady subsonic flow.

The amplitude a is assumed to be small compared with the wave length L:

$$(11.12\text{-}12) \qquad\qquad a \ll L.$$

The fluid, since it is assumed to be perfect, can glide over the plate but cannot penetrate it. Therefore, the velocity vector of the flow must be tangent to the plate. Now the velocity vector has the components

$$(11.12.13) \qquad\qquad U + u', v', w'$$

in the x-, y-, z-direction. On the other hand, the normal vector to the surface

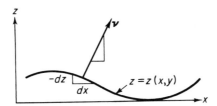

Fig. 11.15 Surface normal and the velocity boundary condition.

(11.12-11) has the following components (see Fig. 11.15):

(11.12-4) $-\dfrac{\partial z}{\partial x},\qquad -\dfrac{\partial z}{\partial y},\qquad 1.$

If the velocity vector, with components given in (11.12-13), is to be tangent to the plate surface, it must be normal to the normal vector (11.12-14). Hence, the condition of nonpenetration can be stated as the orthogonality of the vectors (11.12-13) and (11.12-14), i.e., by the condition that their scalar product vanishes:

$$-(U + u')\frac{\partial z}{\partial x} - v'\frac{\partial z}{\partial y} + w'\cdot 1 = 0.$$

Omitting higher order terms, we obtain the boundary condition

(11.12-15) $w' = U\dfrac{\partial z}{\partial x}$

on the plate. From (11.12-11), this is

(11.12-16) $w' = U\dfrac{a\pi}{L}\cos\dfrac{\pi x}{L}$ $\left(\text{when } z = a\sin\dfrac{\pi z}{L}\right).$

Again, counting on the continuity and differentiability of the function $w'(x, y, z)$, we can write

(11.12-7) $w'(x, y, z) = w'(x, y, 0) + z\left(\dfrac{\partial w'}{\partial z}\right)_{z=0} + \dots.$

For small z, all the terms following the first are higher order terms. Consistently neglecting these terms, we can simplify the boundary condition into

(11.12-18) $w' = U\dfrac{a\pi}{L}\cos\dfrac{\pi x}{L}$ (when $z = 0$).

This boundary condition is not sufficient to determine the solution to our problem, which is governed by either Eq. (11.12-9) or Eq. (11.12-10), depending on whether the flow is subsonic or supersonic. In addition, the conditions at infinity must be specified. There is a great difference between the elliptic and hyperbolic equations with respect to the appropriate types of boundary conditions that may be specified, and we must consider them in some detail.

For the elliptic equation, Eq. (11.12-9), the influence of disturbance is spread out in all directions; and it is reasonable to assume that, for any finite body, the disturbances tend toward zero at distances infinitely far away from the body. A rigorous argument may be based on the total energy that may be imparted to the fluid. If the fluid velocity is distributed in a certain fashion and if it does not tend toward zero at a certain rate as the distance from the body increases toward infinity, an infinitely large energy would have to be imparted to the fluid in order to create the motion, which is impossible. (For further details, see texts on partial differential equations or aerodynamics.) Accordingly, we impose the following conditions for our problem:

(a) The flow is two-dimensional, parallel to the xz-plane, and there is no dependence on the y-coordinate.

(b) All disturbances tend toward zero as $z \to \pm\infty$. In particular,

$$(11.12\text{-}19) \qquad u', v', w' \to 0; \quad \text{i.e.,} \quad \Phi \to \text{const. as } z \to \pm\infty.$$

Turning now to the hyperbolic equation (11.12-10), we find that the disturbances can be carried away along waves of limited dimension. The

Fig. 11.16(a) Flow past a flat plate with a beveled, sharp, leading edge, the top surface being aligned with the free stream of Mach number 8. On the top side of the plate, a laminar boundary layer is revealed by the lighter line. A shock wave is induced by the displacement effect of the boundary layer. Similar features are seen on the lower side. Schlieren system. Flow left to right. Courtesy of Toshi Kubota, California Institute of Technology.

argument of decreasing amplitude does not apply. Instead, the boundary condition must be replaced by the *radiating condition:* that the plate is the only source of disturbance and that the disturbances radiate from the source, not toward it.

This description of the radiation condition is easy to apply when we are concerned with a single source. For example, of the two solutions on the right-hand side of Eq. (11.11-13), the term $f(r - ct)/r$ represents a wave radiating from the origin; hence, for a source at the origin it is the only term admissible under the radiation condition. The condition becomes somewhat confounded when applied to a two-dimensional steady flow. Perhaps the matter can be clarified by examining some photographs of supersonic flow about stationary models in a wind tunnel, such as those shown in Fig. 11.16(a) and (b). Here the flows are from the left to right. We see that the lines of disturbances, which are contours of density of the fluid as revealed by these Schlieren photographs, incline to the right. This direction of inclination of the strong (shock) and weak (Mach) waves is determined by the radiation condition.

Fig. 11.16(b) Scale mode of the Nimbus spacecraft in a 50-in. hypersonic tunnel, at Mach number 8 and Reynolds number of 0.42×10^6/ft. Schlieren system. Flow left to right. Courtesy of Von Karman Gas Dynamics Facility, ARO, Inc.

Now we can return to our problem. It is easily verified by direct substitution that, in the subsonic case, Eq. (11.12-9) can be satisfied by a function of the form

(11.12-20)
$$\Phi = A e^{\mu z} \cos \frac{\pi x}{L}.$$

On substituting Eq. (11.12-20) into Eq. (11.12-9) we obtain

$$-(1 - M^2)\left(\frac{\pi}{L}\right)^2 A e^{\mu z} \cos \frac{\pi x}{L} + A \mu^2 e^{\mu z} \cos \frac{\pi x}{L} = 0,$$

or

(11.12-21)
$$\mu = \pm \left(\frac{\pi}{L}\right)\sqrt{1 - M^2}.$$

If the plus sign is used in Eq. (11.12-21), the function Φ in Eq. (11.12-20) will grow exponentially without limit as $z \longrightarrow \infty$. On the other hand, if the minus sign is used, the condition (11.12-19) can be satisfied. Hence, we may try

(11.12-22)
$$\Phi = A e^{-(\pi/L)\sqrt{1-M^2}\,z} \cos \frac{\pi x}{L}.$$

The vertical velocity w' computed from Φ is

(11.12-23) $$w' = \frac{\partial \Phi}{\partial z} = -\frac{\pi}{L}\sqrt{1 - M^2}\,A e^{-(\pi/L)\sqrt{1-M^2}\,z} \cos \frac{\pi x}{L}.$$

On setting $z = 0$ in Eq. (11.12-23) and applying the boundary condition (11.12-18) we obtain

(11.12-24)
$$A = -\frac{Ua}{\sqrt{1 - M^2}}.$$

Now all the boundary conditions for the subsonic case are satisfied. Hence, the solution for the subsonic case is

(11.12-25)
$$\Phi = -\frac{Ua}{\sqrt{1 - M^2}} e^{-(\pi/L)\sqrt{1-M^2}\,z} \cos \frac{\pi x}{L}.$$

We see that the disturbances decrease exponentially with increasing z. From this solution we can deduce the velocity field, the pressure field, and the density field. In particular, since

(11.12-26)
$$U \frac{\partial u'}{\partial x} = -\frac{1}{\rho} \frac{\partial p'}{\partial x},$$

we have

(11.12-27) $p' = -\rho U u' = -\rho U \dfrac{\partial \Phi}{\partial x}.$

The streamlines for such a flow is plotted in Fig. 11.14.

Turning now to the supersonic case, Eq. (11.12-10), we see that it can be satisfied by the function

(11.12-28) $\Phi = f(x - \sqrt{M^2 - 1}\, z) + g(x + \sqrt{M^2 - 1}\, z),$

where f and g are arbitrary functions, because if we set

(11.12-29) $\xi = x - \sqrt{M^2 - 1}\, z,$

then

$$\frac{\partial f}{\partial x} = \frac{df}{d\xi}, \qquad \frac{\partial f}{\partial z} = -\sqrt{M^2 - 1}\,\frac{df}{d\xi};$$

hence,

$$(M^2 - 1)\frac{\partial^2 f}{\partial x^2} - \frac{\partial^2 f}{\partial z^2} = (M^2 - 1)\frac{d^2 f}{d\xi^2} - (M^2 - 1)\frac{d^2 f}{d\xi^2} = 0,$$

and Eq. (11.12-10) is satisfied. The lines

(11.12-30) $\xi = $ const.,   i.e., $x - \sqrt{M^2 - 1}\, z = $ const.,

are the Mach waves, along which the disturbances are propagated with undiminished intensity. These lines are inclined in the correct direction as revealed by the wind-tunnel photographs. On the other hand, the Mach lines for the function $g(x + \sqrt{M^2 - 1}\, z)$ are inclined in the wrong direction. Hence, the function g must be rejected on the basis of the radiation condition. Therefore, we may try

(11.12-31) $\Phi = f(x - \sqrt{M^2 - 1}\, z).$

From Eq. (11.12-31), we obtain

(11.12-32) $w' = \dfrac{\partial \Phi}{\partial z} = -\sqrt{M^2 - 1}\,\dfrac{df}{d\xi}.$

Comparing Eq. (11.12-32) with the boundary condition (11.12-18), we obtain, when $z = 0$,

(11.12-33) $-\sqrt{M^2 - 1}\left(\dfrac{df}{d\xi}\right)_{z=0} = \dfrac{U a\pi}{L}\cos\dfrac{\pi x}{L} = \dfrac{U a\pi}{L}\cos\dfrac{\pi \xi}{L}\bigg|_{z=0}.$

Hence, on integrating and returning to (11.12-29), we have

$$(11.12\text{-}34) \qquad \Phi = f = -\frac{Ua}{\sqrt{M^2 - 1}} \sin \frac{\pi}{L}(x - \sqrt{M^2 - 1}\, z),$$

which solves the problem. The plot of the streamlines is shown in Fig. 11.13. The contrast of the two cases is dramatic. Whereas in the subsonic case the pressure disturbance is diminished as the distance from the plate increases; in the supersonic case it does not. This is, of course, the reason why a sonic boom hits us with all its fury from a supersonic aircraft but not from a subsonic one.

PROBLEMS

11.9 Derive the Navier-Stokes equation for an incompressible fluid in cylindrical polar coordinates.

Solution: The left-hand side of the Navier-Stokes equation represents acceleration. In polar coordinates the components are a_r, a_θ, a_z, which are given in Eq. (10.9-9) on p. 258. The right-hand side is the vector divergence of the stress tensor. In polar coordinates they are given by Eq. (10.9-11). It remains to write down the stresses in terms of the velocities u, v, w along the radial, circumferential, and axial directions. On p. 138, we have e_{rr}, $e_{\theta\theta}$, etc., expressed in terms of u_r, u_θ, u_z. The relationship between the strain rates $\dot{e}_{rr}$, $\dot{e}_{\theta\theta}$, etc., to the velocities u, v, w are the same. Hence

$$\dot{e}_{rr} = \frac{\partial u}{\partial r}, \qquad \dot{e}_{\theta\theta} = \frac{u}{r} + \frac{1}{r}\frac{\partial v}{\partial \theta}, \text{ etc.}$$

Therefore, from Eq. (7.3-6) and for an incompressible fluid, we have

$$\sigma_{rr} = -p + 2\mu\dot{e}_{rr} = -p + 2\mu\frac{\partial u}{\partial r},$$

$$\sigma_{\theta\theta} = -p + 2\mu\dot{e}_{\theta\theta} = -p + 2\mu\left(\frac{u}{r} + \frac{1}{r}\frac{\partial v}{\partial \theta}\right),$$

$$\sigma_{zz} = -p + 2\mu\dot{e}_{zz} = -p + 2\mu\frac{\partial w}{\partial z},$$

$$\sigma_{r\theta} = 2\mu\dot{e}_{r\theta} = \mu\left(r\frac{\partial(v/r)}{\partial r} + \frac{1}{r}\frac{\partial u}{\partial \theta}\right),$$

$$\sigma_{\theta z} = 2\mu\dot{e}_{\theta z} = \mu\left(\frac{1}{r}\frac{\partial w}{\partial \theta} + \frac{\partial v}{\partial z}\right),$$

$$\sigma_{zr} = 2\mu\dot{e}_{zr} = \mu\left(\frac{\partial u}{\partial z} + \frac{\partial w}{\partial r}\right).$$

A substitution into Eq. (10.9-11) yields the Navier-Stokes equation:

$$\frac{\partial u}{\partial t} + u\frac{\partial u}{\partial r} + \frac{v}{r}\frac{\partial u}{\partial \theta} + w\frac{\partial u}{\partial z} - \frac{v^2}{r} = -\frac{1}{\rho}\frac{\partial p}{\partial r} + \nu\left(\nabla^2 u - \frac{u}{r^2} - \frac{2}{r^2}\frac{\partial v}{\partial \theta}\right) + F_r,$$

$$\frac{\partial v}{\partial t} + u\frac{\partial v}{\partial r} + \frac{v}{r}\frac{\partial v}{\partial \theta} + w\frac{\partial v}{\partial z} + \frac{uv}{r} = -\frac{1}{\rho}\frac{1}{r}\frac{\partial p}{\partial \theta} + \nu\left(\nabla^2 v + \frac{2}{r^2}\frac{\partial u}{\partial \theta} - \frac{v}{r^2}\right) + F_\theta,$$

$$\frac{\partial w}{\partial t} + u\frac{\partial w}{\partial r} + \frac{v}{r}\frac{\partial w}{\partial \theta} + w\frac{\partial w}{\partial z} = -\frac{1}{\rho}\frac{\partial p}{\partial z} + \nu\nabla^2 w + F_z,$$

where

$$\nabla^2 \equiv \frac{\partial^2}{\partial r^2} + \frac{1}{r}\frac{\partial}{\partial r} + \frac{1}{r^2}\frac{\partial^2}{\partial \theta^2} + \frac{\partial^2}{\partial z^2}.$$

The equation of continuity is

$$\frac{1}{r}\frac{\partial}{\partial r}(ru) + \frac{1}{r}\frac{\partial v}{\partial \theta} + \frac{\partial w}{\partial z} = 0.$$

11.10 Blood is a non-Newtonian fluid whose viscosity varies with the strain rate. See Fig. 9.19 and Prob. 9.3. Derive the equation of motion of blood in a form analogous to the Navier-Stokes equations. Formulate mathematically the problem of blood flow in a living heart.

11.11 If air is truly nonviscous, would an airplane be able to fly? What about birds and insects?

11.12 If water is nonviscous, would fishes be able to swim? What are the differences in the arguments for fishes in water and birds in the air?

11.13 Formulate the problem of tides induced on the earth under the influence of the moon. [*Ref.* Horace Lamb, *Hydrodynamics*, New York: Dover Publications, pp. 358–362.]

11.14 Waves are generated in water in a long channel of rectangular cross section. What are the equations with which the wavelength and frequency can be determined?

11.15 Ripples are generated on the surface of water in a deep pond. Does the wave speed depend on the wavelength? Even though the full solution is rather complicated, whether or not the waves are dispersive (i.e., whether the speed depends on wavelength) can be detected when all the basic equations are written down. If the free surface of the pond is taken as the xy-plane and the z-axis points downward, try a two-dimensional solution with velocity components

$$v \equiv 0, \quad u = ae^{-kz}\sin kx \sin \omega t, \quad \text{and} \quad w = -ae^{-kz}\cos kx \sin \omega t.$$

11.16 Consider a ground-effect machine, which uses one or more reaction jets and hovers above the ground. Sketch the streamlines of the flow and write the equations and boundary conditions that govern the machine when it is hovering.

11.17 Analyze the motion in a cumulus cloud in a summer thunderstorm.

What are the variables relevant to this problem? If temperature is an important consideration, how would it be incorporated into the basic equations? Gravity must not be neglected. Present the basic equations. Make a dimensional analysis to determine fundamental dimensionless parameters.

11.18 Water waves run up a sloping beach and create all the panorama on the seashore: surf, riptides, waves, ripples, and foam. Analyze the phenomena mathematically. Give an appropriate choice of variables. Write down the differential equations and boundary conditions. Make simplifying assumptions if you think they are appropriate, but state your assumptions clearly.

11.19 On the beach there are riptides which are fast-moving narrow streams of water which move toward the ocean in a direction perpendicular to the shoreline and are dangerous to swimmers. Now this is an anomaly. For a two-dimensional sloping beach and a two-dimensional water wave, we obtain a three-dimensional solution. Is there any basic objection to this situation (from the mathematician's point of view, not the swimmer's)? Can you name another example of such a phenomenon in nature?

11.20 When wind blows over (perpendicular to) long cylindrical pipes, vortices are shed in the wake. These vortices induce vibrations in the pipe. A trans-Arabic oil line (the above-ground part) was reported to have suffered severe vibrations due to wind. Smokestacks, large rockets, etc., are subjected to these disturbances. Vortex shedding over a long cylinder is three-dimensional; in other words, the shedding is nonuniform along the length of the cylinder, even if the wind and the cylinder are both uniform. Formulate the aerodynamic problem for a fixed, rigid cylinder. Furnish all the differential equations and boundary conditions. Make a dimensional analysis to determine all the dimensionless parameters involved.

11.21 Generalize the problem above to consider the vortex shedding over a flexible, vibrating cylinder.

11.22 Using the equations derived in Prob. 11.9, find the velocity field in a Couette flowmeter (Fig. P3.22, p. 88).

Answer: Let $v = \omega_1 a$ at $r = a$ and $v = \omega_2 b$ at $r = b$. Then

$$v = (a^2 - b^2)^{-1}[(\omega_1 a^2 - \omega_2 b^2)r - a^2 b^2(\omega_2 - \omega_1)/r].$$

11.23 Using the Navier-Stokes equation, find the velocity distribution of a flow in a long cylindrical pipe of rectangular cross section.

FURTHER READING

GOLDSTEIN, S. (ed.), *Modern Development in Fluid Dynamics* (2 vol.), London: Oxford University Press (1938).

LAMB, HORACE, *Hydrodynamics*, 1st ed. 1879, 6th ed. 1932, New York: Dover Publications (1945).

LIEPMANN, H. W. AND A. ROSHKO, *Elements of Gasdynamics*, New York: Wiley (1957).

PRANDTL, L., *The Essentials of Fluid Dynamics*, London: Blackie (1953).

SCHLICHTING, H., *Boundary Layer Theory*, 4th ed., New York: McGraw-Hill (1960).

YIH, CHIA-SHUN, *Fluid Mechanics, a Concise Introduction to the Theory*, New York: McGraw-Hill (1969).

CHAPTER TWELVE

Some Simple Problems in Elasticity

The simplest solids are those obeying Hooke's law. The statics and dynamics of Hookean solids is the theme of the classical theory of elasticity. In this chapter we present the derivation of the basic equations of the theory and show how the theory is simplified under various conditions of symmetry and idealizations. Several types of boundary-value problems will be considered, including those concerned with the propagation of elastic waves.

An important practical problem in engineering is the transmission of forces and moments from one place to another. For this purpose beams and shafts are used. St.-Venant's brilliant success in solving the problems of torsion and bending in the theory of elasticity is an important event in the history of mechanics. We shall consider his theory in some detail in this chapter.

12.1 BASIC EQUATIONS OF ELASTICITY FOR HOMOGENEOUS ISOTROPIC BODIES

In the preceding chapter we have discussed the equations governing the flow of fluids. In this chapter we shall consider the motion of solids which obey the Hooke's law. A Hookean body has a unique natural state, to which the body returns when all external loads are removed. All strains and particle displacements are measured from this natural state: Their values are counted as zero in that state.

The basic equations can be gleaned from the preceding chapters. Let $u_i(x_1, x_2, x_3, t)$, $i = 1, 2, 3$, describe the displacement of a particle located at x_1, x_2, x_3 at time t from its position in the natural state. Various strain measures can be defined for the displacement field. The Almansi strain tensor is expressed in terms of $u_i(x_1, x_2, x_3, t)$ according to Eq. (5.3-4):

$$(12.1\text{-}1) \qquad e_{ij} = \frac{1}{2}\left[\frac{\partial u_j}{\partial x_i} + \frac{\partial u_i}{\partial x_j} - \frac{\partial u_k}{\partial x_i}\frac{\partial u_k}{\partial x_j}\right].$$

The particle velocity v_i is given by the material derivative of the displacement,

$$(12.1\text{-}2) \qquad v_i = \frac{\partial u_i}{\partial t} + v_j\frac{\partial u_i}{\partial x_j}.$$

The particle acceleration α_i is given by the material derivative of the velocity, Eq. (10.3-7),

$$(12.1\text{-}3) \qquad \alpha_i = \frac{\partial v_i}{\partial t} + v_j\frac{\partial v_i}{\partial x_j}.$$

The conservation of mass is expressed by the equation of continuity (10.5-3),

$$(12.1\text{-}4) \qquad \frac{\partial \rho}{\partial t} + \frac{\partial(\rho v_i)}{\partial x_i} = 0.$$

The conservation of momentum is expressed by the Eulerian equation of motion (10.6-7),

$$(12.1\text{-}5) \qquad \rho\alpha_i = \frac{\partial \sigma_{ij}}{\partial x_j} + X_i.$$

Hooke's law for a homogeneous isotropic material is

$$(12.1\text{-}6) \qquad \sigma_{ij} = \lambda e_{kk}\delta_{ij} + 2Ge_{ij},$$

where λ and G are Lamé constants.

Equations (12.1-1) through (12.1-6) together describe a theory of elasticity. If we compare these equations with the corresponding equations for a viscous fluid as shown in Sec. 11.1, we see that the theoretical structure is similar except that here we have a nonlinear strain-and-displacement-gradient relation (12.1-1), in contrast to the linear rate-of-deformation-and-velocity-gradient relation (6.1-3) for the fluid. Hence, the theory of elasticity is more deeply involved in nonlinearity than the theory of viscous fluids.

The nonlinear problem is so wrought with mathematical complexities that a general theory is not available. The solutions of even relatively simple problems usually require extensive numerical treatment. For this reason it is common to simplify the theory by introducing a severe restriction that *the displacements and velocities are infinitesimal* so that Eq. (12.1-1) to (12.1-3) can be linearized. Such a linearized theory is amenable to mathematical treatment. A phenomenon described by a set of linear equations is easier to understand; the principle of superposition applies, and more complex solutions can be obtained by superposition of the simpler ones. In any case, the linearized theory provides a foundation on which the nonlinear theory can be built. One tries to learn as much as possible about the linearized theory and then proceed to discover what special features are introduced by the nonlinearities.

Little can be said about the nonlinear theory at the elementary level. Hence, we shall *linearize* the equations by restricting ourselves to values of u_i, v_i so small that the nonlinear terms in Eq. (12.1-1) through (12.1-3) may be neglected. Thus,

$$(12.1\text{-}7) \qquad e_{ij} = \frac{1}{2}\left(\frac{\partial u_i}{\partial x_j} + \frac{\partial u_j}{\partial x_i}\right),$$

$$(12.1\text{-}8) \qquad v_i = \frac{\partial u_i}{\partial t}, \qquad \alpha_i = \frac{\partial v_i}{\partial t}.$$

Equations (12.1-4) through (12.1-8) together are 22 equations for the 22 unknowns ρ, u_i, v_i, α_i, e_{ij}, σ_{ij}. We may eliminate σ_{ij} by substituting Eq. (12.1-6) into Eq. (12.1–5) and using Eq. (12.1-7) to obtain the well-known *Navier's equation*,

$$(12.1\text{-}9) \qquad G\nabla^2 u_i + (\lambda + G)\frac{\partial e}{\partial x_i} + X_i = \rho\frac{\partial^2 u_i}{\partial t^2},$$

where e is the divergence of the displacement vector $\mathbf{u}$,

$$(12.1\text{-}10) \qquad e = \frac{\partial u_j}{\partial x_j} = \frac{\partial u_1}{\partial x_1} + \frac{\partial u_2}{\partial x_2} + \frac{\partial u_3}{\partial x_3}.$$

∇^2 is the *Laplace operator*. If we write x, y, z instead of x_1, x_2, x_3, we have

$$(12.1\text{-}11) \qquad \nabla^2 = \frac{\partial^2}{\partial x^2} + \frac{\partial^2}{\partial y^2} + \frac{\partial^2}{\partial z^2}.$$

If we introduce the Poisson's ratio v as in Eq. (9.6-9), we can write Navier's equation (12.1-9) as

$$(12.1\text{-}12) \qquad G\left(\nabla^2 u_i + \frac{1}{1-2v}\frac{\partial e}{\partial x_i}\right) + X_i = \rho\frac{\partial^2 u_i}{\partial t^2}.$$

This is the basic field equation of the linearized theory of elasticity.

Navier's equation (12.1-9) must be solved for appropriate boundary conditions, which are usually one of two kinds:

(1) *Specified displacements.* The components of displacement u_i are prescribed on the boundary.

(2) *Specified surface tractions.* The components of surface traction $\overset{v}{T}_i$ are assigned on the boundary.

In most problems of elasticity, the boundary conditions are such that over part of the boundary the displacements are specified, whereas over another part the surface tractions are specified. In the latter case Hooke's law may be used to convert the boundary condition into prescribed values of a certain combination of the first derivatives of u_i.

12.2 PLANE ELASTIC WAVES

To illustrate the use of the linearized equations, let us consider a simple harmonic wave train in an elastic medium. Let us assume that the displacement components u_1, u_2, u_3 (or, in unabridged notations, u, v, w) are infinitesimal and that the body force X_i vanishes. Then it is easy to verify that a solution of the Navier's equation (12.1-9) is

$$(12.2\text{-}1) \qquad u = A\sin\frac{2\pi}{l}(x \pm c_L t), \qquad v = w = 0,$$

where A, l, c_L are constants, provided that the constant c_L is chosen to be

$$(12.2\text{-}2) \qquad c_L = \sqrt{\frac{\lambda + 2G}{\rho}} = \sqrt{\frac{E(1-v)}{(1+v)(1-2v)\rho}}.$$

The pattern of motion expressed by Eq. (12.2-1) is unchanged when $x \pm c_L t$ remains constant. Hence, if the negative sign were taken, the pattern would move to the right with a velocity c_L as the time t increases. The constant c_L is called the *phase velocity* of the wave motion. In Eq. (12.2-1), l is the *wavelength*, as can be seen from the sinusoidal pattern of u as a function of x, at any instant of time. The particle velocity computed from Eq. (12.2-1) is in the same direction as that of the wave propagation (namely, the x-axis). Such a motion is said to constitute a train of *longtudinal waves*. Since at any instant of time the wave crests lie in parallel planes, the motion represented by Eq. (12.2-1) is called a train of *plane waves*.

Next, let us consider the motion

$$(12.2\text{-}3) \qquad u = 0, \qquad v = A \sin \frac{2\pi}{l}(x \pm ct), \qquad w = 0,$$

which represents a train of plane waves of wavelength l propagating in the x-axis direction with a phase velocity c. When Eq. (12.2-3) are substituted into Eq. (12.1-9), it is seen that c must assume the value c_T,

$$(12.2\text{-}4) \qquad c_T = \sqrt{\frac{G}{\rho}}.$$

The particle velocity (in the y-direction) computed from Eq. (12.2-3) is perpendicular to the direction of wave propagation (x-direction). Hence, it is said to be a *transverse wave*. The speeds c_L and c_T are called the characteristic *longitudinal wave speed* and *transverse wave speed*, respectively. They depend on the elastic constants and the density of the material. The ratio c_T/c_L depends on Poisson's ratio only,

$$(12.2\text{-}5) \qquad c_T = c_L \sqrt{\frac{1 - 2\nu}{2(1 - \nu)}}.$$

If $\nu = 0.25$, then $c_L = \sqrt{3}\, c_T$.

Similar to Eq. (12.2-3), the following example represents a transverse wave in which the particles move in the z-axis direction.

$$(12.2\text{-}6) \qquad u = 0, \qquad v = 0, \qquad w = A \sin \frac{2\pi}{l}(x \pm c_T t).$$

The plane parallel to which the particles move [such as the xy-plane in Eq. (12.2-3) or the xz-plane in Eq. (12.2-6)] is called the *plane of polarization*.

Table 9.7, p. 218, gives a brief list of the longitudinal wave velocities of some common media. It is interesting to see that most metals and alloys have approximately the same wave velocities.

Plane waves as described above may exist only in an unbounded elastic continuum. In a finite body, a plane wave will be reflected when it hits a boundary. If there is another elastic medium beyond the boundary, refracted waves occur in the second medium. The features of reflection and refraction are similar to those in acoustics and optics; the main difference is that, in elasticity, an incident longitudinal wave will be reflected and refracted in a combination of longitudinal and transverse waves, and an incident transverse wave will also be reflected in a combination of both types of waves. The details can be worked out by the proper combination of these waves so that the boundary conditions are satisfied.

12.3 SIMPLIFICATIONS

Important simplifications to the equation of the theory of elasticity may come from
(1) Homogeneity and isotropy.
(2) Absence of inertial forces.
(3) High degree of symmetry in geometry.
(4) Plane stress and plane strain.
(5) Thin-walled structures—plates and shells.

Clearly, a simplification is obtained if the number of independent variables is reduced. Thus, if nothing changes with time, the variable t will be suppressed. Homogeneity of materials makes the coefficients of the differential equations constant. Isotropy reduces the number of independent material constants. High degree of symmetry reduces geometric parameters in a problem. Reduction of the general field equations to two diemensions or one dimension reduces the number of independent variables. These simplifications make the physical problems easier to grasp.

Example 1. A Plane State of Stress

A *plane-stress* state depending on x, y only may be visualized as one that exists in a thin membrane stressed in its own plane. Figure 4.1, p. 93, shows an example of such a case. Analytically, a plane-stress state is defined by the condition that the stress components $\sigma_{zz}, \sigma_{zx}, \sigma_{zy}$ vanish everywhere,

$$(12.3\text{-}1) \qquad \sigma_{zz} = \sigma_{zx} = \sigma_{zy} = 0,$$

whereas the stress components $\sigma_{xx}, \sigma_{xy}, \sigma_{yy}$ are independent of the coordinate z. The strain components can be calculated from Hooke's law:

$$e_{xx} = \frac{1}{E}(\sigma_{xx} - v\sigma_{yy}),$$

$$e_{yy} = \frac{1}{E}(\sigma_{yy} - v\sigma_{xx}),$$

$$(12.3\text{-}2) \qquad e_{zz} = -\frac{v}{E}(\sigma_{xx} + \sigma_{yy}),$$

$$e_{xy} = \frac{1}{2G}\sigma_{xy},$$

$$e_{xz} = e_{yz} = 0.$$

On solving these equations for σ_{xx}, σ_{yy}, and σ_{xy}, we obtain

$$(12.3\text{-}3) \qquad \sigma_{xx} = \frac{E}{1 - v^2}(e_{xx} + v e_{yy}), \qquad \sigma_{xy} = \frac{E}{1 + v}e_{xy}.$$

σ_{yy} can be obtained from (12.3-3) by interchanging the index x and y.

Substituting Eq. (12.3-3) into the equation of motion (12.1-5) and using (12.1-7), we obtain the basic equation for plane stress expressed in terms of the displacement components u, v:

$$(12.3\text{-}4) \qquad \begin{aligned} G\left(\frac{\partial^2 u}{\partial x^2} + \frac{\partial^2 u}{\partial y^2}\right) + G\frac{1 + v}{1 - v}\frac{\partial}{\partial x}\left(\frac{\partial u}{\partial x} + \frac{\partial v}{\partial y}\right) + X &= \rho\frac{\partial^2 u}{\partial t^2}, \\ G\left(\frac{\partial^2 v}{\partial x^2} + \frac{\partial^2 v}{\partial y^2}\right) + G\frac{1 + v}{1 - v}\frac{\partial}{\partial y}\left(\frac{\partial u}{\partial x} + \frac{\partial v}{\partial y}\right) + Y &= \rho\frac{\partial^2 v}{\partial t^2}. \end{aligned}$$

This set of equations is similar to but simpler than Navier's Eqs. (12.1-12).

For a plate without forces acting on its lateral surfaces, Eq. (12.3-1) are true on the lateral surfaces. Since the plate is thin one may assume (12.3-1) to be valid throughout the plate. Although this does not directly imply that the stresses σ_{xx}, σ_{xy}, σ_{yy} are independent of the coordinate z, detailed investigations on stress distribution in thin plates under the hypotheses (12.3-1) show that the variations of σ_{xx}, σ_{xy}, σ_{yy} and u, v, w with z are small: The smaller the ratio of the thickness of the plate h to its typical dimension L in the xy-plane, the more σ_{xx}, σ_{xy}, σ_{yy}, u, v, w can be regarded as constant throughout the plate thickness. In the limit $h/L \to 0$ the variations of σ_{xx}, σ_{yy} with z can be made as small as we please. In this sense the plane-stress state may be regarded as representing the state in a thin plate deforming without bending.

Example 2. A Plane State of Strain

If the z-component of displacement w vanishes everywhere, and if the displacements u, v are functions of x, y only, and not of z, the body is said to be in a *plane-strain* state depending on x, y only. Such a state may be visualized as one that exists in a long cylindrical body loaded uniformly along the axis. With a plane-strain state we must have

$$(12.3\text{-}5) \qquad \frac{\partial u}{\partial z} = \frac{\partial v}{\partial z} = w = 0.$$

Substituting the conditions (12.3-5) into Navier's equation (12.1-12), we obtain the basic equation for plane strain:

(12.3-6)
$$G\left(\frac{\partial^2 u}{\partial x^2} + \frac{\partial^2 u}{\partial y^2}\right) + \frac{1}{1-2v}G\frac{\partial}{\partial x}\left(\frac{\partial u}{\partial x} + \frac{\partial v}{\partial y}\right) + X = \rho\frac{\partial^2 u}{\partial t^2},$$
$$G\left(\frac{\partial^2 v}{\partial x^2} + \frac{\partial^2 v}{\partial y^2}\right) + \frac{1}{1-2v}G\frac{\partial}{\partial y}\left(\frac{\partial u}{\partial x} + \frac{\partial v}{\partial y}\right) + Y = \rho\frac{\partial^2 v}{\partial t^2}.$$

This equation is very similar to that governing plane stress, Eq. (12.3-4).

12.4* THE AIRY STRESS FUNCTION

To solve problems in either plane stress or plane strain, one may begin by finding stresses that satisfy the equations of equilibrium. Let us assume that the body forces are absent. Then the equations of equilibrium

(12.4-1)
$$\frac{\partial \sigma_{ij}}{\partial x_j} = 0, \qquad (i = 1, 2)$$

reduce to

(12.4-2)
$$\frac{\partial \sigma_{xx}}{\partial x} + \frac{\partial \sigma_{xy}}{\partial y} = 0,$$

(12.4-3)
$$\frac{\partial \sigma_{xy}}{\partial x} + \frac{\partial \sigma_{yy}}{\partial y} = 0.$$

Equations (12.4-2) and (12.4-3) are two equations with three unknowns σ_{xx}, σ_{xy}, σ_{yy}; and it can be expected that an infinite number of solutions can be obtained.

Indeed, let $\Phi(x, y)$ be an arbitrary function and let

(12.4-4)
$$\frac{\partial^2 \Phi}{\partial y^2} = \sigma_{xx}, \qquad \frac{\partial^2 \Phi}{\partial x^2} = \sigma_{yy}, \qquad -\frac{\partial^2 \Phi}{\partial x\,\partial y} = \sigma_{xy}.$$

We can immediately verify that Eq. (12.4-2) and (12.4-3) are identically satisfied. Therefore, as far as the equations of equilibrium are concerned, Eq. (12.4-4) constitutes a general solution. The problem is reduced to finding a function Φ that satisfies the condition of compatibility and the boundary conditions.

To choose among the infinite number of solutions that satisfy the equation of equilibrium, we use the equations of compatibility (6.3-4). The only compatibility equation not identically satisfied is

(12.4-5)
$$\frac{\partial^2}{\partial y^2}(\sigma_{xx} - v\sigma_{yy}) + \frac{\partial^2}{\partial x^2}(\sigma_{yy} - v\sigma_{xx}) = 2(1 + v)\frac{\partial^2\sigma_{xy}}{\partial x\,\partial y}.$$

Differentiating Eq. (12.4-2) with respect to x and Eq. (12.4-3) with respect to y and adding, we obtain

(12.4-6)
$$\frac{\partial^2 \sigma_{xx}}{\partial x^2} + \frac{\partial^2 \sigma_{yy}}{\partial y^2} = -2 \frac{\partial^2 \sigma_{xy}}{\partial x\, \partial y}.$$

Eliminating σ_{xy} between Eq. (12.4-6) and (12.4-5), we obtain

(12.4-7)
$$\left(\frac{\partial^2}{\partial x^2} + \frac{\partial^2}{\partial y^2}\right)(\sigma_{xx} + \sigma_{yy}) = 0.$$

Substituting the stress components derived from Φ according to Eq. (12.4-4) into Eq. (12.4-7), we obtain finally the equation that governs the function Φ:

(12.4-8)
$$\frac{\partial^4 \Phi}{\partial x^4} + 2\frac{\partial^4 \Phi}{\partial x^2 \partial y^2} + \frac{\partial^4 \Phi}{\partial y^4} = 0.$$

Thus, we conclude that *an arbitrary function $\Phi(x, y)$ that satisfies Eq. (12.4-8) can generate according to Eq. (12.4-4) a stress system $\sigma_{xx}, \sigma_{xy}, \sigma_{yy}$, that automatically satisfies all the equations of equilibrium and compatibiiity.*

This marvelous approach was discovered by the English astronomer Sir George Biddell Airy (1801–1892). The function $\Phi(x, y)$ is called the *Airy stress function*.

Once the problem is reduced to the solution of boundary-value problems associated with the biharmonic equation (12.4-8), the entire machinery of the theory of functions of a complex variable can be brought to bear on the problem. The beautiful mathematical theory that evolved, however, is too extensive to be reviewed even cursorily. We shall be content to refer the reader to more advanced text books.

12.5* TORSION OF A CIRCULAR CYLINDRICAL SHAFT

We shall now illustrate an application of the linearized elasticity theory by considering the problem of torsion. To transmit a torque from one place to another a shaft is employed. In engineering, therefore, it is important to understand the stress distribution in a shaft.

The solution of the problem of torsion by Barre de Saint-Venant in 1855 was an important event in the history of engineering science. The degree of difficulty for this problem depends on the geometry of the shaft. If the shaft is a circular cylinder, the solution is simple. If it is a cylinder of a noncircular cross section or if the shaft has variable cross sections, then it is difficult.

Let us consider the simple problem of torsion of a cylindrical shaft of circular cross section. See Fig. 12.1, which shows the notations and the co-

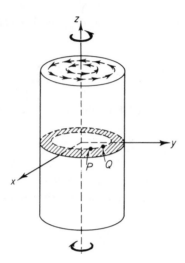

Fig. 12.1   Torsion of a circular shaft.

ordinate axes to be used. Before tackling the problem analytically, let us look at the physical conditions. Under the torque, the shaft twists. Let the cross section at $z = 0$ be fixed. Since the shaft as well as the loading are homogeneous along the z-axis, the twist must be uniform along the z-axis. Hence, the deformation must be expressible in terms of twist per unit length α, which is a constant independent of z. The quantity α represents the rotation of a section at $z = 1$ relative to that at $z = 0$.

By reason of symmetry, it is obvious that a circular cross section of the shaft remains circular when a torque is applied. What about the axial displacements of such a section? Consider a plane cross section such as that at $z = 0$ before the torque is applied. When the torque T is applied, would a particle P in this plane move in the z-direction? Suppose that it did; then by symmetry a particle Q located on the same circle with P would have been displaced by the same amount. Therefore, the axial displacement, if any, would cause a symmetric distortion of the plane. But which way could it go? Could it be bulging up? Let us assume that it does bulge up. Then if we turn the shaft upside down, we would see the plane bulge down. But the loading and the shaft look identical before and after you turned it upside down. In other words, it would seem that it bulges both downward and upward. Since a unique displacement must exist, the only possibility is that the plane section remains plane.

Summarizing the discussion above, we see that the distortion of a circular shaft under a torque must be a relative rotation of the cross sections at a uniform rate of twist. Therefore, the displacement of a particle located at (x, y, z) would appear to be, in polar coordinates,

$$(12.5\text{-}1) \qquad u_r = 0, \qquad u_\theta = \alpha z r, \qquad u_z = 0,$$

or, in rectangular Cartesian coordinates,

$$(12.5\text{-}2) \qquad u_x = -\alpha z y, \qquad u_y = \alpha z x, \qquad u_z = 0,$$

as is shown in Fig. 12.2.

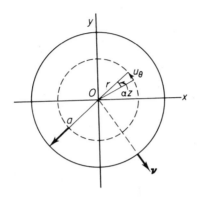

Fig. 12.2   Notations.

We shall now show that this is indeed correct. Since the displacement are given, there is no need to check the compatibility conditions. We must, however, check the equations of equilibrium and the boundary conditions.

From Eq. (12.5.2) we have the strain components

$$e_{xx} = 0, \qquad e_{yy} = 0, \qquad e_{zz} = 0,$$

$$e_{xy} = \frac{1}{2}\left(\frac{\partial u_x}{\partial y} + \frac{\partial u_y}{\partial x}\right) = \frac{1}{2}(\alpha z - \alpha z) = 0,$$

$$(12.5\text{-}3)$$

$$e_{xz} = \frac{1}{2}\left(\frac{\partial u_x}{\partial z} + \frac{\partial u_z}{\partial x}\right) = -\frac{1}{2}\alpha y,$$

$$e_{yz} = \frac{1}{2}\left(\frac{\partial u_y}{\partial z} + \frac{\partial u_z}{\partial y}\right) = \frac{1}{2}\alpha x.$$

The stress-strain relation yields the corresponding stress components:

$$\sigma_{xx} = \sigma_{yy} = \sigma_{zz} = \sigma_{xy} = 0,$$

$$(12.5\text{-}4) \qquad \sigma_{xz} = -G\alpha y,$$

$$\sigma_{yz} = G\alpha x,$$

where G is the shear modulus of the shaft material.

The equations of equilibrium are obtained by omitting α_i and X_i in Eq. (12.1-5):

(12.5-5)
$$\frac{\partial \sigma_{xx}}{\partial x} + \frac{\partial \sigma_{xy}}{\partial y} + \frac{\partial \sigma_{xz}}{\partial z} = 0,$$
$$\frac{\partial \sigma_{yx}}{\partial x} + \frac{\partial \sigma_{yy}}{\partial y} + \frac{\partial \sigma_{yz}}{\partial z} = 0,$$
$$\frac{\partial \sigma_{zx}}{\partial x} + \frac{\partial \sigma_{zy}}{\partial y} + \frac{\partial \sigma_{zz}}{\partial z} = 0,$$

which are obviously satisfied by the stress components given in Eq. (12.5-4).

The boundary conditions of our problem consists of the fact that the lateral surfaces are stress free and that the ends are acted on by a torque. Since there is no tension or compression on the ends, we have

(12.5-6) $\sigma_{zz} = 0$ (on $z = -L$ and $z = L$).

This is satisfied by Eq. (12.5-4).

The stress vector acting on the lateral surface is given by $\overset{v}{T_i}$, where $\mathbf{v}$ denotes the vector normal to the lateral surface. By Cauchy's formula,

(12.5-7) $\overset{v}{T_i} = v_j \sigma_{ij}.$

Setting $i = 1, 2, 3$, we have the three equations

(12.5-8)
$$\sigma_{xx} v_x + \sigma_{xy} v_y + \sigma_{xz} v_z = 0,$$
$$\sigma_{yx} v_x + \sigma_{yy} v_y + \sigma_{yz} v_z = 0,$$
$$\sigma_{zx} v_x + \sigma_{zy} v_y + \sigma_{zz} v_z = 0,$$

where v_x, v_y, v_z are the direction cosines of the normal vector to the lateral surface. Now on the lateral surface it is evident from Fig. 12.2 that the normal vector $\mathbf{v}$ coincides with the radius vector. Hence, the components of $\mathbf{v}$ are

(12.5-9) $v_x = \dfrac{x}{a},$ $v_y = \dfrac{y}{a},$ $v_z = 0.$

Hence, the boundary conditions on the circumference C are

(12.5-10)
$$x\sigma_{xx} + y\sigma_{xy} = 0,$$
$$x\sigma_{yx} + y\sigma_{yy} = 0,$$
$$x\sigma_{zx} + y\sigma_{zy} = 0,$$

which are again satisfied by the Eq. (12.5-4).

There is left to check only the condition that the stresses acting on the ends $z = -L$ and $z = L$ are equipollent to a torque. Referring to Fig. 12.3 and using Eq. (12.5-4), we see that the resultant of the stresses acting on the end cross sections are

$$\iint \sigma_{xz}\, dx\, dy = -G\alpha \iint y\, dx\, dy = -G\alpha \int_{-a}^{a} dx \int_{-\sqrt{a^2-x^2}}^{\sqrt{a^2-x^2}} y\, dy = 0,$$

(12.5-11) $\quad \iint \sigma_{yz}\, dx\, dy = G\alpha \iint x\, dx\, dy = 0,$

$$\iint \sigma_{zz}\, dx\, dy = 0.$$

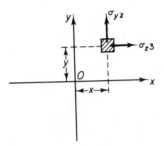

Fig. 12.3 Stresses in a twisted shaft.

Hence, the resultant force vanishes as desired. The resultant moment about the z-axis is, however,

(12.5-12) $\qquad \iint (x\sigma_{yz} - y\sigma_{xz})\, dx\, dy.$

On substituting from Eq. (12.5-4), we have

$$\text{moment} = G\alpha \iint (x^2 + y^2)\, dx\, dy$$

$$= G\alpha \int_{0}^{2\pi} d\theta \int_{0}^{a} r^3\, dr$$

$$= \frac{2\pi G\alpha a^4}{4}.$$

Thus, we see that the resultant moment is indeed a torque of magnitude T:

(12.5-13) $\qquad T = \dfrac{\pi a^4 G\alpha}{2}$

The checking is now complete. All equations of equilibrium and boundary conditions are satisfied. The solution contained in Eq. (12.5-1) through (12.5-4) are exact.

PROBLEM 12.1 Consider the torsion of a shaft of square cross section. Write down all the boundary conditions. Show that the solution contained in Eq. (12.5-1) through (12.5-4) no longer satisfy all the boundary conditions.

12.6* SAINT-VENANT'S PRINCIPLE

The solution of the problem of torsion is exact only if the shear stress distribution acting on the ends of a shaft happens to be exactly as described in Eq. (12.5-4). In reality, we can hardly guarantee such a distribution if the torque is applied in some mechanical manner, for example, by a wrench or by a crankshaft. Therefore, the danger exists that the solution we have found has academic interest only, without practical applications.

St.-Venant felt intuitively that such is not the case. He thought that the applicability of his solution was much greater than the mathematics justified. In 1855 he announced a principle which states that for a slender elongated body the local distribution of forces acting on a small portion of the body may be changed without changing the stress field in the body at some distance from the point where the redistribution was made. The new set of forces must be statically equivalent to the old set of forces. For example, for an elongated body shown in Fig. 12.4, let a set of forces act in a small area A.

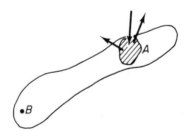

Fig. 12.4 Illustration of St-Venant's principle.

Then at a point B sufficiently distant from A the stress depends only on the resultant force and the resultant moment over A and not on their exact distribution. If the resultant force and moment of the forces acting over A vanish, then at a large distance from A this group of forces induce only negligible stresses.

Applying this principle to our torsion problem, we see that the shear stress distribution in the shaft depends only on the resultant torque and not on how the troque is generated, except near the ends of the shaft.

St.-Venant's principle is famous and convenient, but is it true? How can it be proved? The answer is that the principle has worked remarkably well. A number of precise statements of the principle and a number of sufficient conditions have been found for it. The reader is referred to the author's *Foundations of Solid Mechanics*, pp. 300–309 and 495.

12.7* BEAMS

When a structural member is used to transmit bending moment and transverse shear, it is called a *beam*. Beams are used constantly in engineering and, therefore, are important objects for research. The floor we stand on is resting on beams. An airplane wing is a beam. Bridges are made of beams, and so on. An engineer should know the stress and deformation in a beam, how to choose the materials for a beam, how to use the material efficiently by a proper geometric design, how to minimize the weight, how to maximize the stiffness and stability of beams, how to utilize supports to minimize vibrations, how to calculate the loads that act on the beam (static and moving loads, wind loads on a building, aerodynamic load on an airplane, etc.), how to analyze aeroelastic or hydroelastic interactions in case a beam is used in a fluid flow (such as an airplane wing, a ship structure), etc.

Beams are classified according to the condition of support at the ends. An end is called *simply supported* when it is free to rotate but restrained from lateral translation. An end is said to be *free* when it is free to rotate and deflect. An end is said to be *clamped* when translation and rotation are both prevented.

In Sec. 7.7 we have considered the pure bending of a prismatic beam of a homogeneous isotropic Hookean material. We deduced certain results, but we did not check all the field and boundary conditions. We shall now show that all these conditions are satisfied.

Consider the pure bending of a prismatic beam as shown in Fig. 7.1, p. 171. Let the beam be subjected to two equal and opposite couples M acting in a plane of symmetry of the cross sections of the beam. Let the x, y, z-axes of reference be chosen as in Sec. 7.7, with the origin located at the centroid of a cross section. In Sec. 7.7 we were led to the conclusion that the stress distribution in the beam is

$$(12.7\text{-}1) \qquad \sigma_{xx} = \frac{Ey}{R}, \qquad \sigma_{yy} = \sigma_{zz} = \sigma_{xy} = \sigma_{yz} = \sigma_{zx} = 0,$$

$$(12.7\text{-}2) \qquad \frac{M}{EI} = \frac{1}{R}, \qquad \sigma_{xx} = \sigma_o \frac{y}{c}, \qquad \sigma_o = \frac{Mc}{I},$$

where c is the distance from the neutral surface to the "outer fiber" of the cross section, M is the bending moment, E is the Young's modulus, I is the area moment of inertia of the cross section, and σ_o is the outer fiber stress.

The strains are, therefore,

$$(12.7\text{-}3) \quad e_{xx} = \frac{y}{R}, \qquad e_{yy} = -\nu\frac{y}{R} = e_{zz}, \qquad e_{xy} = e_{yz} = e_{zx} = 0.$$

From these we see that the equations of equilibrium, (12.5-5), are satisfied. The equations of compatibility (6.3-4) are also satisfied. The boundary conditions on the lateral surface of the beam are $\overset{\nu}{T}_i = 0$. Since any normal to lateral surface is perpendicular to the longitudinal axis x, the direction cosine ν_x vanishes; i.e., $\nu_x = 0$ on the lateral surface. Thus the following boundary conditions are satisfied:

$$\overset{\nu}{T}_x = 0 = \sigma_{xx}\nu_x + \sigma_{xy}\nu_y + \sigma_{xz}\nu_z,$$

$$(12.7\text{-}4) \qquad \overset{\nu}{T}_y = 0 = \sigma_{yx}\nu_x + \sigma_{yy}\nu_y + \sigma_{yz}\nu_z,$$

$$\overset{\nu}{T}_z = 0 = \sigma_{zx}\nu_x + \sigma_{yz}\nu_z + \sigma_{zz}\nu_z.$$

The boundary conditions at the ends of the beam are that the stress system must correspond to a pure bending moment, and without a resultant force. The stress system (12.7-1) does that, as discussed in Sec. 7.7. Hence the solution is exact if the boundary stresses on the ends of the beam are distributed precisely in the manner specified by Eq. (12.7-1), because then all the differential equations and boundary conditions are satisfied.

One of the restrictions imposed in the derivation given in Sec. 7.7, namely, that the cross section of the beam has a plane of symmetry, can be removed. Let us consider, then, a prismatic beam with an arbitrary cross section such as the one shown in Fig. 12.5. Assume the same stress and strain distribution as in Eq. (12.7-1), (12.7-2), and (12.7-3). The boundary conditions (12.7-4) are also satisfied. The resultant axial force is zero again when the orgin is

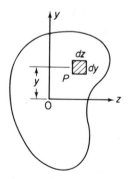

Fig. 12.5   An unsymmetric cross section.

taken at the centroid of the cross section. The resultant moment about the z-axis by the traction acting on the end section is given by the surface integral over the cross-section A:

$$M_z = \int_A \sigma_{xx} y \, dA = \frac{E}{R} \int_A y^2 \, dA = \frac{EI_z}{R}.$$

It is the same as before, except that we have added the subscript z to show that the bending moment and the area moment of inertia of the cross section are both taken about the z-axis. The resultant moment about the y-axis, however, is a new element to consider. It is given by an integration of the traction $\sigma_{xx} \, dA$ acting on an element of area dA situated at a distance z from the y-axis. Hence

$$(12.7\text{-}5) \qquad\qquad M_y = \int_A \sigma_{xx} z \, dA.$$

On substituting Eq. (12.7-1) into the above, we obtain

$$(12.7\text{-}6) \qquad\qquad M_y = \frac{E}{R} \int_A yz \, dA.$$

The last integral is the negative of the *product of inertia of the cross-sectional area*:

$$(12.7\text{-}7) \qquad\qquad P_{yz} = -\int_A yz \, dA \equiv -\iint_A yz \, dy \, dz.$$

Hence

$$(12.7\text{-}8) \qquad\qquad M_y = \frac{-EP_{yz}}{R}.$$

In case the beam cross section has a plane of symmetry in which the bending moment acts, we choose the xy-plane as that plane of symmetry; then $P_{yz} = 0$. It follows that $M_y = 0$, which shows that our solution in Sec. 7.7 is satisfactory. In the general case, we now choose the coordinate axes in such a way that the product of inertia vanishes. Then

$$(12.7\text{-}9) \qquad\qquad P_{yz} = 0, \qquad M_y = 0,$$

and the moment vector is parallel to the z-axis with a magnitude equal to M_z.

The product of inertia vanishes if the y- and z-axes are the principal axes of inertia. Hence in order that a moment acting in a plane produces bending in the same plane, it is necessary that the plane be a principal plane, i.e., one contain-

ing a principal axis of inertia of every cross section. Combining the require-
ments (7.7-4) and (12.7-9), we see that *the coordinate axes y, z must be chosen
as centroidal axes in the direction of the principal axes of the area moments of
inertia.*

Our verification is now complete. We have found that the stress system
(12.7-1) is exact if y is measured from the neutral axis in the direction of a
principal axis. The stress system satisfies the equations of equilibrium, the
equations of compatibility, and the boundary conditions. Our intuitive
assumption that plane sections remain plane is verified in this case.

More refined theories of bending can be found in many books, e.g.,
Sokolnikoff's *Mathematical Theory of Elasticity.* The history of the devolp-
ment of man's understanding of beam action is a fascinating subject well-
recorded in Timoshenko's book, *History of the Strength of Materials.* Modern
investigation began with Galileo, but it is again to the credit of St.-Venant
that the problem was solved within the general theory of elasticity.

12.8 CONCLUDING REMARKS

We have given in this chapter a few examples of the theory of elasticity.
It is evident that only very special boundary-value problems can be solved
with ease. Practically all practical problems require the solution of a rather
difficult set of equations and boundary conditions. Thus we see that the problem of torsion of a circular shaft is easy to solve, but the problem of the torsion of a cylindrical shaft of non-circular cross section is much more difficult
to solve. Similar comments can be made with respect to other problems:
the bending of a curved beam, the torsion of a tapered shaft, the propagation of seismic waves in the earth which is nonuniform in material, the reflection of seismic waves on the foundation of our building, etc. Realistic
boundary conditions may not be analytically simple!

Much remains to be learned about fluid and solid mechanics. The present
book is merely an introduction.

PROBLEMS

12.2 Compute the location of the centroid and the centroidal moments of
inertia of the areas of the sections in Fig. P12.2. Use the following procedure
if you wish. Let x and y be horizontal and vertical axes, respectively. Compute
both coordinates $\bar{x}$, $\bar{y}$ of the centroids and the moments of inertia I_x, I_y about
the x-axis and the y-axis, respectively.

$$I_x = \oiint y^2 \, dx \, dy, \qquad I_y = \oiint x^2 \, dx \, dy.$$

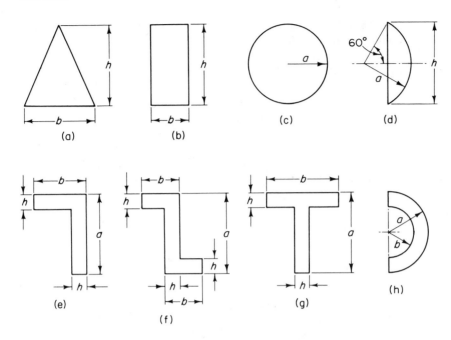

Fig. P12.2   Various beam cross sections.

Then compute the centroidal moments of inertia of the area by the formulas

$$\bar{I}_x = \oiint (y - \bar{y})^2 \, dx \, dy = I_x - \bar{y}^2 A,$$

$$\bar{I}_y = \oiint (x - \bar{x})^2 \, dx \, dy = I_y - \bar{x}^2 A.$$

12.3 Compute the products of inertia P_{xy} of the areas named in Prob. 12.2:

$$P_{xy} = - \oiint xy \, dx \, dy.$$

Compute the centroidal product of inertia:

$$\bar{P}_{xy} = - \oiint (x - \bar{x})(y - \bar{y}) \, dx \, dy = P_{xy} + \bar{x}\bar{y}A.$$

Answer: The product of inertia vanishes whenever x or y is a line of symmetry. Hence $\bar{P}_{xy} = 0$ in Cases a, b, c, d, g, and h. For Cases e and f, we can break the area up into small rectangles, for each of which P_{xy} about its centroid is zero. Therefore the contribution of each small rectangle to the product of inertia is given by the term $\bar{x}\bar{y}A$.

(e) Thus $\bar{P}_{xy} = -\frac{1}{4}(a^2h^2 + b^2h^2 - h^4) - \bar{x}\bar{y}h(a + b - h)$ where $\bar{x}$, $\bar{y}$ are given in Prob. 12.2(e).

(f) $\bar{P}_{xy} = -\frac{1}{2}bh(a - h)(b - h)$.

12.4 Show that the symmetric matrix of inertia

$$\begin{pmatrix} I_x & P_{xy} \\ P_{yx} & I_y \end{pmatrix}$$

transforms according to the tensor transformation rule with respect to rotation of coordinates in the xy-plane. Hence, Mohr's circle method may be used to compute the moments and product of inertia with respect to new axes obtained by the rotation of coordinates.

Note 1: If we define an inertia tensor

$$I_{ij} = \int (\delta_{ij} x_k x_k - x_i x_j)\, dA,$$

in which the Latin indices range over 1, 2, then its components can be identified as the *moments* $(i = j)$ and *products* $(i \neq j)$ of inertia of the area.

Note 2: In the theory of dynamics of rigid bodies in a three-dimensional space, the *inertia tensor* is defined as

$$I_{ij} = \int (\delta_{ij} x_k x_k - x_i x_j)\, dm$$

whose components are called the *moments* $(i = j)$ and *products* $(i \neq j)$ of inertia of the mass.

12.5 Compute the centroidal moments of inertia $\bar{I}_x$, $\bar{I}_y$ and the centroidal product of inertia $\bar{P}_{xy}$ of the L section as shown in Fig. P12.2(e). The matrix

$$\begin{pmatrix} \bar{I}_x & \bar{P}_{xy} \\ \bar{P}_{xy} & \bar{I}_y \end{pmatrix}$$

is symmetric but not diagonal. Consider a rotation of coordinates from xy to $x'y'$:

$$x' = x \cos \theta + y \sin \theta,$$
$$y' = -x \sin \theta + y \cos \theta.$$

Compute the inertia matrix with respect to the new axes $x'y'$:

$$\begin{pmatrix} \bar{I}_{x'} & \bar{P}_{x'y'} \\ \bar{P}_{x'y'} & \bar{I}_{y'} \end{pmatrix}.$$

Determine the angle θ which is required to make $\bar{P}_{x'y'}$ vanish. The x'-, y'-axes are then called the principal axes.

Note: Use results of Prob. 12.2, 12.3, and 12.4 and Mohr's circle.

12.6 A 2 in. $\times$ 6 in. rectangular beam is used to resist a bending moment which acts in a vertical plane (see Fig. P12.6). Compare the differences in the extreme fiber stresses, the bending stiffness, and the deflections in the following cases:

(a) The 6-in. side is placed vertically.

(b) The 6-in. side is placed horizontally.

Answer: Configuration a is nine times stiffer and deflects nine times less under the same load, and the extreme fiber stress is one-third that of configuration b.

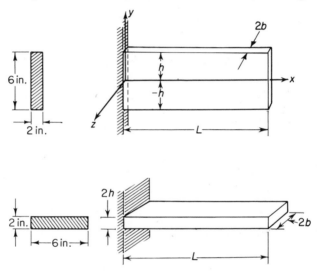

Fig. P12.6   Proper use of beams.

12.7 Plastic torsion of a circular cylinder. Let a circular cylinder be made of a material such as mild steel which obeys Hooke's law approximately when the maximum shear stress is less than the yield stress τ_0 corresponding to a shearing strain $e_{xy}^{(0)} = (1/2G)\tau_0$. If the shearing strain continues to increase beyond $e_{xy}^{(0)}$, the shear stress would remain constant and equal to τ_0. Such a material is said to be perfectly plastic. Deduce a relationship between the torque T and the rate of twist α when the yield point is exceeded. Assume that the strain distribution in the shaft is the same as in the case of a perfectly elastic shaft. Determine the maximum torque that can be sustained by the shaft. See Fig. P12.7.

Solution: If torque exceeds the maximum elastic limit, the outer part of the shaft would yield plastically while the shearing stress remains at the constant value τ_0, see Fig. P12.7(b). The maximum torque a shaft can carry corresponds to Case c. Figure P12.7(a) corresponds to the fully elastic case.

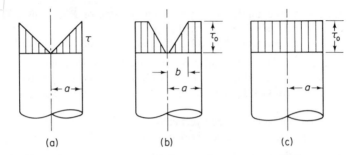

Fig. P12.7 Stress distribution in plastic torsion of a circular cylinder.

In the elastic range, if the outer fiber shear stress is τ, the torque is

$$T = \int_0^a 2\pi r^2 \frac{r}{a}\,\tau\,dr = \frac{2\pi\tau}{a}\frac{r^4}{4}\Big|_0^a = \frac{\pi a^3}{2}\tau. \tag{1}$$

In the plastic range, if the plastic zone extends from $r = b$ to $r = a$,

$$T = \frac{\pi b^3}{2}\tau_0 + \int_b^a 2\pi r^2\tau_0\,dr = \frac{\pi b^3}{2}\tau_0 + \frac{2\pi}{3}(a^3 - b^3)\tau_0. \tag{2}$$

The maximum torque possible is, when $b = 0$,

$$T_{\max} = \frac{2\pi}{3}a^3\tau_0. \tag{3}$$

The stress τ_0 is reached when

$$e_{xy} \geq e_{xy}^{(0)} = \frac{1}{2G}\tau_0. \tag{4}$$

Let α be the angle of twist per unit length. Consider a segment of unit length. The upper end of the segment is rotated by an angle α relative to the lower end. At a radius a the displacement is $a\alpha$. This induces a local shear strain of $\frac{1}{2}\alpha a$. If $\frac{1}{2}\alpha a < \tau_0/2G$, everywhere is elastic. The yield condition is reached when

$$\frac{1}{2}\alpha a = \frac{\tau_0}{2G}, \quad \text{or} \quad \alpha = \frac{\tau_0}{Ga}. \tag{5}$$

If $\alpha > \tau_0/Ga$, yielding occurs, Fig. 12.7(b), and Eq. (2) apply. Figure 12.7(c) and Eq. (3) apply when $b \longrightarrow 0$, $\alpha \longrightarrow \infty$.

12.8 Formulate the problem of a hollow sphere subjected to an internal pressure p_i and an external pressure p_e. If the sphere is made of homogeneous elastic material obeying Hooke's law, the solution can be obtained obviously by superposition of two solutions: one for a uniform compression and one for the pressure differential $p_i - p_e$.

Can the same be done if the material is perfectly plastic? And if the material is nonlinearly elastic as all biological tissues are?

Solution: By symmetry, the deformation of the spherical shell is such that only one component of displacement does not vanish, namely, the radial component u. Hence if we use spherical coordinates (r, Φ, θ) as a frame of reference, we have $u_r = u \neq 0$, $u_\Phi = u_\theta = 0$. By symmetry, the shear stresses and strains all vanish:

$$\sigma_{r\Phi} = \sigma_{r\theta} = \sigma_{\Phi\theta} = 0, \qquad e_{r\Phi} = e_{r\theta} = e_{\Phi\theta} = 0. \tag{1}$$

The normal strains are

$$e_{rr} = \frac{\partial u}{\partial r}, \qquad e_{\theta\theta} = \frac{u}{r}, \qquad e_{\Phi\Phi} = \frac{u}{r}. \tag{2}$$

See Sec. 5.9 for a similar derivation for cylindrical polar coordinates.

To derive the equation of equilibrium, note that on account of the symmetry we need to consider only the balance of radial forces. Let us consider a small conical element extending small angles $d\theta$, $d\Phi$ at the origin, as shown in Fig. P12.8. The area of the surface $ABCD$ at radius r subtended by this cone is $r^2 \, d\theta \, d\Phi$. The radial force normal to the surface $ABCD$ is $\sigma_{rr} r^2 \, d\theta \, d\Phi$. There is a corresponding radial force acting on the surface $EFGB$. These two opposite forces have a net radial resultant:

$$\frac{\partial}{\partial r}(r^2\sigma_{rr}) \, d\theta \, d\Phi \, dr.$$

The stresses $\sigma_{\theta\theta}$ and $\sigma_{\Phi\Phi}$ each contributes a radial resultant:

$$-\sigma_{\theta\theta} \, r \, d\Phi \, d\theta \, dr, \qquad -\sigma_{\Phi\Phi} \, r \, d\theta \, dr \, d\Phi.$$

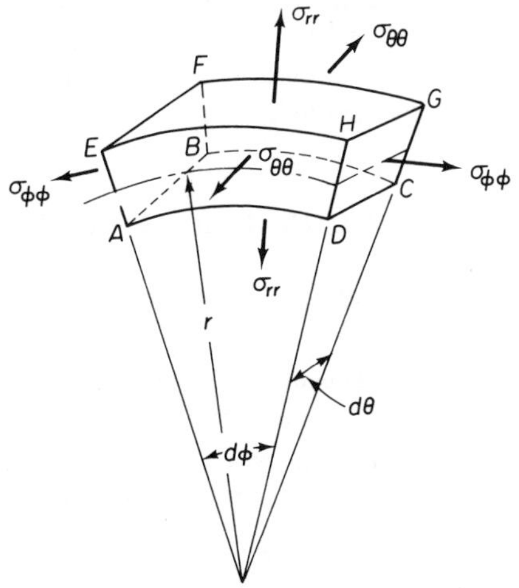

Fig. P12.8 An element of a spherical shell.

Adding, we obtain the equation of equilibrium

$$\frac{\partial \sigma_{rr}}{\partial r} + \frac{2}{r}\sigma_{rr} - \frac{\sigma_{\theta\theta}}{r} - \frac{\sigma_{\Phi\Phi}}{r} = 0. \tag{3}$$

The boundary conditions are that the radial stress σ_{rr} is equal to $-p_i$ at the inner radius R_i and $-p_e$ at the external radius R_e.

Now let us consider the homogeneous Hookean material that obeys the stress-strain relationship:

$$\sigma_{rr} = \lambda(e_{rr} + e_{\theta\theta} + e_{\Phi\Phi}) + 2\mu e_{rr} = (\lambda + 2\mu)\frac{\partial u}{\partial r} + \frac{2\lambda u}{r},$$

$$\sigma_{\theta\theta} = \sigma_{\Phi\Phi} = (2\lambda + 2\mu)\frac{u}{r} + \lambda \frac{\partial u}{\partial r}. \tag{4}$$

On substituting Eq. (4) into (3), we obtain the differential equation

$$\frac{\partial^2 u}{\partial r^2} + \frac{2}{r}\frac{\partial u}{\partial r} - \frac{2u}{r^2} = 0; \tag{5}$$

whereas the boundary conditions read

$$(\lambda + 2\mu)\frac{\partial u}{\partial r} + 2\lambda \frac{u}{r} = -p_i \text{ at } r = R_i, \text{ and } = -p_e \text{ at } r = R_e. \tag{6}$$

These equations are linear. The principle of superposition applies. The sum of two solutions is itself a solution. Hence we can resolve the solution into two subsolutions as noted in the problem.

If the material is either perfectly plastic or nonlinearly elastic, the stress-strain relationship is nonlinear. Equations (1), (2), and (3) remain valid, but Eq. (4), (5), and (6) must be modified; and they become nonlinear in u. For a set of nonlinear equations the sum of two solutions is no longer a solution. The principle of superposition is not valid, and the device of separating the given problem into two component problems cannot be used.

FURTHER READING

FUNG, Y. C., *Foundations of Solid Mechanics*, Englewood Cliffs, N.J.: Prentice-Hall (1965).

LOVE, A. E. H., *The Mathematical Theory of Elasticity*, 1st ed. 1892, 4th ed. 1927, New York: Dover Publications (1944).

MUSKHELISHVILI, N. I., *Some Basic Problems of the Theory of Elasticity*, translated from Russian by J. R. M. Radok, Groningen, Netherlands: Nordhoff.

RAYLEIGH, J. W. STRUTT, *The Theory of Sound*, 1st ed. 1877, 2nd ed. revised, New York: Dover Publications (1945).

SOKOLNIKOFF, I. S., *Mathematical Theory of Elasticity*, 2nd ed., New York: McGraw-Hill (1956).

TIMOSHENKO, S. P., *History of Strength of Materials*, New York: McGraw-Hill (1953).

TIMOSHENKO, S. AND N. GOODIER, *Theory of Elasticity*, New York: McGraw-Hill, 1st ed. 1934, 2nd ed. 1951.

General Problems

G.1 Consider a string of uniform density and material, stretched tight between two posts (e.g., a violin string, a piano wire). It is struck at a point. Vibration ensues. Formulate the problem mathematically. Give both the differential equations and the boundary conditions.

G.2 Consider a gong used in the orchestra. Formulate the mathematical problem of gong vibration.

G.3 Formulate a mathematical description of the clouds floating in the sky. How do they move about? Include enough parameters so that the great variety of things you see daily can be described and deduced.

G.4 An airplane flies in the air at a forward speed V relative to the ground. In a steady level flight it is pulled by a 4000 horsepower engine. How does the wing maintain this flight? To answer this question we need to write down the field equations for the air, that of the airplane wing, and the boundary conditions at the interface between the air and the wing. Present a full set of equations which would be sufficient to furnish a mathematical theory in principle.

G.5 The elastic waves in the rails as a train approaches are typical of many dynamics problems. We can easily hear the impact of the wheels of the train (if we put our ears to the rail) long before the train can be seen. Then as the train comes by, we can see the deflection of the rails under the wheels. Formulate the problem mathematically so that both of these features can be exhibited.

G.6 Feel the pulse on your wrist. It is a composite elastic wave in your artery. The most important component is undoubtedly the elastic response of the artery to the pressure wave in the blood. To a lesser extent there must be other waves which are propagated along the arterial wall and caused by disturbances further upstream or downstream. Our arteries are elastic. Formulate a mathematical theory for our pulse propagation. Leonhard Euler (1707–1783) formulated the problem and presented an analysis as early as 1775.

G.7 Galileo (1564–1642) proposed the following method for measuring the frequency of vibration of a gong. Attach a small, sharp, pointed knife to a slender rod. Pull the rod over the gong at a constant speed. The vibration of the gong would cause the rod to chatter. Examine the metal surface of the gong and measure the spacing of the marks, from which the frequency may be calculated.

Explain whether this method would work. How would you compute the frequency? Formulate the problem mathematically from the point of view of the theory of elasticity. Assume a good musical gong to assure that the material is a linear elastic solid which obeys Hooke's law.

G.8 The phenomenon of chatter in machine tools is not unlike Galileo's gong experiment. Consider the problem of a high-speed lathe. Formulate the problem of chatter which ruins a good machine operation. Propose ways to alleviate the problem.

G.9 A beam vibrates. Write down the differential equation and boundary conditions for a vibrating beam and a method of determining the frequency of vibration.

G.10 A circular cylindrical shaft spins about its longitudinal axis at an angular speed ω radians/sec. The shaft is simply supported at both ends. Lateral vibrations are always possible when the shaft spins. However, if the rate of spin reaches a critical value, the lateral deformation becomes excessive, and so-called "whirling" sets in. Describe the phenomenon mathematically. Formulate the equations with which the critical whirling speed can be determined.

G.11 The shaft of an airplane propeller is subjected to both a tension and a torque. How would you propose to measure the stresses in the shaft in flight? How would you measure the power delivered to the propeller in flight?

Index

A

Acceleration, convective, 246, 247
  components in polar coordinates, 258
  material, 246, 247
Acoustics:
  basic equation of, 292
  velocity of sound, 293
Adkins, J. E., 168, 185
Airy, Sir George Biddell, 315
Airy stress function, 314, 315
Almansi, E., 127
Almansi strain tensor, 127, 308
Aris, R., 61
Avogadro, Amedeo, 201
  number, 202

B

Barotropic fluid, 288, 291
Beams:
  bending moment, 15
  classification, simply-supported, clamped,
    free, 174, 321
  curvature of, 173, 175
  deflection, 173–76
  the largest stress in, 173
  moment diagram, 16
  neutral surface of, 172
  outer fiber stress, 173
  simply-supported, 14
Bending of beams, 171–76, 321–24
Bingham plastic, 234, 235
Blasius, H., 283–85
Blood, non-Newtonian viscosity, 232, 233
Body, 11
  free, 11
Body force, 64, 68, 69

Body moment, 63
Boltzmann, L., 202, 231
Boundary conditions:
  fluid, free surface, 270
  solid-fluid interface, 261, 262, 268
  solids, 116, 310
  stress, 79–82
  subsonic flow, 298
  supersonic flow, 299
  velocity, 297
Boundary layer, 279–85
  Prandtl's equation, 282
  thickness, 282
Bradshaw, A., 215
Brenner, S., 223
Bridgman, P. W., 213, 214
Briggs, L. J., 205
Brillouin, L., 61
Buckling, 89
Bulk modulus:
  of air, 212
  of liquid, 215
  relation to other moduli, 217, 218
  of solid, 216, 217

C

Caloric equation of state, 291
Cartesian tensors, analytical definition, 52, 53
Cauchy, A. L., 65, 127
  formula, 70–72
  strain tensors, 128, 141
Cavitation, 205
  damage, 206
Cesaro, E., 158
Chapman, S., 209
Chien, S., 232, 233
Circulation, 287–89

Clements, J. A., 273
Compatibility condition, 154–58
  equation of, 156
  plane strain, 156
  of St.-Venant, 158
  in three-dimensions, 157
Compressibility:
  of air, 212
  of liquid, 213
Compressible flow, 291–302
  basic equations, 292, 295, 296
Conservation laws:
  of angular momentum, 252
  of energy, 253–56
  of mass, 250, 251, 260
  of momentum, 250–52, 257
  in polar coordinates, 256–60
Constants:
  Avogadro's number, 202
  Boltzmann, L., 202
  Lamé, G., 168, 217, 218
  Loschmidt's, 202
Constitutive equations, 5, 7, 163–70
  Bingham material, 234, 235
  gas, 202
  Hookean elastic solid, 167, 216–19
  Hookean solid with temperature
    variation, 170
  liquids, 213
  living tissue, 223–27
  Newtonian fluid, 129
  non-Newtonian fluids, 232–33
  nonviscous fluids, 164
  plasticity of metals, 219–20
  seawater, 213–15
  van der Waals' equation, 202–204
  viscoelastic material, 227–32
  viscoplastic materials, 233–35
Continuum, concept of, 2–4
Continuity, equation of, 250, 260
  compressible fluid, 250, 291
  in polar coordinates, 260
Contraction (tensor), 58
Convective acceleration, 247
Convention:
  strain notation, 121, 128–29
  stress notation, 18, 20, 65–67
  summation of indices, 44
Coordinates, transformation, 47, 51
  curvilinear, 77
  cylindrical polar, 77–79, 135, 136
Cottrell, A. H., 210, 211, 212, 222, 223, 238

Couette, M., 88, 270, 304
Coulomb, C. A., 270
Couple, 8
Couple-stress, 63, 65
Creep functions solid, 228–31
Critical points, liquid-vapor-gas, 201
Curl, 59
Curvature:
  of space curve, 43

D

D'Alembert's principle, 7
Deformation, analysis, 121–52 (*see also* Strain)
Deformation gradients, 130, 141
Detrusions, 129 (*see also* Shear strain)
Dirac-delta function, 229
Dislocations, 223
Displacement field, infinitesimal, 124, 130
Displacement vector, 123, 124
  in polar coordinates, 136, 139
  radial and tangential, 138
  relation to velocity, 154
Divergence, 59
Dorsey, N. F., 214
Duhamel-Neumann thermoelasticity law, 170

E

e-δ identity, 45, 47
Earthquake, 116, 147
Eirich, F., 238
Elastic stability, 89
Elasticity, of solids, 167, 169, 216–19
  basic equations, 308–10
  effect of temperature, 169
  nonlinear, 223–27
  theory of, 307–31
Elliptic equation, 296
Ellis, A. T., 206
Energy, conservation of, 253–56
  equation, 254, 255, 256
Enskog, D., 209
Equations of:
  continuity (*see* Continuity)
  energy (*see* Energy)
  equilibrium (*see* Equilibrium)
  motion (*see* Motion)
  state gas, 202–04
    water, 204, 213–15
Equilibrium, 8
  necessary conditions, 8, 10

Equilibrium, equations of, 72–74, 257, 314, 318
Eringen, A. C., 185
Euclidean metric space, 188
Euler, L., 65, 68, 334
Eulerian equation of motion, 251, 291
Eulerian strain tensor, 127, 132–34 (*see also* Strain tensor)

F

Fage, A., 270
Finger's strain tensor, 142
Finite strain components, 122, 127–132
  geometric interpretation, 132–134
Flow, Hagen-Poiseuille, 278
Fluid:
  critical points, 200, 201
  isotropic viscous, 165–67, 195
  rate-of-deformation-and-velocity-gradient relation, 154
Force, body and surface, 64, 68, 69
Free-body diagram, 5, 11, 67, 69
Freundlich, H., 236
Fung, Y. C. B., 135, 158, 161, 170, 223, 224, 239, 321, 330

G

Galileo Galilei, 181, 324, 334
Gauss, K., 243
  theorem, 72, 242, 243, 244
Gel, 235
Glaser, D. A., 205
Goldstein, S., 304
Goodier, N., 180, 331
Goursat, E., 52
Gradient, 59
Gray, D. E., 205
Green, A. E., 127, 158, 161, 168, 185
Green's (George) strain tensor, 127
Green's (George) theorem, 243
Gregersen, M. I., 232, 233

H

Hagen-Poiseuille flow, 278
Hamel, G., 127
Heat flux vector, 254
Hedrick, E. R., 52
Helmholtz's theorem, 288, 289
Hohenemser, K., 234

Hooke, R., 125
Hooke's law, 167, 168, 216, 308
Hyperbolic equation, 296

I

Ideal gas, 164
Incompressible fluid, 166
  constitutive equation, 164
  equation of continuity, 266, 274
  field equations, 274
Index, 43
  dummy and free, 44
Indicial notation, 43
Inertia force, 6
Inertial frame of reference, 6
Infinitesimal strain components, 129, 130
  geometric interpretation, 129, 130
  polar coordinates, 135
Integrability condition, 155
Interface condition, 81
Internal energy, 254, 255
International System of Units, 18
Invariants, 99
  isotropy, 197
  strain, 135
  stress, 99
  stress deviations, 103
Irrotational flow, 289–91
Isentropic flow, 291
Isotropic materials, 165, 166, 168, 187, 195
Isotropic tensor, 188
  rank 1, nonexistence, 189, 190
  rank 2, 190, 191
  rank 3, 191–92
  rank 4, 192–95
  higher rank, 198
Isotropy, 5, 187–98

J

Jacobian determinant, 52, 125, 246
Jaeger, J. C., 142
Jeans, J., 202, 239
Jeffreys, H., 198

K

Kaplan, W., 52
Kaye, G. W. C., 215
Kê, T. S., 227
Kelvin, Lord (William Thomson), 227, 230

Kelvin's model of viscoelasticity, 227–31
  (*see also* Standard linear solid)
Kelvin's theorem, 288
  fluid line, 289
Kestin, J., 284
Kinematic viscosity, 207, 208
Kinney, G. F., 239
Kronecker delta, 45
Kubota, T., 298

L

Laby, T. H., 215
Lagrange, J. L., 243
  strain tensor, 127
  stress, 225
Lamb, H., 291, 304
Lamé, G., 114, 217
  constants, 168, 217
  ellipsoid, 107, 135
  relation to other elastic moduli, 217
  shear modulus, 168
Laminar boundary layer, 282
Laplace, Pierre Simon M. de, 271, 283
  operator, 59, 267
Laplacian, 59
Levich, V. G., 239
Li, Yuan-Hui, 213
Liepmann, H. W., 305
Lodge, A. S., 238
Longitudinal waves, 310
Loschmidt's number, 202
Love, A. E. H., 66, 168, 185, 330
Lundberg, J. L., 233

M

Mach number, 213, 295
Mach waves, lines, 301
Mass, conservation of, 246, 250, 308
Material derivative, 247, 248–50, 308
Material description, 245
Material isotropy, 187
Matrix, 48
  orthogonal, 48
Maxwell, J. C., 207, 227, 228–31, 235
Maxwell solid, 228–30
  creep function and relaxation function,
    228, 229
Mean curvature, 272
Mean free path, 4, 25, 209
Mechanics, 2
Membrane, thin, 92

Michal, A. D., 61
Millikan, Robert, 113, 275
Minimal surface, 272
Modulus, bulk, 212, 217
  elasticity, 217
  relaxed, 230
  rigidity, 217
  shear, 217
Mohr, Otto, 95, 108
Mohr's circle, 95–97, 108–12, 134
  proof of construction, 96, 108
  special sign convention, 95
  three-dimensional states, 108–12
Moment, 9
  bending,
Momentum, conservation of, 250, 308
  linear, 67, 72
  moment of, 68, 252, 253
Morrey, C. B., 52
Motion, equations of, 69, 76, 250, 258, 267
  polar coordinates, 256–60
Muskhelishvili, N.I., 330

N

Nadai, A., 239
Navier's equation, 309, 310
Navier-Stokes equation, 265, 267, 277, 282
  dimensionless, 274, 279
Neutral plane, 172
Newton, Isaac:
  law of gravitation, 42
  laws of motion, 6, 7, 68, 241, 250, 251
  law of viscosity, 207
  velocity of sound, 293
Newtonian fluid, 163, 165, 167
  static pressure in, 165, 166
Nikuradse, J., 284, 285
Nimbus spacecraft, 299
Non-Newtonian fluids, 163, 167, 232, 233
Nonviscous fluid, 164, 285
  constitutive equation, 166
  equation of motion, 286
Normal modes of vibration, 92
Normal strain (*see* Strain)
Normal stress (*see* Stress)
No-slip condition, 268–70
Numbers:
  Avogadro, 202
  Loschmidt, 202
  Mach, 213, 295
  Reynolds, 273–75, 278, 279, 284

O

Octahedral planes, 87
Orthogonal matrix, 48
Orthogonal transformation, 48,
Oseen, 270
Ostrogradsky, M., 243

P

Parabolic velocity profile, 277, 278
Pauling, L., 239
Pearson, K., 66
Peptization, 235
Perfect gas law, 202
Permutation, tensor, 45
    connection with Kronecker delta, 45
    symbol, 45
Phase velocity, 310
Photomicrograph of a stainless steel, 206
Plane elastic waves, 310
Plane strain, 313
Plane stress, 93, 312, 313
    Mohr's circle, 95
Plane waves, 310
Plasma, blood, 232
Plasticity of metals, 219–20
Poise, 18, 207
Poiseuille, J. L. M., 207, 270, 278
Poisson, S. D., 218
Poisson's ratio, 147, 217, 218, 219
Polarization, the plane of, 311
Potential equation, 290
Potential flow, 290, 295–302
Prager, W., 234
Prandtl, L., 280, 282, 305
Principal axes, 92, 94, 98
    of strain, 134, 195
    of stress, 98, 100, 195
Principal coordinates, 92
Principal directions, 94
Principal planes, 92, 98, 134
Principal strains, 134
Principal stress, 91, 92, 94, 102
    main theorem, 98
    proof of existence, 100
    proof of real-valuedness, 100
    stress invariants, 99
Principal stress deviation, 104
Proportional limit, 220
Protter, M. H., 52

Pryce-Jones, J., 236
Pure shear, 142

R

Radiating condition, 299
Radius vector, 50, 54
Radok, J. R. M., 330
Rayleigh, Lord (J. W. Strutt), 147, 330
Rayleigh wave, 147
Relaxation function, 229, 230, 231, 232
Relaxation time, 230
Reynolds, Osborne, 267, 278, 279
Reynolds number, 273, 274, 275, 279
    boundary-layer thickness, 282
Rheology, 233
Rockets, 8
Roshko, A., 305
Rotation, infinitesimal, 130

S

Saint-Venant, Barre de, 127, 315, 320, 324
Saint-Venant's principle, 320
Scalar product, 38
Scalars, analytical definition, 53, 54
Scalar triple product, 40
Schleicher, K. E., 215
Schlichting, H., 284, 305
Schlieren photographs, 298
Sechler, E. E., 139
Shear modulus, 168, 217
    for iron and steel alloys, 221
Shear strain, notation, warning, 128, 130
    maximum, 222
    pure, 142
    simple, 129, 142
Shear strength of metals, 222
    ideal, 222
Shear stress, 66, 101, 207
    maximum, 91, 95, 102, 108
SI units, 18
Simple shear, 142
Skin friction coefficient, 279
Sokolnikoff, I. S., 61, 324, 331
Sol, 235
    sol-gel transformation, 235
Solenoidal vector field, 286
Solid, Hookean elastic, 167, 168
Sound, speed, 184, 218, 291
Spectrum of relaxation, 231, 232
Standard linear solid, 228–30
Statics, 8

Stokes, George G., 113, 166, 270
Stokes' fluid, 166
Stokes' problem, sphere in viscous fluid, 113
Stokes' theorem, 244, 261
Strain, 121–52
    Cartesian coordinates, 127
    finite, 126, 127–29, 132–34
    infinitesimal, 127, 128, 129–31
    invariants, 135, 197
    plane state, 313, 314
    polar coordinates, 135–40
    principal, 134
    shear, 129
Strain deviation, 135, 216
Strain-energy function, 196
Strain-rate tensor, 153, 154, 256
Strain tensor, 126, 127
    Almansi's, 127, 308
    Cauchy's, 127, 128, 141
    Eulerian, 127
    Finger's, 142
    Green's, 127
    Lagrangian, 127
Stream function, 286, 287
Stress, 16, 63
    boundary conditions, 79–82
    components, notation, 65, 66
        sense of positive, 66, 67
        transformations of coordinates, 76, 77
    compressive, 18
    couple-stress, 65
    definition, 64
    ellipsoid, (see also Lamé, G.) 106–7
    equation of motion and equilibrium, 72,
        74, 76, 250, 256
    invariants, 87, 91, 99, 104, 105
    matrix, 63, 91, 101
    normal, 16, 18, 66
    notations, 20
    plane state, 92–97, 312
    polar coordinates, 79, 256
    principal (see Principal stress)
    shear, 18, 66
    sign convention, 66, 67
    tension, 18
    tensor transformation, 63
    vector, 65 (see also Stress vector)
Stress concentration, 114
Stress-deviation tensor, 91, 103, 216
Stress function, Airy, 314–315
Stress-strain-rate relationship, 165, 266
        (see also Constitutive equations)

Stress-strain relationship, 167, 168, 216
Stress tensor, 72, 77
    components, 65, 70, 77
    curvilinear coordinates, 77
    invariants, 99
    symmetry, 74–75
    transformation of coordinates, 76, 77
Stress vector, 65, 70
Structure
    statically indeterminate, 27
Subsonic flow, 294
Summation convention, 44
Superheated liquid, 203
Supersonic flow, 294
Surface, minimal, 272
Surface force, 64, 68, 69
Surface tension, 270–73

T

Tait, P. G., 213
Tait equation, 213, 215
Taylor, H. M., 233
Tensile strength:
    iron and steel alloys, 221
    liquid, 204, 205
Tensor, 18, 37, 55
    Cartesian, 53
    contraction, 58
    definition, 53, 54
    dual, 131
    isotropic, 188 (see also Isotropic materials,
        Isotropic tensor)
    notations, 56
    partial derivatives, 57
    quotient rule, 56
    rank, 53
    rotation, 131
    spin, 154
Thermodynamics, the first law, 241, 253
Thixotropy, 235
Thomas, T. Y., 198
Timoshenko, S. P., 180, 331
Todhunter, I., 66
Torsion, 315–20
    of space curve, 43
Townsend, A. A., 270
Traction, 65 (see also Stress vector)
Transformation of coordinates, 47, 51
    admissible, 52, 55
    Jacobian, 52, 246
    orthogonal, 48, 188

Transformation of coordinates (Cont.):
proper and improper, 52
rotation, 47
vector, 50
Transverse wave speed, 311
Truss, 11
statically indeterminate, 27
Turbulence, 278, 279

U

Unit-impulse function (Dirac-delta
function), 229
Unit-step function, 228
Usami, S., 232, 233

V

van der Waals, Johannes D., 202, 203, 204
Vectors, 37
analytical definition, 53, 54
base, 49
cross product, 39, 59
dot product, 38, 59
dual, 131
field, 53
solenoidal, 286
notation, 37
rules of operation, 38
scalar product, 38
scalar triple product, 40, 51
transformation, 50
unit, 16, 37
vector product, 39, 46
vector triple product, 46
zero, 37
Velocity field, 153, 154
Velocity of sound, 184, 218, 291–93
Velocity potential, 158, 290
Vibration, modes and frequencies, 92
Vincent, R. S., 205

Viscoelasticity, 227–32
Visco-plastic material, 233–35
Viscosity:
air, 208
atomic interpretation, 209–12
blood, 233
coefficient, 207
liquid, 210
Newtonian concept, 207
solid, 212
water, 208
Voigt, W., 227, 228, 229
Voigt solid, 228, 229
von Kármán, Theodore, 220
Vorticity, 120, 154, 287–89

W

Wave equation, 292
Waves:
Acoustic, 292–94
Longitudinal, 310
Polarization, 311
Rayleigh, 147, 148
Shear, 311
Surface, 148
transverse, 311
Whetham, 270

Y

Yield function, 234
Yield strength, for iron and steel alloys, 221
Yield stress, 219, 233
Yih, Chia-shun, 305
Young, Thomas, 217, 221, 271

Z

Zener, C., 227, 239
Zerna, W., 158, 161